Synthesis Lectures on Engineering, Science, and Technology

The focus of this series is general topics, and applications about, and for, engineers and scientists on a wide array of applications, methods and advances. Most titles cover subjects such as professional development, education, and study skills, as well as basic introductory undergraduate material and other topics appropriate for a broader and less technical audience.

Shuangming Yang · Badong Chen

Neuromorphic Intelligence

Learning, Architectures and Large-Scale Systems

 Springer

Shuangming Yang
School of Electrical and Information
Engineering
Tianjin University
Tianjin, China

Badong Chen
National Key Laboratory of Human-Machine
Hybrid Augmented Intelligence, National
Engineering Research Center for Visual
Information and Applications, and Institute
of Artificial Intelligence and Robotics
Xi'an Jiaotong University
Xi'an, China

ISSN 2690-0300   ISSN 2690-0327 (electronic)
Synthesis Lectures on Engineering, Science, and Technology
ISBN 978-3-031-57872-4   ISBN 978-3-031-57873-1 (eBook)
https://doi.org/10.1007/978-3-031-57873-1

This Springer imprint is published by the registered company Springer Nature Switzerland AG
The registered company address is: Gewerbestrasse 11, 6330 Cham, Switzerland

Paper in this product is recyclable.

Preface

Artificial intelligence has been in development for over 60 years, with many researchers dedicated to achieving machine intelligence that can match or even surpass human intelligence. However, existing artificial intelligence systems can only achieve human-level intelligence in specific tasks, and there remains a significant gap between machine and human intelligence. As a result, artificial intelligence research requires novel ideas, technologies, and methods. Neuromorphic intelligence refers to a new computing paradigm inspired by the neural information processing in the human brain. Neuromorphic intelligence is meaningful for understanding the neural information processing principles, as well as aiding super-computing systems with stronger computing power and lower power consumption towards artificial general intelligence (AGI). Thus, it is necessary to revisit and explore the field of neuromorphic intelligence.

Neuromorphic intelligence requires a highly cross-disciplinary integration of neuroscience, computer science, and information technology. It results in the development of new types of learning algorithms, system architectures, and large-scale systems for neuromorphic intelligence that should be applied to the new generation of artificial intelligence and supercomputers. For neuromorphic learning, it refers to the research to improve the learning capability of spiking neural networks, including generalization ability, robustness, and learning speed. It is expected to produce neuromorphic intelligence that is comparable to human intelligence. For neuromorphic architecture, the neural system in the human brain has features of high parallelism, computational complexity, speed, and efficiency that the current general computer based on von Neumann architecture cannot match. Neuromorphic architectures aim to exploit the parallelism of neuromorphic hardware, leading to faster and more efficient execution. For large-scale systems, they can provide substantial energy efficiency and speed advantages compared to traditional processors, enabling real-time, large-scale implementations of brain-inspired models and algorithms. It successfully leads to significant improvements in computing energy consumption, scale, and efficiency by changing the design paradigms of the existing computing systems. In summary, it is believed that the significant progress made in this field will provide an excellent view of the development and revolution of artificial intelligence, which will bring significant changes to social life and industrial society.

Over the past few years, the authors have systematically investigated various aspects of neuromorphic intelligence, including computational models, learning algorithms, computing architectures, and system implementations. They have achieved a series of breakthrough results with significant potential for both scientific research and practical applications. The purpose of this book is to integrate all these research perspectives and achievements into an academic book, serving as a crucial reference for learners and researchers in the relevant application fields.

To this end, the style and approach of the book will be tailored to senior undergraduates or graduate students with a foundational understanding of artificial intelligence theory, computer architecture basics, and knowledge of digital circuit implementations. It is also suitable for researchers or practitioners with experience in neuromorphic computing, including algorithms and hardware implementations. Each chapter will begin with the necessary research background, followed by a focus on cutting-edge work, and conclude with a summary. In summary, this book will serve as a significant reference in the field of neuromorphic intelligence.

Tianjin, China   Shuangming Yang
Xi'an, China   Badong Chen

Contents

Acronyms

τ	Decay time constant
V	Membrane potential
R	Resistance constant factor
V_{th}	Threshold membrane potential
V_{reset}	Resting potential value
I_{app}	External stimulus input current
u	Recovery variable
a, b, c, d	Important parameters of Izhikevich neuron model
g_K	Maximum conductance of potassium channels
g_{Na}	Maximum conductance of sodium channels
g_L	Conductance of leaky channels
E_{Na}, E_K	Reversal potential of different ion channels
E_L	Reversal potential of leaky channels
m, n, h	Activation gating variables of the corresponding ion channels
ΔT	Integration step
C_m	Membrane capacitance
V^0	Membrane potential of soma
V^{0a}	Membrane potential of apical dendrite
V^{0b}	Membrane potential of basal dendrite
W^0	Synaptic weight in the input layer
Y	Synaptic weight of feedback synapses
b^0	Bias term
s^{input}	Filtered spiking activities in the input layer
s^1	Filtered spiking activities in the output layer
t^{input}	Spiking time of the input neuron
κ	Response kernel
τ_L, τ_s	Long- and short-time constants
Θ	Heaviside step function
$\delta(.)$	Nonlinear sigmoid function
ϕ_{max}	Maximum firing rate of neurons

L^0	Loss function based on basal dendrites
ϕ^{0*}	Target firing rate
a_s	Slope of the modified PWL function
b_i	Intercept of the modified PWL function
CF_{RE}	Error evaluation criterion
p	Image histogram
μ	Mean of cluster
σ^2	Variance of cluster
B	Bit width
si	Sign bit
FL	Fractional length
mb	Mantissa bit
R^e	Membrane resistance of the excitatory dendrite
R^i	Membrane resistance of the inhibitory dendrite
g^{inh}	Synaptic conductance of inhibitory dendrite
g^{exc}	Synaptic conductance of excitatory dendrite
$V^{den,i}$	Membrane potentials of inhibition dendrite
$V^{den,e}$	Membrane potentials of excitatory dendrite
V^{exc}	Reverse membrane potential of excitatory dendrite
V^{inh}	Reverse membrane potential of inhibitory dendrite
W^{in}	Synaptic weight of soma
W^{iin}	Synaptic weight of inhibitory dendrite
W^{ein}	Synaptic weight of excitatory dendrite
d^{input}	Delay of input synapses for soma
d^{iinput}	Delay of input synapses for inhibitory dendrite
d^{einput}	Delay of input synapses for inhibitory dendrite
d^{rec}	Delay of recurrent synapses for soma
$direc$	Delay of recurrent synapses for inhibitory dendrite
d^{erec}	Delay of recurrent synapses for inhibitory dendrite
Γ	Firing rate of neuron
τ_a	Adaptation time constant
$N(0,1)$	Zero average unit variance Gaussian distribution
w_0	Weight scaling factor
z	Neural spike train
O^{PPO}	Clipped surrogate objective of PPO
μ_f	Regularized hyperparameter
k	Total number of epochs
θ	Current policy parameter
θ_{old}	Fixed parameter
$L(\cdot)$	Loss function
s_e	Environment state

r_w	Reward
ξ	Real coordinate
ζ	Action vector
β	Sum of Dirac pulses
$E[\cdot]$	Expectation operator
$k_\sigma(\cdot,\cdot)$	Shift-invariant kernel function with bandwidth σ
$f(\cdot,\cdot)$	Joint probability density function
CT	Correntropy
$C\hat{T}$	Sample estimator of correntropy
$\hat{f}_E$	Kernel density estimation of probability density function
G_Σ	Gaussian function
H_2	Renyi's quadratic entropy
V_{ip}	Information potential
J_1	Instantaneous information potential
W_P	Length of the Parzen window
$Q\lceil\cdot\rceil$	Quantization operator
ρ_E	Desired distribution
ζ_d	Corresponding density for each peak
E^R	Restricted minimum error entropy
ϕ_N	Number for each quantization word
C	Quantization word
J_{R2}	Gradient based on entropy
X	Source variable
H	Compressive variable
$d(\cdot,\cdot)$	Distortion measurement function
$p(\cdot)$	Probability distribution
$p(\cdot|\cdot)$	Conditional probability distribution
$I(\cdot;\cdot)$	Mutual information
$R(\cdot)$	Rate distortion theory function
$Z(\cdot,\cdot)$	Probability normalization function
β	Lagrange parameter
$H_{SE}(\cdot)$	Shannon entropy
L_{ce}	Cross-entropy loss function
Ψ	Influence function of MMCC
$E(\cdot)$	Mathematical expectations
f_{target}	A predefined target firing rate
f	Mean firing rate of all the neurons
N_{total}	Total duration on a specific task
λ_f	Hyperparameter of the regularization of the firing rate
a_{th}	Adaptive firing threshold
A_{th}	Resulting threshold voltage

λ_v	Hyperparameter measuring membrane potential regularization
h	Hidden variable
e	Dynamics of the eligibility trace
ε	Eligibility vector
Ψ	Pseudo-derivative
τ_{out}	Timing constant of readout neurons
t_{refrac}	Refractory period duration
H_{ea}	Non-differentiable Heaviside function
η_{out}	Learning rate of outer loop
λ_f	Spike rate regularization
L_{HS}	Hilbert-Schmidt independence criterion
L_{TO}	TOIB principle
λ_v	Voltage regularization
t_{img}	Number of time steps per image
η	Learning rate
N_{HSNN}	Network size of HSNN
q_{ada}	Neuron fraction using adaptation
N_{batch}	Batch size for outer loop optimization
N_{PSNN}	Network size of the PSNN
τ_{LS}	Timing constant learning signals of readouts
$f_{tarPSNN}$	Target firing rate for PSNN
O^{PPO}	Clipped surrogate objective of PPO
ϑ	Current policy parameter
Ω	Channel number
$\dagger$	Pattern height
$\varkappa$	Pattern width
$O(\cdot)$	Time complexity
$[\Xi, \Upsilon]$	Hilbert space
$\mathbf{Z}^*$	Original learning objective of IB
υ	Learnable parameter
$\delta_D(x)$	Dirac delta function
g_b	Batch size
τ_{index}	Batch index
μ	Mean value
σ	Variance value
$\bar{\mu}$	Normalization of mean value
$\bar{\sigma}$	Normalization of variance value
w_{noise}	Noisy weight update value
Δw	Weight change
σ_{SN}	Standard variance of the Gaussian noise on synapses
σ_{MP}	Standard variance of the Gaussian noise on membrane

β_{uni}	Uniform noise density
z_{ray}	Rayleigh noise
$\alpha_{ray}, \beta_{ray}$	Critical parameters of the Rayleigh noise
f_{gs}	Formulation of the Gaussian noise
n_{KS}	Kernel size
n_{OFM}	Output feature map size
n_{IC}	Input channel number
n_{OC}	Output channel number
Q_α	Joint distribution
s_R	Firing rate
O	Representation of output spike information
g_W	Mapping result of network parameter
y_m	Probability of classification result
H^l	Input height
W^l	Input width
k^l	Filter height
C_i^l	Input channel size
C_o^l	Output channel size
E_{AC}	Power consumption of accumulating
E_{MAC}	Power consumption of multiplication-and-accumulate
L_{IB}	IB loss
L_{SNIB}	SNIB loss
L_{SO}	SOIB loss
x_{SPN}	Neural spikes over all time steps
x_{NEU}	Neuron number
$dest$	Destination node
S	Node address of the current router
CF_{BUR}	Cost function of buffer utilization
N_{VC}	Number of virtual channels per buffer
N_{PPB}	Number of packets per buffer
N_{PL}	Total number of physical links
t_{tot}	Total time from the first packet to the last packet
N_R	Total number of routers in NoC architecture
N_{BSS}	Buffer used in the node
N_{BSN}	Buffer used in the router
Q	Cost function for computational efficiency
t_{exp}	Experimental calculation time of the simulated system
t_{bio}	Biological activity time of the simulated system
$ceil(\cdot)$	Upward value of a number
CF_{ratio}	Ratio of IBFT to BFT two different NoC characteristics
CF_{upd}	Number of layers determines the parallelism

f_{PCI}	Path cost of the intermediate block
f_{hop}	Number of nodes to pass through during transmission
f_R	Average of all the time a packet spends in the router
f_{FIFO}	Time the packet spends in the FIFO in the transmission path
f_{PCSD}	Path cost for the source and target nodes
f_{src}	Time when the synaptic information is converted to a packet
f_{dst}	Time when the packet is restored to synaptic information
f_{PC}	Path cost on the 3D multicast network NoC
Y_{max}	Network length of 3D NoC
X_{max}	Network width of 3D NoC
Z_{max}	Network height of 3D NoC
C	Current node
D	Destination node
β_{grc}	Silicon synaptic unit
x_{lr}	Linear response criterion
ω	Angular frequency of the oscillation
$xsof$	Software computation result
$xhar$	Hardware computation result
x_{sof}	Average value of software computation result
x_{har}	Average value of hardware computation result
E_p	Mean spike events of the PKJ neurons
ω	Angular frequency of the oscillation
$A+$	Time difference between pre- and post-synaptic neuron firing
$\tau+$	Time window parameters in the process of synaptic strength
N_{Npipe}	Pipeline depth of the neuron calculation unit
N_{Spipe}	Pipeline depth of the synaptic current calculation unit
f_{max}	Maximum calculation frequency
τ_{ahp}	Conductance delay time
u^x	Membrane recovery variable
I^{syn}	Total synaptic current received by the neuron
τ_{Re}	Decay constant of the synaptic receptor
E_{Re}	Synaptic potential of the relevant receptor
g_x	Synaptic conductance
G_x	Maximum synaptic conductance
AC	Accumulation
AER	Address event representation
AGI	Artificial general intelligence
ANN	Artificial neuron network
ARM	Advanced reduced instruction set computer machine
ASC	ANN-to-SNN conversion
BFT	Butterfly fat tree

BG	Basal ganglia
BNTT	Batch normalization through time
BPTT	Back-propagation through time
CF	Climbing fiber
CIP	Cross-information potential
CMAC	Cerebellar model articulation controller
CMOS	Complementary metal-oxide-semiconductor
CN	Counter number
CNN	Convolutional neural network
CNU	Cerebellar neuron unit
CORDIC	Coordinate rotation digital computer
CPG	Central pattern generator
CPU	Central processing unit
CR	Conditioned response
DBS	Deep brain stimulation
DEP	Dendritic event-driven processing
DLC	Dedicated logic circuit
DVS	Dynamic vision sensor
EBC	Event-based camera
EC	Energy consumption
EEG	Electroencephalogram
EM9E	Embedded multiplier 9-bit element
EOMN	Extra-ocular motor neuron
FIFO	First in first out
FLOPs	Floating point operations
FPGA	Field programmable gate array
GAN	Generative adversarial network
GPe	Globus pallidus externus
GPi	Globus pallidus internus
GrC	Granular cell
HBP	Human brain project
HESFOL	Heterogeneous ensemble-based spike-driven few-shot learning
HH	Hodgkin-and-Huxley
HOSIB	High-order spike-based information bottleneck
HSNN	Hippocampus-inspired SNN
HSTC	High-speed Terasic connector
IB	Information bottleneck
IBFT	Improved butterfly fat tree
IF	Integrate-and-fire
IMP	Improved multi-information processor
ISI	Interspike interval

ITL	Information theoretic learning
KDE	Kernel density estimation
KL	Kullback-Leibler
LE	Logic element
LIB	Local information bottleneck
LIF	Leaky integrate-and-fire
LMS	Least mean square
LSM	Liquid state machine
LSTM	Long-term short-term memory
LTD	Long-term depression
LTP	Long-term potentiation
LUT	Lookup table
MAC	Multiplication-and-accumulation
MCC	Maximum correntropy criterion
MeMEE	Meta-learning with minimum error entropy
MF	Mossy fiber
MMCC	Mixture maximum correntropy criterion
MMSE	Minimum mean square error
MNIST	Modified national institute of standards and technology
MNP	Multi-compartment neuron processor
MoE	Mixture of experts
MP	Membrane potential
MSE	Mean squared error
MSIP	Multiple synaptic information processing
MSynOps	Million synaptic operations per second
MTSO	Multi-timescale optimization
MWM	Matrix-vector multiplication
NNI	Normalized noise intensity
NoC	Network on chip
PC	Purkinje cell
PDF	Probability density function
PF	Parallel fiber
PFC	Prefrontal cortex
PKJ	Purkinje
PLA	Piecewise linear approximation
PLL	Phase locked loop
PP	Plateau potential
PPO	Proximal policy optimization
PSNN	PFC-inspired SNN
P-STDP	Pair-based STDP
RAM	Random access memory

ReLU	Rectified linear unit
RKHS	Reproducing kernel Hilbert space
RMCC	Restricted maximum correntropy criterion
RMEE	Restricted minimum error entropy
RNN	Recurrent neural network
ROM	Read-only memory
RSNN	Recurrent spiking neural network
SAM	Self-adaptive multi-compartment model
SBP	Surrogate back-propagation
SCCU	Synaptic current computation units
SCP	Synaptic current processor
SDL	Segregated dendritic learning
SDRAM	Synchronous dynamic random access memory
SFOL	Spike-driven few-shot online learning
SGD	Stochastic gradient descent
SIBoLS	Spike-based information bottleneck with learnable state
SLM	Shift logic multiplier
sMNIST	Sequential modified national institute of standards and technology
SNIC	Saddle node on invariant circle
SNN	Spiking neural network
SoC	System-on-chip
SOIB	Second-order information bottleneck
SoPC	System on a programmable chip
SP	Salt-and-pepper
SpiNNaker	Spiking neural network architecture
STDP	Spike timing-dependent plasticity
STN	Subthalamic nucleus
Str	Striatum
SVM	Support vector machine
SynOps	Synaptic operations per second
TC	Thalamocortical
tdBN	Threshold-dependent batch normalization
TLIF	Two-compartment LIF
TMB	Total memory bit
TOIB	Third-order information bottleneck
T-STDP	Triplet-based STDP
VAE	Variational autoencoders
VIB	Variational information bottleneck
VN	Vestibular nucleus
VNN	Visual-pathway-inspired neural network
WIB	Without information bottleneck
WTA	Winner-take-all

Introduction of Neuromorphic Intelligence

1.1   Neuromorphic Intelligence

Neuromorphic intelligence is a rapidly evolving field of research that draws inspiration from the structure and function of the brain to develop novel computational models and systems. This objective relies on the innovation of computational models and hardware systems. In terms of computational models, neuromorphic intelligence utilizes spiking neural networks (SNNs), which encode information through the precise timing of electrical spikes [1]. The learning algorithms employed by SNNs enable networks to dynamically adjust connections and synaptic strengths in response to input, facilitating the recognition of patterns in data [2, 3]. In terms of hardware systems, neuromorphic architectures differ from traditional computing systems that depend on von Neumann architecture and sequential processing. Neuromorphic systems, on the other hand, leverage parallel processing and event-driven communication [4, 5]. This approach is designed to execute tasks commonly associated with human intelligence, including perception, reasoning, decision-making, and learning, in a more human-like and energy-efficient manner. Neuromorphic systems allows neuromorphic computing to perform neuromorphic intelligence in a highly parallel and energy-efficient manner. As illustrated in Fig. 1.1, neuromorphic systems extracts the inspiration from human brain, and can enhance the level of neuromorphic intelligence. Furthermore, neuromorphic intelligence will facilitate the understanding of human brain intelligence. Neuromorphic intelligence has the potential to revolutionize many fields, including artificial intelligence, robotics, and cognitive neuroscience [6–9]. It could provide new insights into the workings of the brain and lead to the development of more efficient and robust systems with lower power consumption towards artificial general intelligence (AGI).

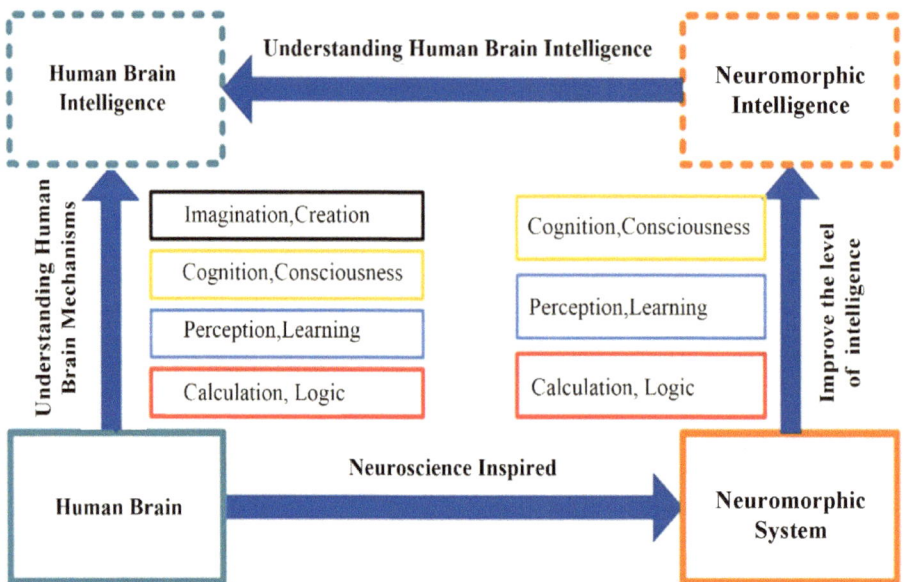

Fig. 1.1 Human brain intelligence and neuromorphic intelligence

1.2   Spike-Based Learning for Neuromorphic Computing

1.2.1   Spiking Neuron Models

SNNs are a kind of brain-inspired neural networks considered as the third generation of artificial neural networks (ANNs) [10–13]. Unlike traditional ANNs, the core concept of SNNs is to emulate the way neurons generate spikes in the biological nervous system. The fundamental units of SNNs are spiking neurons, which mimic the behavior of biological neurons in generating spikes. In SNNs, information is transmitted in the form of discrete spikes instead of continuous activations, with the timing and intensity of each spike regarded as crucial computational elements. This event-driven transmission approach makes SNNs closer to the functioning of the biological neural system, where neurons also transmit information in the form of spikes. SNNs provide a novel paradigm that closely related to the working mechanism of the brain, offering a new perspective for interdisciplinary research in neuroscience and artificial intelligence.

Spiking neuron models constitute the basic components of SNNs, describing how neurons generate output spikes based on the temporal patterns of input spikes. Common spiking neuron models include the integrate-and-fire (IF) model, Izhikevich model [14] and the leaky integrate-and-fire (LIF) model [15]. These models take into account the changes in membrane potential of neurons and how input spikes lead to neuron activation. The design of spiking neuron models focuses on simulating the electrophysiological characteristics of biological neurons, aiming to achieve more biologically plausible neural network simulations. We briefly introduce some representative kinds of spiking neuron models below.

(1) IF neuron model

The IF model as a typical neuron model can be described by the following formulation as

$$\tau_m \frac{dV}{dt} = -V + I(t)$$
$$if\ V \geq V_{th},\ then\ V = V_{reset}$$

(1.1)

where τ_m is the membrane time constant for the voltage decay, and $I(t)$ represents the input current. When the membrane potential is larger than the threshold voltage, the neuron omits a spike and restores the membrane potential to its resting value. When the input external stimulus current is large enough, the membrane potential accumulates until it reaches the membrane potential threshold, while the membrane potential gradually decays in the absence of external current stimulus and finally reaches the resting potential value.

(2) LIF neuron model

The LIF model as a simple neuron model can be described by the following mathematical expression as

$$C_m \frac{dV}{dt} = -G_L(V - E_L) + I_{app}$$
$$if\ V \geq V_{th},\ then\ V = V_{reset}$$

(1.2)

where C_m is the membrane surface capacitance, and G_L is the conductance. E_L is the equilibrium potential, and I_{app} is the external stimulus input current.

(3) Izhikevich neuron model

The Izhikevich model is based on neuronal dynamics theory and it is proposed by E.M. Izhikevich, which is able to simulate different complex firing patterns in cortical neurons, such as regular spiking and bursting [14]. The dynamic properties of the Izhikevich model are controlled by two variables by the following expressions as

$$\begin{cases} \dfrac{dV}{dt} = 0.04V^2 + 5V + 140 - u + I_{app} \\ \dfrac{du}{dt} = a(bV - u) \end{cases} \tag{1.3}$$

$$if \ V \geq 30mV, \ then \begin{cases} V \leftarrow c \\ u \leftarrow u + d \end{cases}$$

where V represents the neuronal membrane potential, and u is the recovery variable, which plays a negative feedback role on the membrane potential. I_{app} is the external stimulus input current. a, b, c, and d are the four important parameters of the model, which determine the firing frequency and firing patterns of the neuron.

In addition to the models mentioned above, other models include the Hodgkin-Huxley (HH) model [16], Morris-Lecar model [17], Fitzhugh-Nagumo model [18], and Hindmarsh-Rose model [19].

It is worth mentioning that the aforementioned models can be used to construct spiking neurons with a single compartment as well as spiking neurons with multiple compartments, incorporating structures such as dendrites and axons [20]. The detailed introduction of multi-compartment neuron models with dendrites and their learning methods will be provided in Chap. 2.

1.2.2   Typical Learning Approaches of SNNs

In order to realize efficient learning of SNNs for neuromorphic intelligence, there are several learning approaches, such as ANN-to-SNN conversion (ASC) method and direct training method of SNNs. The learning methods are further introduced in detail as below.

(1) ASC method

ASC [21–25] refers to the process of converting ANNs to SNNs. ANNs are typically designed to work with continuous signals and are composed of simple computational units (neurons) that perform weighted sum and non-linear activation functions. ANNs are typically trained using back-propagation, which is a supervised learning algorithm that requires the use of continuous, differentiable activation functions. However, SNNs use discrete, non-differentiable spike trains to represent information, which makes them difficult to train using back-propagation. The conversion process involves mapping the weights and activations of an ANN to the parameters of an SNN. There are several methods proposed for ASC, including: ***Weight Normalization***. In this method, the weights of the ANN are normalized such that they fall within a specific range, and then they are used to define the synaptic strengths in the SNN. The spiking threshold for each neuron is also normalized based on the distribution of weights. This method has been shown to produce accurate SNNs with relatively low computational cost. ***ReLU to Spike***. This method involves replacing the rectified linear unit (ReLU) activation function in the ANN with a spiking function that generates spike events when the input is positive. The synaptic strengths in the SNN are then defined based on the weights of the ANN. This method has been shown to produce SNNs with high accuracy and good energy efficiency. ***Temporal Coding***. In this method, the time intervals between the spike events are used to encode information in the SNN. The weights of the ANN are mapped to the synaptic strengths between the spiking neurons, and the spike timings are adjusted based on the input signal. This method has been shown to produce SNNs that are robust to noise and can handle complex temporal patterns. ***Binarization***. This method involves binarizing the weights and activations of the ANN to 0 and 1 values, and then mapping them to the spiking weights and firing events in the SNN. This method has been shown to produce SNNs that are highly efficient in terms of memory and computation.

(2) Direct training of SNNs

Direct training methods for SNNs are a class of learning algorithms that directly optimize the network weights using gradient descent, without requiring the conversion of an ANN to an SNN. These methods typically use surrogate gradients, which are differentiable approximations of the non-differentiable spiking activation functions. Direct training methods have the advantage of being able to train SNNs in an end-to-end manner, without the need for separate training of ANNs and conversion to SNNs. However, they can be computationally expensive, and may require the use of specialized hardware to efficiently simulate spiking neurons.

Direct training methods of SNNs involve optimizing the parameters of the network directly using spike-based learning rules [26–30]. There are several direct training methods for SNNs. ***Spike-Timing-Dependent Plasticity (STDP)***. STDP is a biologically-inspired learning rule that adjusts the synaptic strengths between neurons based on the timing of their spikes. When a presynaptic neuron fires before a postsynaptic neuron, the strength of the synapse is increased, and when the postsynaptic neuron fires before the presynaptic neuron, the strength of the synapse is decreased. STDP has been used to train SNNs for various tasks, including image classification and speech recognition. ***Surrogate Gradient Learning***. Surrogate gradient learning is a family of learning algorithms that allow back-propagation training of SNNs. In surrogate gradient learning, a differentiable function is used to approximate the non-differentiable spiking function of the neurons. This allows the use of back-propagation techniques to train the network. Surrogate gradient learning has been used to train SNNs for tasks such as object recognition and speech recognition. ***Temporal Contrastive Learning***. Temporal contrastive learning is a spike-based learning rule that optimizes the parameters of the network to maximize the contrast

between the responses of the network to different input stimuli. This is achieved by training the network to recognize the order of events in the input data. Temporal contrastive
learning has been used to train SNNs for various tasks, including speech recognition
and robot control. ***Feedback Alignment***. Feedback alignment is a learning algorithm that
replaces the weight updates that would normally be provided by back-propagation with
fixed random matrices. This allows the network to be trained using only feedforward
and lateral connections. Feedback alignment has been used to train SNNs for tasks such
as object recognition and motor control. ***Direct Feedback Alignment***. Direct feedback
alignment is an extension of feedback alignment that allows the network to receive error
signals directly from the output layer, rather than relying on random feedback weights.
This can improve the performance of the network and make it easier to train. Direct feedback alignment has been used to train SNNs for tasks such as image classification and
speech recognition.

1.3   Neuromorphic Architectures

Neuromorphic architectures [31–33] are inspired by the way neurons in the brain work,
and they aim to create AGI systems that are more energy-efficient, faster, and capable
of performing tasks that traditional computer architectures cannot. Neuromorphic architectures can perform tasks such as object recognition, pattern recognition, and speech
recognition. Another important aspect of neuromorphic architectures is to simulate the
behavior of neurons and synapses. The mesh and torus architectures [34, 35] are two classic types of neuromorphic architectures in terms of network-on-chip (NoC) design. The
mesh architecture is a two-dimensional grid of processing nodes, where each node is connected to its four nearest neighbors by communication channels. The nodes in the mesh
can be designed to simulate neurons, synapses, or both, and they can be programmed to
perform various types of computations. The mesh architecture is relatively simple. However, its regular architecture can limit its flexibility and scalability. The torus architecture
is similar to the mesh architecture, but it is arranged in a three-dimensional grid. This
architecture allows for more efficient communication between nodes, since each node is
connected to its six nearest neighbors in the same layer, as well as to its four neighbors
in the adjacent layers. The torus architecture is well-suited to simulating large-scale neural networks, since it can be easily scaled up by adding more layers or nodes. However,
its implementation can be more complex than the mesh architecture, and it may require
specialized hardware to achieve high performance. Both the mesh and torus architectures
have been used in a variety of neuromorphic systems, including the Neurogrid system [36]
developed at Stanford University, which uses a mesh architecture to simulate millions of
neurons in real-time, and the SpiNNaker system [37], which uses a torus architecture to
simulate up to a million neurons in real-time. It's worth noting that there are many other
types of neuromorphic architectures beyond the mesh and torus architectures, including

more complex architectures like hierarchical networks and non-regular networks. Each architecture has its own advantages and disadvantages, and the choice of architecture depends on the specific requirements of the application.

Neuromorphic architectures are still an active area of research, and there are many challenges that need to be addressed before they can be widely used. With continued research and development, neuromorphic architectures have the potential to revolutionize AGI and neuroscience.

1.4   Neuromorphic Systems and Event-Based Sensors

1.4.1   Neuromorphic Systems

Neuromorphic systems represent a paradigm shift in the field of artificial intelligence, drawing inspiration from the intricate architecture and dynamic functionality of the human brain [6, 36–38]. Rooted in the burgeoning field of neuromorphic engineering, these systems leverage the emulation of neural circuits, synapses, and neuronal dynamics to enable machines to process information in a manner akin to biological brain. Unlike traditional von Neumann architecture, neuromorphic systems embrace parallelism and distributed processing, mirroring the brain's ability to perform complex tasks simultaneously. This departure from conventional computing paradigms offers the potential for enhanced efficiency, adaptability, and robustness in handling cognitive tasks. The hardware implementations of neuromorphic systems span various platforms, including analog circuits, digital circuits, and emerging technologies such as memristive devices. These platforms aim to replicate the behavior of biological neurons and synapses, fostering the creation of efficient and scalable neuromorphic architectures. The applications of neuromorphic systems are multifaceted, encompassing domains such as robotics, image and speech recognition, cognitive computing, and neuroscience research. With their unique blend of neurobiological inspiration and cutting-edge hardware implementations, these systems hold great promise for addressing complex cognitive tasks, pushing the boundaries of computational capabilities, and contributing to a deeper comprehension of the intricate workings of the human brain. Generally, there are four types of hardware implementation types for neuromorphic systems, including analog, digital, digital-analog, and emerging technology implementations.

(1) Implementation methods
(a) Analog method

Analog implementation in neuromorphic systems involves utilizing analog circuits to emulate the behavior of biological neurons and synapses [39–41]. This approach leverages

continuous signals and analog components to mimic the dynamics of neural processing. Analog circuits use continuous voltage signals to represent information, contrasting
with digital systems relying on discrete voltage levels. Neurons in analog neuromorphic
systems are often modeled using differential equations describing the analog behavior
of membrane potentials and synaptic currents. Capacitors or other analog components
are often employed to implement memory-like properties in synapses, enabling them
to retain information over time. Transistors and resistors play crucial roles in shaping
the characteristics of analog circuits, including the dynamics of neurons and synapses.
Analog implementation in neuromorphic systems offers several advantages. Firstly, analog circuits provide a closer approximation to the continuous and dynamic behavior of
biological neurons, contributing to the biological realism of neuromorphic systems. Secondly, analog circuits often exhibit lower power consumption compared to their digital
counterparts, making them suitable for power-critical applications. Thirdly, analog circuits
naturally support parallel processing and continuous computation, potentially leading to
faster and more efficient computations. Fourthly, analog circuits can operate in real-time,
essential for applications requiring immediate and dynamic responses, such as robotics or
sensory processing. Fifthly, analog neuromorphic systems integrate well with sensor data,
enhancing their applicability to sensory-driven tasks. Lastly, analog circuits can implement synaptic plasticity, crucial for developing neuromorphic systems that can adapt and
learn from experience. While analog implementation offers these advantages, it's essential
to note challenges such as sensitivity to noise, precision limitations, and potential difficulties in scaling to large systems. Ongoing research aims to address these challenges and
further enhance the capabilities of analog neuromorphic systems.

(b) Digital method

Digital implementation in neuromorphic systems utilizes digital circuits and algorithms to
replicate the behavior of biological neurons and synapses [42–44]. In contrast to analog
implementations that utilize continuous signals, digital neuromorphic systems operate on
discrete signals represented by binary values. Neurons in digital neuromorphic systems
are frequently modeled using SNNs, and synaptic connections are typically represented
by digital weights, indicating the strength of connections between neurons. There are
several advantages associated with digital neuromorphic systems. Firstly, digital circuits
enable precise computation by operating with discrete signals. This precision is crucial
for tasks requiring accurate numerical representation and calculation, contributing to the
overall accuracy of the neuromorphic system. Secondly, digital circuits provide flexibility in logic design and programming, allowing for easier implementation of changes to
the system's architecture, parameters, or algorithms compared to analog implementations.
This adaptability is valuable for evolving requirements and research findings. Thirdly,
digital neuromorphic systems are well-suited for scalability, facilitating the design of

large-scale systems to handle complex tasks and process extensive datasets. Fourthly, digital signals exhibit greater resilience to noise compared to analog signals, ensuring signal integrity and promoting more reliable information processing within the neuromorphic system. Fifthly, digital neuromorphic systems can be seamlessly integrated into existing digital platforms and technologies, enhancing overall system functionality through compatibility with other digital systems and components. Lastly, digital circuits can efficiently implement complex algorithms, particularly beneficial for advanced neural network architectures, learning algorithms, and intricate processing tasks in neuromorphic systems. The versatility of digital circuits allows them to handle a wide range of computational tasks, making them advantageous for neuromorphic systems designed to perform diverse functions, from pattern recognition to decision-making. While digital circuits offer advantages in terms of precise computation and scalability, it's important to note that they may consume more power compared to analog circuits. Ongoing research is focused on addressing challenges related to power consumption and developing scalable architectures in the ongoing development of digital neuromorphic systems.

(c) Mixed analog-digital method

A mixed analog and digital implementation method in neuromorphic systems involves combining analog and digital components to efficiently simulate the intricate functions of the brain [45–47]. This approach capitalizes on the strengths of analog circuits for specific tasks and the precision and flexibility of digital circuits for others. Analog circuits can replicate the behavior of biological neurons and synapses, allowing for a more accurate representation of synaptic inputs and outputs. Digital components play a crucial role in the precise control of the overall system. Digital control logic manages parameter configurations, controls information flow, and implements higher-level algorithms. Digital components also handle specific computation tasks requiring high precision, such as complex mathematical operations and sophisticated algorithms. In a mixed implementation, individual processing units may comprise both analog and digital elements, optimizing the strengths of each type and creating a versatile and efficient system. Appropriately designed analog circuits contribute to energy efficiency, surpassing their digital counterparts. The mixed implementation method allows for a flexible architecture that can scale according to application requirements, essential for constructing larger and more intricate neuromorphic networks. The mixed analog and digital approach aims to strike a balance between the benefits of analog and digital technologies, resulting in a neuromorphic system proficient in cognitive tasks while maintaining the precision and adaptability needed for diverse applications. This approach remains an active area of research as scientists and engineers strive to develop more capable and energy-efficient neuromorphic computing systems.

(d) Emerging technology method

Emerging technologies within neuromorphic systems are in a constant state of evolution, encompassing various techniques in this field [48–51]. There are some typical types of emerging technologies for neuromorphic systems. Firstly, memristors emulate the synaptic behavior observed in biological systems, enabling the storage and retrieval of information [52, 53]. Memristors exhibit significant promise in neuromorphic computing, particularly due to their capacity to efficiently replicate synaptic plasticity. The integrated architecture for storage and computation prevents redundant operations, ensuring the avoidance of repeated reading of weight parameters during execution. The compact size, high speed, and low energy consumption of memristors offer substantial advantages in constructing 3D stacks and developing highly efficient and dense neuromorphic computing platforms. Moreover, the commendable conductance plasticity of memristors satisfies the fundamental requirements for constructing neuromorphic computing platforms that support online learning functions. Secondly, optical computing harnesses light signals instead of conventional electronic signals [54, 55]. Within neuromorphic systems, optical computing has the potential to facilitate faster and more energy-efficient processing. Ongoing exploration of photonic synapses and neurons further enhances their viability for integration into neuromorphic systems. Thirdly, the principles of quantum computing are under investigation for their applicability in neuromorphic systems [56, 57]. Quantum neuromorphic computing combines quantum mechanics with neural network architectures, potentially providing advantages in processing intricate patterns and solving specific problems more efficiently. Fourthly, biohybrid systems integrate living cells or biological components with artificial neural networks [58, 59] These systems aspire to merge the adaptability and learning capabilities inherent in biological systems with the computational efficiency offered by artificial neuromorphic circuits. Researchers and engineers are actively engaged in exploring these emerging technologies to augment the capabilities of neuromorphic systems across diverse applications, including artificial intelligence, robotics, and sensor networks.

(2) Representative work

Some representative neuromorphic processors and systems are briefly introduced them as below.

(a) TrueNorth

TrueNorth is a neuromorphic processor developed by IBM and is built to work with SNNs [60]. TrueNorth operates in an event-driven manner, meaning computations are performed only when necessary, leading to energy efficiency. TrueNorth's architecture is composed of a two-dimensional array of cores, each containing a set of neurons and synapses. It has a flexible and scalable structure that allows for the creation of large-scale neuromorphic systems. TrueNorth is designed for a variety of applications, including cognitive computing, machine learning, and artificial intelligence. Its low power consumption makes it suitable for edge computing and Internet of Things (IoT) devices. TrueNorth has been used in various research projects and collaborations, and has contributed to the advancement of neuromorphic computing, offering a platform for researchers and developers to experiment with brain-inspired architectures.

(b) SpiNNaker

SpiNNaker stands as a groundbreaking neuromorphic computing system developed by the University of Manchester. The term "SpiNNaker" itself signifies "Spiking Neural Network Architecture," and its primary objective is to replicate the intricate operations of the human brain through a massively parallel architecture [37]. The system's architecture comprises a two-dimensional array of processing nodes or cores, each adept at modeling thousands of neurons and synapses. Operating in real-time, SpiNNaker empowers researchers to simulate and scrutinize large-scale neural networks with a level of biological fidelity. Designed with flexibility and scalability at its core, SpiNNaker provides researchers with the freedom to create and experiment with expansive neuromorphic systems. Its architecture facilitates the modeling of diverse neural network structures and behaviors. SpiNNaker represents a pioneering effort in the realm of neuromorphic computing, furnishing researchers with a potent platform to delve into the complexities of neural networks. It not only provides insights into the fundamental principles of information processing in the brain but also contributes significantly to the advancement of artificial intelligence. The system has been actively employed in diverse research projects, focusing on the study of neural systems, sensory processing, and the development of models for understanding brain-related disorders. Moreover, SpiNNaker has been instrumental in simulating the behavior of robotic systems, establishing a crucial link between neuromorphic research and practical applications in the fields of robotics and control systems.

(c) Loihi

Loihi, developed by Intel, stands as a neuromorphic chip [61]. Engineered to replicate the structure and function of the human brain, Loihi utilizes SNNs for information processing. The chip comprises thousands of artificial neurons and synapses working in parallel, facilitating the execution of complex computations with minimal power consumption. A standout feature of Loihi is its event-driven architecture, ensuring computations occur only when necessary, thereby enhancing energy efficiency. Loihi is characterized by its programmability and adaptability, providing researchers the flexibility to explore and implement diverse neural network models. Well-suited for tasks like pattern recognition, machine learning, and other cognitive computing applications, the chip is designed to enable real-time learning and adaptation. This attribute makes Loihi a valuable tool for developing intelligent systems capable of learning and responding to dynamic environments. Intel's Loihi chip has found applications in various research and development projects, playing a crucial role in advancing neuromorphic computing and delving into new frontiers of artificial intelligence. With its innovative architecture and capabilities, Loihi stands out as a significant player in the expanding field of neuromorphic computing.

(d) Neurogrid

Neurogrid is a neuromorphic hardware platform developed by researchers at Stanford University that is designed to simulate the activity of large-scale neural networks in real time [36]. It addresses the challenge of efficiently simulating the complex interactions of a large number of neurons, making it a valuable tool for researchers studying neural circuits and seeking to understand the principles of brain function. Neurogrid utilizes custom-designed analog circuits, which are more power-efficient compared to traditional digital circuits when simulating the behavior of neurons and synapses. It has the potential to accelerate progress in the field of neuroscience by providing a powerful tool for investigating the neural basis of cognition and behavior. Additionally, Neurogrid provides a platform for researchers to experiment with neural network models, investigate the principles of neural computation, and potentially develop more efficient neuromorphic algorithms.

(e) Tianjic

Tianjic is developed by researchers at Tsinghua University, and it represents an effort to create a brain-inspired computing architecture [62]. Tianjic's architecture supports a wide range of neural network models, including convolutional neural networks (CNNs), recurrent neural networks (RNNs), and SNNs. It is highly energy-efficient, consuming only a few watts of power per chip. The architecture is also highly configurable, allowing researchers to adapt its behavior to suit different applications and tasks. One of the key advantages of Tianjic is its ability to perform real-time processing of sensory data. Tianjic is intended for a variety of applications, including artificial intelligence, machine learning, and robotics. Its architecture is expected to be particularly efficient for tasks that involve pattern recognition and sensorimotor integration. The energy-efficient design of Tianjic also makes it ideal for use in devices that operate on limited power, such as mobile devices and IoT devices.

(f) BrainScaleS

BrainScaleS constitutes a neuromorphic computing platform that emerged as part of the Human Brain Project, a comprehensive European research initiative devoted to advancing our comprehension of the brain and fostering the development of brain-inspired technologies [63]. Specifically, BrainScaleS directs its focus towards the hardware implementation of neuromorphic systems, offering a robust platform for the simulation of expansive SNNs [64]. The distinctive design of BrainScaleS incorporates a wafer-scale integration strategy, where numerous neuromorphic chips are seamlessly integrated onto a singular wafer. This configuration facilitates parallel processing on a substantial scale, enabling the simultaneous simulation of a vast number of neurons and synapses. The platform leverages analog circuits to replicate the intricate behavior of neurons and synapses, demonstrating a well-suited capability for emulating the continuous and dynamic nature inherent in biological neural processes. An essential objective of BrainScaleS is to realize real-time simulation capabilities for large-scale neural networks. This pursuit proves critical for in-depth examinations of the dynamic aspects within neural circuits and applications necessitating low-latency processing. By providing an innovative and efficient infrastructure, BrainScaleS contributes significantly to the exploration of neuromorphic research, fostering a deeper understanding of neural computation and offering potential insights for the development of brain-inspired computing in artificial intelligence.

(g) Braindrop

Braindrop, developed at Stanford University, is a neuromorphic computing platform engineered to replicate the intricate functionalities of the human brain through the hardware implementation of SNNs [65]. Employing a mixed-signal architecture that seamlessly integrates digital and analog elements, Braindrop signifies a pioneering leap forward as the inaugural neuromorphic system tailored for programming at an elevated level of abstraction. Diverging from earlier neuromorphic systems, which necessitated hardware expertise for neurosynaptic-level programming, Braindrop distinguishes itself with a distinctive approach. Its computations are meticulously articulated as interconnected nonlinear dynamical systems, effortlessly transposed to hardware through an automated process. This innovative methodology not only capitalizes on the dynamic computational primitives inherent in Braindrop's subthreshold analog circuits but also mitigates disparities and temperature-related sensitivities in their responses at the network level. This distinctive approach facilitates the efficient and adaptable execution of SNN computations. Braindrop is engineered to conduct large-scale simulations of SNNs, positioning it as an apt choice for investigating intricate neural circuits and behaviors.

(h) DYNAP

DYNAP (dynamic neuromorphic asynchronous processors) [66] comprises a suite of solutions developed by SynSence, a company associated with the University of Zurich. The company features a patented event-routing technology facilitating communication between cores and has introduced several neuromorphic processors: DYNAP-SE2, DYNAP-SEL, and DYNAP-CNN. While the DYNAP-SE2 and DYNAP-SEL chips are non-commercial projects under development by neuroscientists as research tools, DYNAP-CNN is positioned as a commercial chip designed for the efficient execution of CNNs converted to SNNs. Unlike the analog computing and digital communication employed in the research-focused DYNAP-SE2 and DYNAP-SEL chips, DYNAP-CNN operates as a fully digital solution. Designed for feed-forward, recurrent, and reservoir networks, DYNAP-SE2 is utilized by researchers to explore SNN topologies and communication models. On the other hand, the key features of the DYNAP-SEL chip include on-chip learning support and extensive fan-in/out network connectivity. The DYNAP-SEL chip incorporates five cores, with only one core featuring plastic synapses. Released with a development kit in 2021, the DYNAP-CNN chip is optimally paired with event-based sensors and is suitable for image classification tasks. In inference mode, the chip can execute an SNN converted from a CNN, provided there are no more than nine convolutional or fully connected layers and no more than 16 output classes. Notably, on-chip learning is not supported, and the original CNN must be initially created with PyTorch and trained using classical methods. Additionally, leveraging the Sinabs.ai framework (an open-source PyTorch-based library), the convolutional network can be converted to a spiking form for execution on DYNAP-CNN in the inference mode.

1.4.2 Event-Based Sensors

Event-based sensors are a type of sensors that operate similar to biological systems by detecting changes in the environment and transmitting information through spikes or events. They have gained significant popularity in neuromorphic computing due to their low power consumption, high temporal resolution, and robustness to noise. The types of event-based sensors include dynamic vision sensor (DVS), event-based camera (EBC), neuromorphic audio sensors, and event-based gas sensors. DVS is a type of event-based sensor that detects changes in luminance in a scene and outputs a stream of events indicating the location, polarity, and time of occurrence of each change [67]. The sensor has been widely used in robotic applications, such as object recognition, tracking, and localization. The EBC is a type of camera that captures events in the form of pixel-level changes in brightness, rather than capturing images at fixed intervals like traditional cameras [68]. The camera is useful in applications that require high temporal resolution and low power consumption, such as high-speed tracking, event-based motion detection, and visual servoing. The EBC is also less sensitive to motion blur and has a higher dynamic range than traditional cameras. Neuromorphic audio sensors are sensors that mimic the functioning of the cochlea in the ear, detecting sound waves through a series of filters and generating spikes that represent the frequency and intensity of the sound [69]. Neuromorphic audio sensors have been used in applications such as speech recognition, sound localization, and acoustic event detection. Event-based gas sensors detect changes in the concentration of gases in the environment and generate events that represent the concentration level and time of occurrence [70, 71]. Event-based gas sensors have been used in applications such as air quality monitoring, gas leak detection, and industrial process control. In summary, event-based sensors offer several advantages over traditional sensors in neuromorphic computing, such as low power consumption, high temporal resolution, and robustness to noise [70–72].

1.5 Organization of the Book

The authors have systematically developed novel approaches for neuromorphic intelligence, including algorithms, architectures, and large-scale systems. Algorithms are used to train the SNNs towards better performance, and architectures aim at deploying SNNs in more efficient manners. Large-scale systems are implemented based on the designed architectures. The work is a vital integration of artificial intelligence and neuroscience, and has great practical application potential towards AGI. The purpose of this book is to combine all these perspectives and results in a single academic book that will become an essential reference for researchers and practitioners in relevant application fields. The topics of this book are divided into the following chapters, and are introduced as below.

This chapter reviews the basic aspects of neuromorphic intelligence, including spike-based learning approaches, neuromorphic architectures, neuromorphic systems, and event-based sensors for neuromorphic computing.

Chapter 2 introduces spike-based learning methods with multi-compartment model along with the biological background. Credit assignment by spiking dendrites and spike-based learning with self-adaptive multi-compartment model are introduced in detail.

Chapter 3 focuses on spike-based learning methods with information theory, including maximum correntropy criterion, minimum error entropy criterion, and information bottleneck. Robust spike-driven few-shot learning, robust spike-driven meta-learning, as well as robust deep SNNs' learning approaches, are provided.

Chapter 4 studies the scalable architectures of neuromorphic system. Classical mesh and torus architectures are first introduced, and a series of novel architectures are then provided, including IBFT, 3D mesh, and fault-tolerant mesh architectures.

Chapter 5 summarizes large-scale digital neuromorphic systems, including existing work, as well as LaCSNN, CerebelluMorphic, and BiCoSS neuromorphic systems developed by the authors. System design and experiments are both presented.

References

1. Ghosh-Dastidar S, Adeli H. Spiking neural networks. Int J Neural Syst. 2009;19(04):295–308.
2. Tavanaei A, Ghodrati M, Kheradpisheh SR, et al. Deep learning in spiking neural networks. Neural Netw. 2019;111:47–63.
3. Kasabov N, Dhoble K, Nuntalid N, et al. Dynamic evolving spiking neural networks for on-line spatio- and spectro-temporal pattern recognition. Neural Netw. 2013;41(Complete):188–201
4. Roy K, Jaiswal A, Panda P. Towards spike-based machine intelligence with neuromorphic computing. Nature. 2019;575(7784):607–17.
5. Furber S. Large-scale neuromorphic computing systems. J Neural Eng. 2016;13(5):051001.
6. Ivanov D, Chezhegov A, Kiselev M, et al. Neuromorphic artificial intelligence systems. Front Neurosci. 2022;16:1513.
7. Bartolozzi C, Indiveri G, Donati E. Embodied neuromorphic intelligence. Nat Commun. 2022;13(1):1024.
8. Aitsam M, Davies S, Di Nuovo A. Neuromorphic computing for interactive robotics: a systematic review. IEEE Access. 2022.
9. Davies M, Wild A, Orchard G, et al. Advancing neuromorphic computing with loihi: a survey of results and outlook. Proc IEEE. 2021;109(5):911–34.
10. Yamazaki K, Vo-Ho VK, Bulsara D, et al. Spiking neural networks and their applications: a review. Brain Sci. 2022;12(7):863.
11. Nunes JD, Carvalho M, Carneiro D, et al. Spiking neural networks: a survey. IEEE Access. 2022;10:60738–64.
12. Wang X, Lin X, Dang X. Supervised learning in spiking neural networks: a review of algorithms and evaluations. Neural Netw. 2020;125:258–80.
13. Lobo JL, Del Ser J, Bifet A, et al. Spiking neural networks and online learning: an overview and perspectives. Neural Netw. 2020;121:88–100.
14. Izhikevich EM. Simple model of spiking network. IEEE Trans Neural Netw. 2004;15(5):1063–70.
15. Rast AD, Galluppi F, Jin X, et al. The leaky integrate-and-fire neuron: a platform for synaptic model exploration on the spinnaker chip. In: The 2010 international joint conference on neural networks (IJCNN). IEEE;2010. p. 1–8.
16. Häusser M. The Hodgkin-Huxley theory of the action potential. Nat Neurosci. 2000;3(11):1165–1165.
17. Tsumoto K, Kitajima H, Yoshinaga T, et al. Bifurcations in Morris-Lecar neuron model. Neurocomputing. 2006;69(4–6):293–316.
18. Rocsoreanu C, Georgescu A, Giurgiteanu N. The FitzHugh-Nagumo model: bifurcation and dynamics. Springer Science & Business Media;2012.
19. Storace M, Linaro D, de Lange E. The Hindmarsh–Rose neuron model: bifurcation analysis and piecewise-linear approximations. Chaos: Interdiscip J Nonlinear Sci. 2008;18(3).
20. Halnes G, Augustinaite S, Heggelund P, et al. A multi-compartment model for interneurons in the dorsal lateral geniculate nucleus. PLoS Comput Biol. 2011;7(9): e1002160.
21. Jiang H, Anumasa S, De Masi G, et al. A unified optimization framework of ANN-SNN conversion: towards optimal mapping from activation values to firing rates. In: International conference on machine learning. PMLR;2023. p. 14945–74.
22. Gao H, He J, Wang H, et al. High-accuracy deep ANN-to-SNN conversion using quantization-aware training framework and calcium-gated bipolar leaky integrate and fire neuron. Front Neurosci. 2023;17:1141701.

23. Diehl PU, Zarrella G, Cassidy A, et al. Conversion of artificial recurrent neural networks to spiking neural networks for low-power neuromorphic hardware. In: 2016 IEEE international conference on rebooting computing (ICRC). IEEE;2016.

24. Wang B, Cao J, Chen J, et al. A new ANN-SNN conversion method with high accuracy, low latency and good robustness. In: Proceedings of the thirty-second international joint conference on artificial intelligence, IJCAI-23;2023. p. 3067–75.

25. Diehl S, Cook M. Conversion of artificial neural networks to spiking neural networks with rank-order coding. 2015.

26. Diehl P, Cook M. Unsupervised learning of digit recognition using spike-timing-dependent plasticity. Front Comput Neurosci. 2015;9:1662–5188.

27. Neftci EO, Mostafa H, Zenke F. Surrogate gradient learning in spiking neural networks: bringing the power of gradient-based optimization to spiking neural networks. IEEE Signal Process Mag. 2019;36(6):51–63.

28. Wu Y et al. Temporal contrastive learning for spiking neural networks. 2019.

29. Nøkland A. Direct feedback alignment provides learning in deep neural networks. Adv Neural Inform Process Syst. 2016;29.

30. Pfeiffer M, Pfeil T. Deep learning with spiking neurons: opportunities and challenges. Front Neurosci. 2018;12:774.

31. Sengupta A, Panda P, Roy K. Recent advances in neuromorphic computing architectures: a survey. IEEE Access. 2019;7:147233–77.

32. Fischl I, Martin MA. Neuromorphic computing: a review of emerging materials, devices, and architectures. Adv Mater. 2021;33(25): e2008663.

33. Boahen K. A neuromorphic computer for modeling cortical circuits. In: Proceedings of the IEEE international joint conference on neural networks;2006. p. 2167–72.

34. Zamarreño-Ramos C, Linares-Barranco A, Serrano-Gotarredona T et al. Multicasting mesh AER: a scalable assembly approach for reconfigurable neuromorphic structured AER systems. Application to ConvNets. IEEE Trans Biomed Circ Syst. 2012;7(1):82–102.

35. Akbari N, Modarressi M. A high-performance network-on-chip topology for neuromorphic architectures. In: 2017 IEEE international conference on computational science and engineering (CSE) and IEEE international conference on embedded and ubiquitous computing (EUC), vol. 2. IEEE;2017. p. 9–16.

36. Benjamin BV, Gao P, McQuinn E, et al. Neurogrid: a mixed-analog-digital multichip system for large-scale neural simulations. Proc IEEE. 2014;102(5):699–716.

37. Furber SB, Galluppi F, Temple S, et al. The spinnaker project. Proc IEEE. 2014;102(5):652–65.

38. Christensen DV, Dittmann R, Linares-Barranco B, et al. 2022 roadmap on neuromorphic computing and engineering. Neuromorphic Comput Eng. 2022;2(2): 022501.

39. Petrovici M A, Schmitt S, Klähn J et al. Pattern representation and recognition with accelerated analog neuromorphic systems. In: 2017 IEEE international symposium on circuits and systems (ISCAS). IEEE;2017. p. 1–4.

40. Cramer B, Billaudelle S, Kanya S, et al. Surrogate gradients for analog neuromorphic computing. Proc Natl Acad Sci. 2022;119(4): e2109194119.

41. Koickal TJ, Hamilton A, Tan SL, et al. Analog VLSI circuit implementation of an adaptive neuromorphic olfaction chip. IEEE Trans Circ Syst I: Reg Papers. 2007;54(1):60–73.

42. Kornijcuk V, Jeong DS. Recent progress in real-time adaptable digital neuromorphic hardware. Adv Intell Syst. 2019;1(6):1900030.

43. Yang S, Deng B, Wang J, et al. Scalable digital neuromorphic architecture for large-scale biophysically meaningful neural network with multi-compartment neurons. IEEE Trans Neural Netw Learn Syst. 2019;31(1):148–62.

44. Gutierrez-Galan D, Schoepe T, Dominguez-Morales JP, et al. An event-based digital time differ-
    ence encoder model implementation for neuromorphic systems. IEEE Trans Neural Netw Learn
    Syst. 2021;33(5):1959–73.
45. Milde MB, Blum H, Dietmüller A, et al. Obstacle avoidance and target acquisition for robot
    navigation using a mixed signal analog/digital neuromorphic processing system. Front Neuro-
    robotics. 2017;11:28.
46. Zendrikov D, Solinas S, Indiveri G. Brain-inspired methods for achieving robust computation
    in heterogeneous mixed-signal neuromorphic processing systems. Neuromorphic Comput Eng.
    2023;3(3): 034002.
47. Mostafa H, Corradi F, Stefanini F et al. A hybrid analog/digital spike-timing dependent plastic-
    ity learning circuit for neuromorphic VLSI multi-neuron architectures. In: 2014 IEEE interna-
    tional symposium on circuits and systems (ISCAS). IEEE;2014. p. 854–857.
48. Upadhyay NK, Jiang H, Wang Z, et al. Emerging memory devices for neuromorphic computing.
    Adv Mater Technol. 2019;4(4):1800589.
49. Rajendran B, Alibart F. Neuromorphic computing based on emerging memory technologies.
    IEEE J Emerg Selected Top Circ Syst. 2016;6(2):198–211.
50. Ielmini D, Ambrogio S. Emerging neuromorphic devices. Nanotechnology. 2019;31(9):
    092001.
51. Woo J, Kim JH, Im JP, et al. Recent advancements in emerging neuromorphic device technolo-
    gies. Adv Intell Syst. 2020;2(10):2000111.
52. Li Y, Wang Z, Midya R, et al. Review of memristor devices in neuromorphic computing:
    materials sciences and device challenge. J Phys D: Appl Phys. 2018;51(50): 503002.
53. Prezioso M, Merrikh-Bayat F, Hoskins BD, et al. Training and operation of an integrated neu-
    romorphic network based on metal-oxide memristors. Nature. 2015;521(7550):61–4.
54. Portner K, Schmuck M, Lehmann P, et al. Analog nanoscale electro-optical synapses for neu-
    romorphic computing applications. ACS Nano. 2021;15(9):14776–85.
55. Hou YX, Li Y, Zhang ZC, et al. Large-scale and flexible optical synapses for neuromor-
    phic computing and integrated visible information sensing memory processing. ACS Nano.
    2020;15(1):1497–508.
56. Marković D, Grollier J. Quantum neuromorphic computing. Appl Phys Lett. 2020;117(15).
57. Ghosh S, Nakajima K, Krisnanda T, et al. Quantum neuromorphic computing with reservoir
    computing networks. Adv Quantum Technol. 2021;4(9):2100053.
58. George R, Chiappalone M, Giugliano M et al. Plasticity and adaptation in neuromorphic biohy-
    brid systems. Iscience. 2020; 23(10).
59. Broccard FD, Joshi S, Wang J, et al. Neuromorphic neural interfaces: from neurophysiological
    inspiration to biohybrid coupling with nervous systems. J Neural Eng. 2017;14(4):041002.
60. Merolla PA, Arthur JV, Alvarez-Icaza R, et al. A million spiking-neuron integrated circuit with
    a scalable communication network and interface. Science. 2014;345(6197):668–73.
61. Davies M, Srinivasa N, Lin TH, et al. Loihi: a neuromorphic manycore processor with on-chip
    learning. IEEE Micro. 2018;38(1):82–99.
62. Pei J, Deng L, Song S, et al. Towards artificial general intelligence with hybrid Tianjic chip
    architecture. Nature. 2019;572(7767):106–11.
63. Petrovici MA, Vogginger B, Müller P, et al. Characterization and compensation of network-level
    anomalies in mixed-signal neuromorphic modeling platforms. PloS one. 2014;9(10): e108590.
64. Pehle C, Billaudelle S, Cramer B, et al. The BrainScaleS-2 accelerated neuromorphic system
    with hybrid plasticity. Front Neurosci. 2022;16: 795876.
65. Neckar A, Fok S, Benjamin BV, et al. Braindrop: a mixed-signal neuromorphic architecture with
    a dynamical systems-based programming model. Proc IEEE. 2018;107(1):144–64.

66. Moradi S, Qiao N, Stefanini F, et al. A scalable multicore architecture with heterogeneous memory structures for dynamic neuromorphic asynchronous processors (DYNAPs). IEEE Trans Biomed Circ Syst. 2017;12(1):106–22.
67. Delbruck T, Lang M. Robotic applications of neuromorphic vision sensors. Philos Trans Roy Soc B: Biol Sci. 2015;370(1677):20140377.
68. Lichtsteiner P, Posch C, Delbruck T. A 128 × 128 120 dB 15 μs latency asynchronous temporal contrast vision sensor. IEEE J Solid-State Circ. 2008;43(2):566–76.
69. Linares-Barranco B, Serrano-Gotarredona T. A current-mode cochlea with spike output. IEEE Trans Biomed Circ Syst. 2011;5(3):266–78.
70. Liu SC, Delbruck T. Neuromorphic sensory systems. Curr Opin Neurobiol. 2010;20(3):288–95.
71. Lin H, Li X, Liu Y. Recent advances in event-based sensors: from materials, design, to applications. Front Mater. 2021;8:38.
72. Buhler T, Cavigelli L, Benini L. Neuromorphic sensors: progress and prospects. Proc IEEE. 2017;105(1):43–65.

Spike-Based Learning with Multi-compartment Model

2.1 Multi-compartment Neuron Model

2.1.1 Biological Background

A multi-compartment neuron model [1, 2] is a mathematical representation of a neuron that takes into account the spatially distributed nature of neuronal signaling. Unlike the traditional point neuron models, multi-compartment models divide the neuron into several compartments, each with its own set of electrical properties. The model thus provides a more realistic and detailed picture of how neurons process and transmit information. In a multi-compartment neuron model, the membrane potential and currents are calculated separately for each compartment, and the resulting signals are then integrated to determine the overall response of the neuron. Multi-compartment neuron models have been used in many areas of neuroscience research, including the study of synaptic integration, dendritic processing, and the role of ion channels in neuronal excitability [3, 4]. They have also been used to model various types of neurons, such as pyramidal cells in the cortex and Purkinje cells in the cerebellum. There are several different types of multi-compartment neuron models, each with its own strengths and limitations. Some examples include the cable theory model, the compartmental model, and the morphological model [5–7]. Multi-compartment neuron models are typically implemented using numerical methods, such as the finite difference method or the finite element method. These methods solve the partial differential equations that describe the electrical properties of the neuron in each compartment. Several software tools are available for simulating multi-compartment neuron models to investigate the biological mechanism, such as NEURON [8], GENESIS [9], and MOOSE [10].

In a multi-compartment neuron model, dendrites are commonly depicted as distinct compartments, each endowed with its own unique set of electrical characteristics [13].

© The Author(s), under exclusive license to Springer Nature Switzerland AG 2024   21
S. Yang and B. Chen, *Neuromorphic Intelligence*, Synthesis Lectures on Engineering,
Science, and Technology, https://doi.org/10.1007/978-3-031-57873-1_2

This affords a more precise simulation of the intricate processes inherent in dendrites, such as synaptic integration and the transmission of electrical signals throughout the dendritic tree [14]. The compartmentalization of dendrites within the model facilitates an in-depth examination of dendritic processing, elucidating how distinct regions of dendrites collectively shape the overall response of the neuron. The model has the capacity to integrate intricate details of dendritic morphology and branching patterns, shedding light on the spatial distribution of ion channels and the impact of dendritic geometry on neuronal signaling. Similarly, the axon of a neuron can be incorporated into the multi-compartment model, enabling a detailed exploration of axonal properties and functions [15]. The model can simulate the propagation of action potentials along the axon, taking into account variations in membrane potential and currents along its length. By integrating the axon into the multi-compartment model, researchers can delve into how diverse factors, such as ion channel distribution and axonal geometry, shape the transmission of signals through the axon. Neuronal dendrites play a pivotal role in synaptic integration, processing signals from multiple synapses. The multi-compartment model, by portraying dendrites as distinct compartments, facilitates the investigation into how synapses located at different points on the dendritic tree collectively contribute to the overall response of the neuron. This modeling approach empowers researchers to explore the intricate interactions among dendrites, synapses, and ion channels in a more realistic manner.

2.1.2 Neuronal Dendrites

Dendrites [11, 12] are the branching extensions of neurons that receive inputs from other neurons or sensory stimuli, and play a crucial role in the integration and processing of neural information. Dendrites can range from simple, unbranched structures to highly complex, extensively branched trees that cover a large area. The structure of dendrites varies depending on the type of neuron and its function. Some neurons, such as sensory neurons, have short, simple dendrites, while others, such as pyramidal neurons in the cortex, have highly branched dendritic trees that can span several hundred micrometers. Dendrites are covered with spines, small protrusions that serve as the primary sites of synaptic input. The number and shape of spines can also vary depending on the type of neuron and its function. For example, neurons involved in learning and memory have more and larger spines than those that are not. The electrical properties of dendrites also play an important role in neuronal function. Dendrites contain various types of ion channels that regulate the flow of ions in and out of the neuron, and influence the integration of synaptic inputs. They can also generate action potentials, although this is less common than in the axon. Dendritic processing is a complex process that involves the integration of multiple inputs and the generation of nonlinear responses. It is influenced by the morphology of the dendritic tree, the distribution of ion channels, and the properties of synaptic inputs. Computational models have been developed to simulate dendritic processing and investigate the mechanisms underlying it.

2.2 Credit Assignment by Spiking Dendrites for Efficient Learning

2.2.1 Learning with Dendrites

As described in [20], learning needs neurons to receive signals to assign the credit for behavior. Since the behavioral impact in early network layers is based on downstream synaptic connections, credit assignment in multi-layer networks is challenging. Previous solutions in artificial intelligence use the back-propagation of error algorithm, but this is unrealistic in the neural systems. Rather than requiring weight transport, current biologically-plausible solutions to the credit assignment problem use segregated feed-forward and feedback signals [16, 17]. In fact, the cortico-cortical feedback signals to pyramidal neurons can transmit the necessary error information. These works show how the circuitry needed to integrate error information may exist within each neuron. The idea is that both feed-forward sensory information in the neocortex and the higher-order cortico-cortical feedback are received by different dendritic compartments, including basal and apical dendrites [18]. In a pyramidal cell, distal apical dendrites are distant from the soma, and communicate with the soma based on active propagation using the apical dendritic shaft, driven predominantly by voltage-gated calcium channels. Further, there exist dynamics of plateau potentials (PPs) that generate prolonged upswings in the membrane potential. These are based on the nonlinear dynamics of voltage-gated calcium current, and drive bursting at the soma [19]. The PPs of the apical dendritic activities can induce learning in pyramidal neurons in vivo [20].

Inspired by these phenomena, previous studies have proposed a learning algorithm with segregated dendrites [20, 21]. The idea is that the basal dendritic compartment is coupled to the soma for processing bottom-up sensory information, and the apical dendritic compartment is used to process top-down feedback information to calculate credit assignment and induce learning using PPs. As shown in Fig. 2.1, the simple spiking behaviors of the IF neurons can be triggered by excitatory input spikes. The new state of the neural membrane potential with an input arriving is determined by the last updating time and the previous state. Thus, the event-driven neuron only updates when an input spike is received. Then the membrane potential decay after the last update is retroactively calculated and applied. The synaptic weight is then used to contribute to the resulting membrane potential. A spike event is emitted when the membrane potential exceeds a spike threshold, and then the neural activity is reset and mutual inhibition with coupled neurons is realized. Finally, the membrane potential and spiking event are written to memory to store the network state of the next update of neural activity.

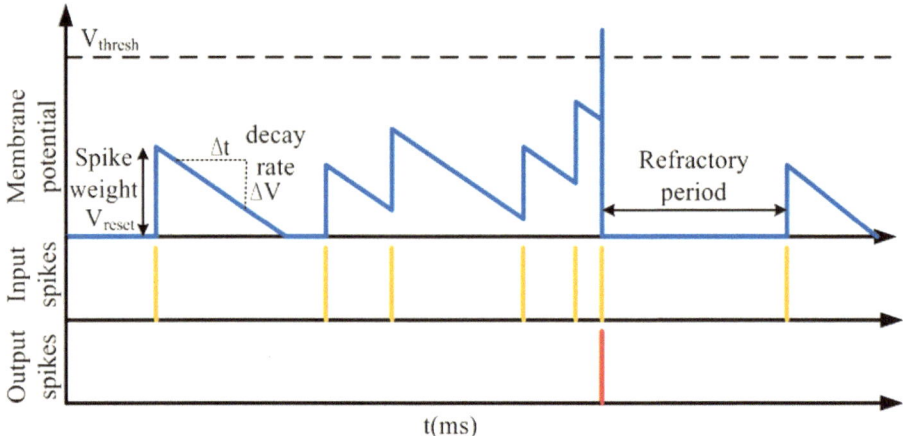

Fig. 2.1 Event-driven neural computing, showing the process of synaptic weights and time affecting the neural membrane potential and the refractory period [20]

2.2.2   Network Architecture and Learning Algorithms

The network diagram, named the dendritic event-driven processing (DEP) algorithm, utilizes the idea in the previous study by Guerguiev et al. as shown in Fig. 2.2 [21], which consists of an input layer with $m = 784$ neurons, a hidden layer with $n = 500$ neurons, and an output layer with $p = 10$ neurons. Since the primary interests are in the realization of neuromorphic networks, the model is restricted to discrete systems based on the Euler method, where N is the time step for discretization. This way of representation is popular in the hardware implementation of spiking neural networks because of its feasibility of implementation and routing. Poisson spiking neurons are used in the input layer, whose firing rate is determined by the intensity of image pixels ranging from 0 to Φ_{max}. In the hidden layer, neurons are modeled using three functional compartments, which are basal dendrites, apical dendrites and soma. The membrane potential of the ith neuron in the hidden layer is updated as follows:

$$\tau \frac{V_i^0(N+1) - V_i^0(N)}{\Delta T} = -V_i^0(N) + \frac{g_b}{g_l}\left(V_i^{0b}(N) - V_i^0(N)\right)$$
$$+ \frac{g_a}{g_l}\left(V_i^{0a}(N) - V_i^0(N)\right) \tag{2.1}$$

where g_l, g_b, and g_a stand for the leak conductance, the basal dendrites conductance, and the apical dendrite conductance, and ΔT is the integration step. The parameter $\tau = C_m/g_l$, is a time constant, where C_m represents the membrane capacitance. The variables V^0, V^{0a}, and V^{0b} represent the membrane potentials of soma, apical dendrite and basal dendrite respectively. The dendritic compartments are defined as weighted sums for the

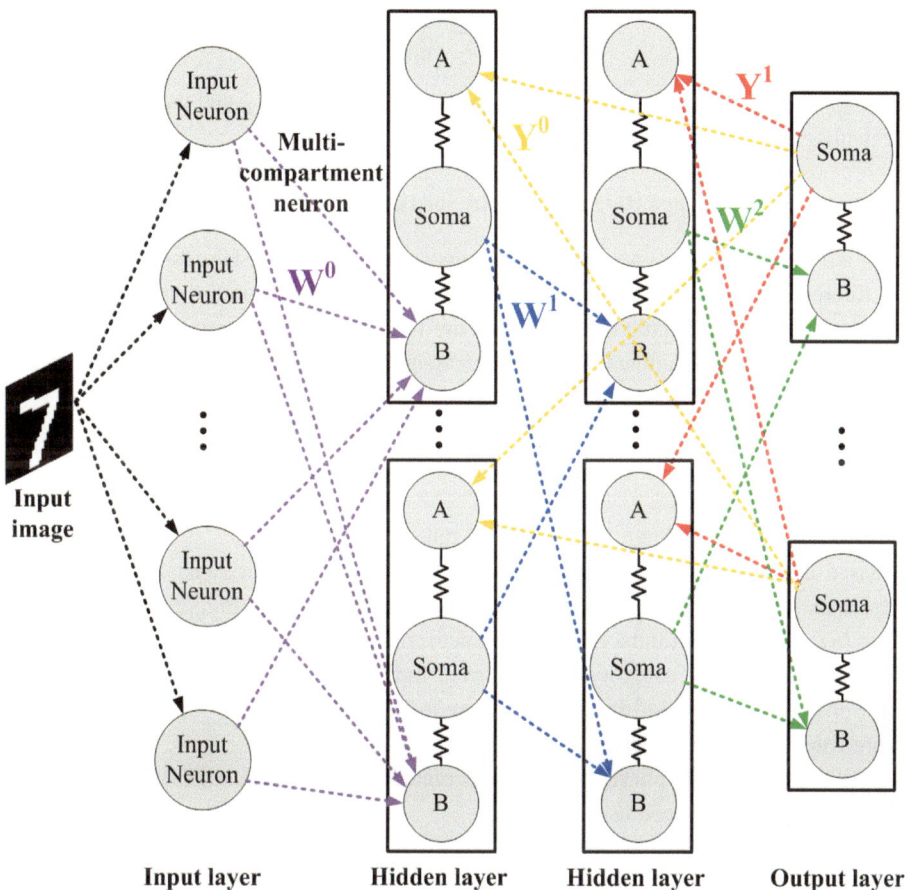

Fig. 2.2 Network architecture. Matrices Y and W represent the feedback and feedforward synaptic weight matrix respectively [20]

ith hidden layer neuron as follows:

$$
\begin{cases}
V_i^{0b}(N) = \sum_{j=1}^{m} W_{ij}^0 s_j^{input}(N) + b_i^0 \\
V_i^{0a}(N) = \sum_{j=1}^{p} Y_j s_j^1(N)
\end{cases}
\tag{2.2}
$$

where W_{ij}^0 and Y_{ij} are synaptic weights in the input layer and feedback synapses respectively. The constant b_i^0 is defined as a bias term, and s^{input} and s^1 are the filtered spiking activities in the input layer and output layer respectively. The variable s^{input} is calculated based on the following equations as

$$s_j^{input}(t) = \sum_k \kappa\left(t - t_{jk}^{input}\right) \tag{2.3}$$

where t_{jk}^{input} represents the kth spiking time of the input neuron j, and the response kernel is calculated as

$$\kappa(t) = \left(e^{-t/\tau_L} - e^{-t/\tau_s}\right)\Theta(t)/(\tau_L - \tau_s) \tag{2.4}$$

where τ_L and τ_s are long- and short-time constants, and Θ is the Heaviside step function. The filtered spike trains at apical synapses s^1 is modeled based on the same method. The spiking activities of somatic compartments are based on Poisson processes, whose firing rates are based on a nonlinear sigmoid function $\delta(.)$ for the ith hidden layer neuron as follows:

$$\Phi_i^0(N) = \phi_{max}\sigma\left(V_i^0(N)\right) = \phi_{max}\frac{1}{1 + e^{-V_i^0(N)}} \tag{2.5}$$

where Φ_{max} represents the maximum firing rates of neurons.

Based on the learning algorithm of Guerguiev et al. [21], two phases are alternated to train the network: the forward and target phases as shown in Fig. 2.3. In the forward phase $I_i(t) = 0$, while it induces any given neuron i to spike at maximum firing rate or be silent according to the category of the current input image when the network undergoes target phase. At the end of the forward phase and the target phase, the set of PPs α_t and α_f are calculated respectively.

At the end of each phase, PPs are calculated for apical dendrites of hidden layer neurons, which are defined as follows

$$\begin{cases} \tau\frac{V_i^1(N+1)-V_i^1(N)}{\Delta T} = -V_i^1(N) + I_i(N) \\ \qquad\qquad + \frac{g_d}{g_l}\left(V_i^{1b}(N) - V_i^1(N)\right) \\ V_i^{1b}(N) = \sum_{i=1}^m W_{ij}^1 s_j^0(N) + b_i^1 \end{cases} \tag{2.6}$$

where t_1 and t_2 represent the end times of the forward and target phases respectively. $\Delta t_s = 30$ ms represents the settling time for the membrane potentials, and Δt_1 and Δt_2 are formulated as follows

$$\begin{cases} \Delta t_1 = t_1 - (t_0 + \Delta t_s) \\ \Delta t_2 = t_2 - (t_1 + \Delta t_s) \end{cases} . \tag{2.7}$$

The temporal intervals between plateaus are sampled based on an inverse Gaussian distribution randomly. Although the system computes in phases, the specific length of the phases is not vital, provided there has been a long enough time to integrate the input currents.

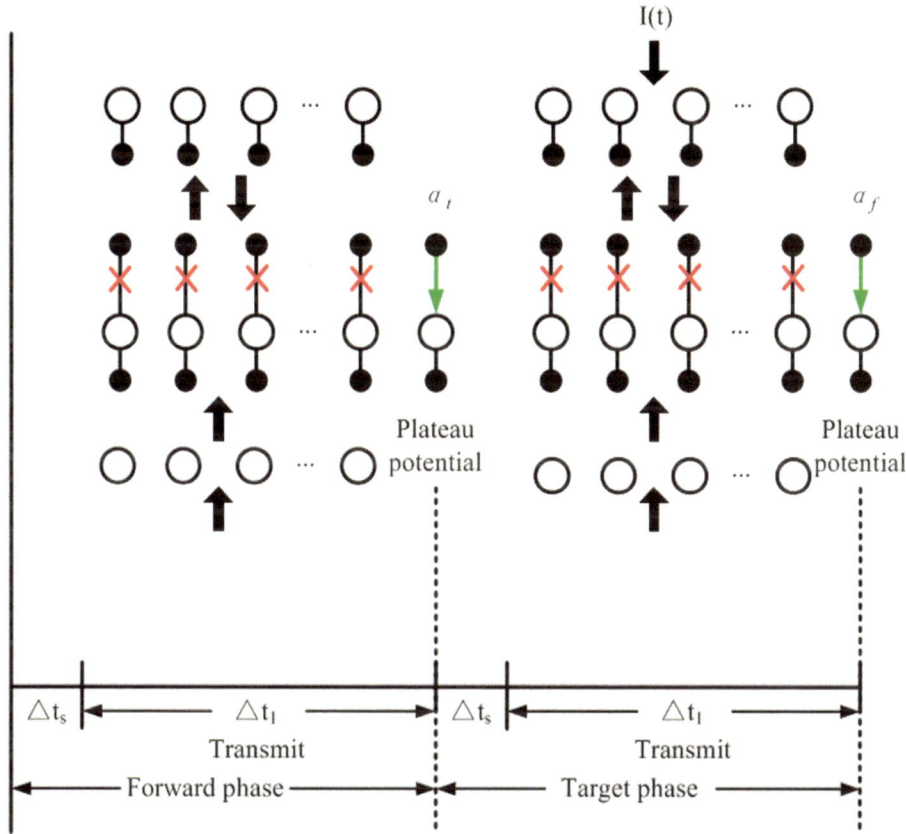

Fig. 2.3 Network computing phases for learning proposed by Yang et al. [20] Guerguiev et al. [21]

During the forward phase, an image is presented to the input layer without teaching current at the output layer between time t_0 to t_1. At t_1 a PP is computed in the hidden layer neurons and the target phase begins. During the target phase the image is also presented into the input layer that also receives teaching current, forcing the correct neuron in the output layer to its maximum firing rate while others are silent. At time t_2 another set of PPs in the hidden layers are computed. Plateau potentials for the end of both the forward and the target phases are calculated as follows

$$\begin{cases} \alpha_i^f = \delta\left(\frac{1}{\Delta t_1} \int_{t_1-\Delta t_1}^{t_1} V_i^{0a}(N)dt\right) \\ \alpha_i^t = \delta\left(\frac{1}{\Delta t_2} \int_{t_2-\Delta t_2}^{t_2} V_i^{0a}(N)dt\right) \end{cases} \tag{2.8}$$

where Δt_s represents a time delay of the network dynamics before integrating the plateau, and $\Delta t_i = t_i - (t_i - 1 + \Delta t_s)$.

The basal dendrites in the hidden layer update the synaptic weights W_0 based on the minimization of the loss function as follows

$$L^0 = \left\| \Phi^{0*} - \Phi_{max}\delta\left(\overline{V^{0^f}}\right) \right\|_2^2. \tag{2.9}$$

And the target firing rate is defined as

$$\Phi_i^{0*} = \overline{\Phi_i^{0^f}} + \alpha_i^t - \alpha_i^f \tag{2.10}$$

where the variable α^f and α^t are PPs in the forward and target phases. Then the formulation can be obtained as

$$L^0 \approx \left\| \alpha^i - \alpha^f \right\|_2^2. \tag{2.11}$$

And the formulation can be described as follows

$$\begin{cases} \frac{\partial L^0}{\partial W^0} \approx -k_b\left(\alpha^t - \alpha^f\right)\Phi_{max}\delta'\left(\overline{V^{0^f}}\right) \circ \overline{s^{input^f}} \\ \frac{\partial L^0}{\partial b^0} \approx -k_b\left(\alpha^t - \alpha^f\right)\Phi_{max}\delta'\left(\overline{V^{0^f}}\right) \end{cases} \tag{2.12}$$

where the constant k_b is given as

$$k_b = g_b/(g_l + g_b + g_a). \tag{2.13}$$

Φ^{0*} is treated as a fixed state for the hidden layer neurons to learn. The synaptic weights of basal dendrites are updated to descend the approximation of the gradient as follows

$$\begin{cases} W^0 \to W^0 - \eta^0 P^0 \frac{\partial L^0}{\partial W^0} \\ b^0 \to b^0 - \eta^0 P^0 \frac{\partial L^0}{\partial b^0} \end{cases}. \tag{2.14}$$

In the target phase the activity is also fixed and no derivatives are used for the membrane potentials and firing rates. The feedback weights are held fixed. In order to avoid the complicated computation induced by nonlinear functions, the PWL approach is used. Both the functions $\delta(x)$ and $\delta'(x)$ are modified based on the PWL method, which can be formulated as follows

$$f_{PLA} = \begin{cases} a_{s1}x + b_{i1}, & \text{when } x \le s_1 \\ a_{s2}x + b_{i2}, & \text{when } s_1 < x \le s_2 \\ \cdots \\ a_{si}x + b_{ii}, & \text{when } x > s_{i-1} \end{cases} \tag{2.15}$$

where a_s and b_i are the slope and intercept of the modified PWL functions respectively ($i = 1, 2 ..., n$). Since the range of the segment points are constrained, an exhaustive search algorithm is used in the determination of the PWL functions. The determination of the coefficient values is based on an error evaluation criterion as follows

$$CF_{RE} = \frac{1}{n} \sqrt{\sum_{i=1}^{n} \frac{(f_{ori}(i) - f_{PLA}(i))^2}{f_{ori}(i)^2}} \qquad (2.16)$$

where n represents the total sampling points. If the modified function cannot meet the accuracy requirement represented by CF_{RE}, its segment number will be added by 1 until it can be guaranteed. Since the multiplication operation is replaced by "adder" and "shifter" in the study, the coefficient value a_s in the PWL functions should be a power of 2 (for example: 1, 2, 4 or 0.5, 0.25, etc.). The parameter values of the PWL methods are listed in Table 2.1. The PWL functions are depicted in Fig. 2.4.

The digital neuromorphic algorithm requires less multiplication operations. The Otsu's thresholding method is used to binarize the filtered spike trains, which can iterate all possible threshold values and compute the expansion measure of each pixel level of the threshold [22]. Therefore, each pixel will fall in either foreground or background. Firstly, separate all the pixels into two clusters based on the threshold as follows

Table 2.1 Parameters values of the PWL methods (From [20])

$\delta(x)$	a_s	b_i	Condition
$i = 1$	0.0078125	0.05	$x \leq -3.4$
$i = 2$	0.0625	0.24	$-3.4 < x \leq -1.3$
$i = 3$	0.25	0.5	$-1.3 < x \leq 1.3$
$i = 4$	0.0625	0.76	$1.3 < x \leq 3.4$
$i = 5$	0.0078125	0.95	$3.4 < x$
$i = 6$	0	0.9999	$\delta(x) \leq 0$
$i = 7$	0	0.0001	$\delta(x) \geq 1$
$\delta'(x)$	a_s	b_i	Condition
$i = 1$	0.0078125	0.05	$x \leq -3.2$
$i = 2$	0.03125	0.15	$-3.2 < x \leq -2$
$i = 3$	0.0625	0.25	$-2 < x \leq 0$
$i = 4$	-0.0625	0.25	$0 < x \leq 2$
$i = 5$	-0.03125	0.15	$2 < x \leq 3.2$
$i = 6$	-0.0078125	0.05	$x > 3.2$
$i = 7$	0	0.0001	$\delta'(x) \leq 0$

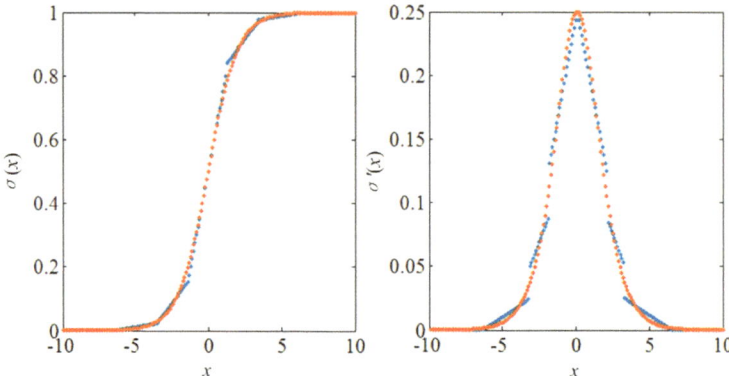

Fig. 2.4 PWL function [20]

$$\begin{cases} q_1(t) = \sum_{i=1}^{t} p(i) \\ q_2(t) = \sum_{i=t+1}^{L} p(i) \end{cases} \tag{2.17}$$

where p represents the image histogram. Secondly, the mean of each cluster is calculated by the formulation as follows

$$\begin{cases} \mu_1(t) = \sum_{j=1}^{t} \frac{i \cdot p(i)}{q_1(t)} \\ \mu_2(t) = \sum_{j=t+1}^{L} \frac{i \cdot p(i)}{q_2(t)} \end{cases} \tag{2.18}$$

Thirdly, calculate the individual class variance as follows

$$\begin{cases} \sigma_1^2(t) = \sum_{i=1}^{t} [i - \mu_1(t)]^2 \frac{p(i)}{q_1(t)} \\ \sigma_2^2(t) = \sum_{i=t+1}^{L} [i - \mu_2(t)]^2 \frac{p(i)}{q_2(t)} \end{cases} \tag{2.19}$$

Fourthly, square the difference between the means formulated as follows

$$\begin{aligned} \sigma_b^2(t) &= \sigma^2 - \sigma_w^2(t) \\ &= q_1(t)\big[1 - q_1(t)\big][\mu_1(t) - \mu_2(t)]^2 \end{aligned} \tag{2.20}$$

Finally, the formulation can be maximized and the solution is t that is maximizing $\sigma_b{}^2(t)$.

Neuromorphic hardware is largely made out of arithmetic elements and memories. Multipliers are the most space and power-hungry arithmetic elements of the digital neuromorphic implementation. The realization of a deep neural network is mainly dependent on matrix multiplications. The key arithmetic operation is the multiply-accumulate operation. The reduction of the precision of the multipliers, especially for the weight matrix, is vital for the efficient realization of deep neural networks. Recent researches have focused on the reduction of model size and computational complexity by using low bit width weights of neural networks [23]. Other neuromorphic hardware systems implement bistable synapses based on a 1-bit weight resolution, which is shown to be sufficient for memory formation [24]. However, the models do not only use spike timings, but also use additional hardware resources to read the postsynaptic membrane potential [25]. Therefore, DEP algorithm is trained using dynamic fixed-point representation. In dynamic fixed point, each number is represented as follows

$$(-1)^{si} \cdot 2^{-FL} \sum_{i=0}^{B-2} 2^i \cdot mb_i \tag{2.21}$$

where B represents the bit-width, si is the sign bit, FL is the fractional length, and mb the mantissa bits.

The algorithm is presented in Table 2.2. In the pseudo code, the synaptic weight matrix W is the input of the algorithm. Total_bit represents the total bit width of the fixed-point number, and IF_bit is the integer bitwidth. The fractional bitwidth is represented by LF_bit. The integer and fractional parts are represented by W_IF and W_LF. The binary integer and fractional parts are represented by W_IF_bit and W_LF_bit respectively. The symbol bit is represented by W_s, and R_max defines the fault-tolerant ratio. The error rate refers to the difference between the binary number and the original decimal number divided by the original decimal number. If the error rate exceeds the defined fault-tolerant rate, a specialized process will be used for the binary number. Since the large error occurs in the situation when the considered number is close to 0, this number will be set to 0 if the error rate exceeds R_max. The term W is an $a*b$ synaptic weight matrix to be trained. The first loop is in the line 2. This loop is in the line 2, which is the row loop of the matrix. The second loop is in the line 3, which is the column loop of the matrix. There are two judgments in the algorithm. The first judgement is to determine the symbol bit. If it is negative, then the symbol is 1. If it is positive, then the symbol is 0. The second judgement is to determine the positive and negative when the binary number is converted to decimal number. If the sign bit is 1, it is negative. And it is positive when the sign bit is 0. The third judgment is to consider the error rate between the newly converted number and the original number. If the error rate exceeds the fault-tolerant ratio, the newly converted number will be replaced by 0 for efficient calculation on neuromorphic systems. Finally, the updated synaptic weight matrix W_new is output by the processing of the algorithm.

Table 2.2 Training strategy with dynamic fixed point representation (From [20])

Input: W, Output, W_new;
W is a*b matrix.
1: **for** i = 1:a
2:     **for** j = 1:b
3:         **if** W(i, j) <0
4:             W_s = 1
5:         **else**
6:             W_s = 0
7:         **end if**
8:         W_IF_bit = W_IF→bit;
9:         IF_bit = length(W_IF_bit)
10:        LF_bit = Total_bit − IF_bit -1
11:        W_LF_bit = (W_LF, LF_bit)→bit;
12:        W_new(i , j) = W_IF_bit→dec + W_LF_bit→dec;
13:        **if** W_s = 1
14:            W_new(i , j) = - W_new(i, j) ;
15:        **end if**
16:        **if** (W_new(i, j) − W(i, j)) / W(i, j) > R_max
17:            W_new(i, j) = 0;
18:        **end if**
19:    **end for**
20: **end for**

By using the algorithm, the memory usage on hardware can be optimized and the energy efficiency of neuromorphic systems can be further improved.

2.2.3   Performance Evaluation

To demonstrate the effectiveness of the learning algorithm, the standard modified national institute of standards and technology (MNIST) database is employed. The MNIST dataset contains 70,000 28×28 images of handwritten digits. The image number in the training and testing sets are 60,000 and 10,000 respectively. The dataset is divided into 10 categories for ten integers 0–9, and each image has an associated label. The network is trained with no hidden layer, with one hidden layer and two hidden layers on the 60,000 MNIST training images for 10 epochs, and tested the classification accuracy using the 10,000-image test set. As shown in Fig. 2.5a, the two-layer network with no hidden layer has poor classification performance of 62.1%. In contrast, the three-layer network with hidden layer has an accuracy of 95.1% by the 10th epoch. The network can take advantage of the multi-layer architecture to enhance the learning performance, which is the critical characteristics of deep learning [26]. Another critical characteristic of deep learning is the capability to generate representations, which obtains task-related information and ignores irrelevant sensory details [27, 28]. The t-distributed stochastic neighbor embedding (t-SNE) algorithm is used to investigate the information abstraction of the algorithm. The t-SNE algorithm can reduce the dimensionality of data with the preservation of local structure and nonlinear manifolds in high-dimensional space. It is a powerful approach to visualize the structure of high-dimensional data [29]. The t-SNE algorithm is applied to the hidden layer, which shows that the categories are better segregated with only a small amount of splitting or merging of category clusters as shown in Fig. 2.5b. Therefore, the algorithm has the capability of learning the developing representations in the hidden layer, in which the categories are quite distinct. It reveals that the algorithm can be applied in a deep learning framework. In addition, the algorithm relies on the phenomenon of feedback alignment, in which the feed-forward system comes to align with the feedback weights so that a useful error signal is provided.

The DEP learning algorithm in a network with one hidden layer trained on permutation invariant MNIST is explored, although it can be generalized to other datasets in

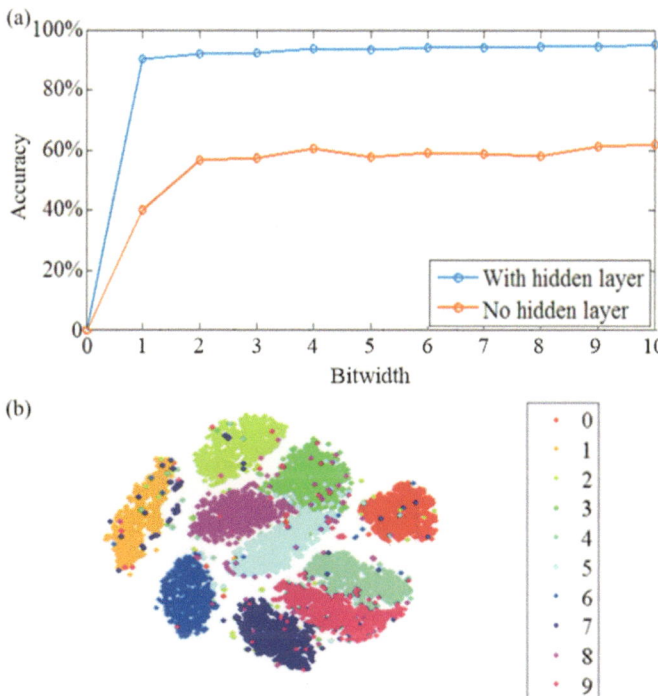

Fig. 2.5 Learning performance of the DEP algorithm. **a** Learning accuracy across 10 epochs of training. **b** Results of t-SNE dimensionality reduction applied to the activity patterns of the hidden layer after 10 epochs of training [20]

theory. Rather than seeking for the optimized classification performance, the equivalent non-spiking neural networks trained by standard BP and random BP are compared with the algorithm, with the parameters tuned to obtain the highest classification accuracy in the current classification task. Weight updates are conducted during each digit input into the spiking network, which is different from the batch gradient descent that performs weight updates once per the entire dataset. As shown in Fig. 2.6, the DEP algorithm requires fewer iterations of the dataset to obtain the peak classification performance in comparison with the two alternative methods. The reason is that the spiking neural network with DEP algorithm can be updated multiple times during each input, which results in faster convergence of learning. In addition, for the equivalent computational resources, online learning with gradient descent strategy has the capability to deal with more data samples and requires less on-chip memory for implementation [30]. Therefore, for the same number of calculation operations per unit time, online gradient-descent-based learning converges faster than batch learning. Since potential applications of neuromorphic hardware is with real-time streaming data, it is essential for the online learning with DEP algorithm.

As shown in Fig. 2.7, learning on neuromorphic system can be energy efficient by using the DEP algorithm, because only active connections in the network induce synaptic operations per second (SynOps) operation. In order to show the learning efficiency, the

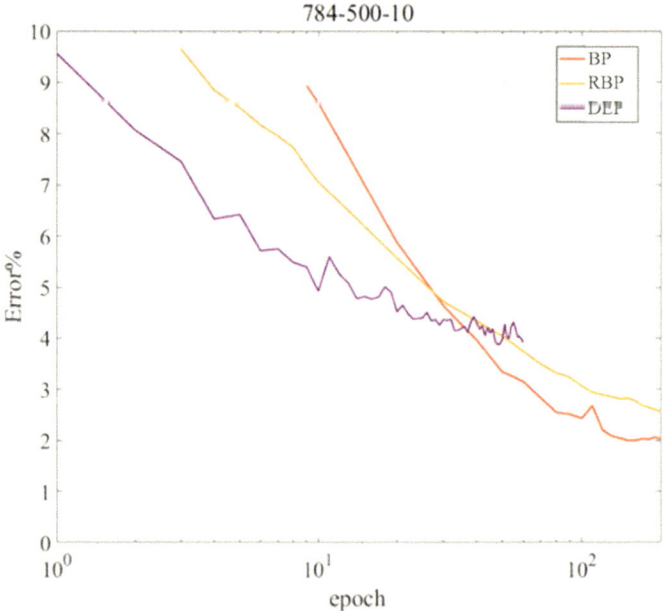

Fig. 2.6 MNIST classification error based on spiking neural network with DEP learning rule and fully connected artificial neural networks with BP and RBP learning rules [20]

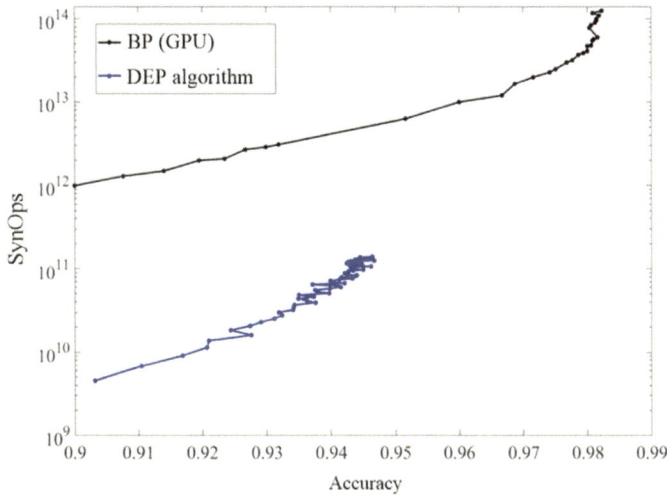

Fig. 2.7 Comparisons of accuracy and SynOps between the DEP algorithm and BP algorithm on GPU [20]

number of multiplication-and-accumulation (MAC) operations using the BP algorithm is compared with SynOps number with the algorithm. This advantage is critical and promising for neuromorphic computing because SynOps in a dedicated neuromorphic system use much less power than MAC operations on a GPU platform. The learning accuracy of the algorithm increases quickly but the final accuracy is lower than an ANN.

As shown in Fig. 2.8, the response of the DEP algorithm after stimulus onset is one synaptic time constant. It leads to 11% error and improves as the spikes number of the neurons in the output layer increases. Classification using the first spike induced less than 20kSynOps events, most of which exist between the input and hidden layer. In the state-of-the-art neuromorphic system, the energy consumption of a synaptic operation is around 20 pJ [31, 32]. On such neuromorphic system, single spike classification based on the network can potentially induce 400 pJ, which is superior to the state-of-the-art work in digital neuromorphic hardware (≈ 2 µJ) at this accuracy [33] and potentially 50 thousand more efficient than current GPU technology. The reasons for the low energy cost can be divided into three aspects. Firstly, the segregated dendrite can generate a PP within 50–60 ms, which determines the training time of the network. The training time can be thereby reduced in this way, which can cut down the number of spikes with the decreasing of the training time for each image. Secondly, the conventional BP algorithm induces a trend to make neurons spike with maximum firing rate, and induces synchronization within layers. This means a larger number of spikes. Thirdly, the communication between layers in the algorithm uses a Poisson filter, and Φ_{max} is set to be 0.2. These results suggest that the DEP learning algorithm can take full use of the spiking dynamics, with

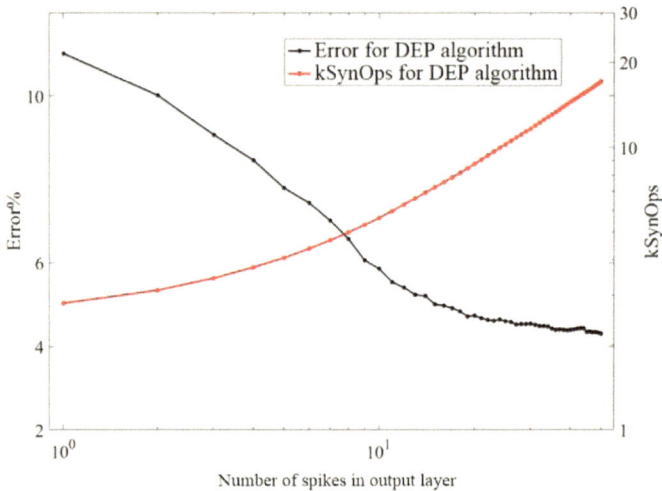

Fig. 2.8 Classification error in the DEP network as a function of the number of spikes in the output layer, and total number of synaptic operations incurred up to each output spike [20]

the learning accuracy comparable to the spiking network that is trained specifically for single spike recognition in previous study [33].

Figure 2.9 shows the distribution of spike times in the output layer, which is the times at which the SNN makes a decision for all the 10,000 test set images. The SNN with DEP algorithm makes a decision after most of the hidden layer neurons have spiked. The network is thus able to make more accurate and robust decisions about the input images, based on the PPs generated by the dendrites in the DEP algorithm.

As shown in Fig. 2.10, 30 neurons in the hidden layer are selected randomly to explore the selectivity for 10 categories of MNIST data set. The negative log probability for each of the 30 neurons to spike for each of the ten categories is explored, which means the negative log probability for a neuron to participate in the classification of a specified category. Probability is calculated from the response of the SNN to the ten thousand test digits. It reveals that some neurons are highly selective, while most of the neurons are more broadly tuned. Some of the neurons are mostly silent, but all the neurons in the SNN model contribute to at least one category of classification with the ten thousand test digits. In other word, neurons are typically broadly tuned and contribute to the classification of more than one category. The bit width of the integer part using the fixed-point calculation is set to 8 to avoid the overflow problem during computation. In contrast, the dynamic fixed-point is not required to determine the bit width of either integer or fractional part. As shown in Fig. 2.11, the fixed-point representation requires 22 bits to obtain a satisfied learning performance that exceeds 90%. The dynamic fixed-point representation just needs 16 bits to realize high-performance learning. Therefore, the dynamic fixed-point

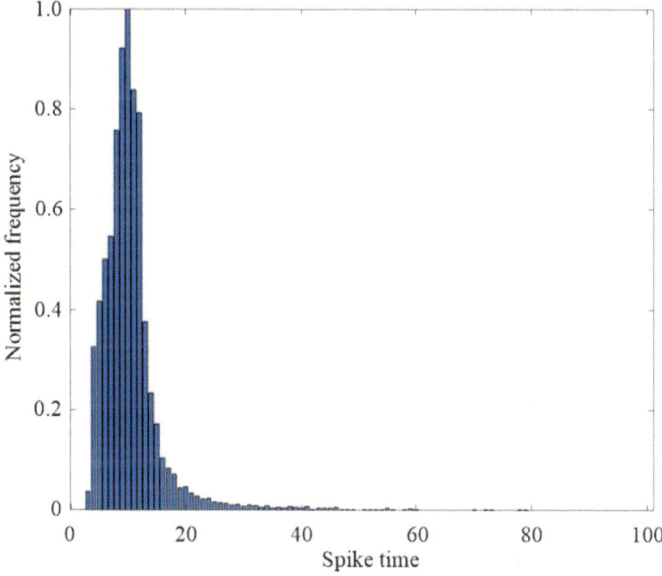

Fig. 2.9 Histograms of spike times in the output layer spike across the 10 000 test set images [20]

representation in the algorithm provides an efficient approach to reduce the computational hardware resource cost and power consumption for neuromorphic computing.

2.3 Spike-Based Learning with Self-adaptive Multi-compartment Model

2.3.1 Self-adaptive Multi-compartment Neuron Model

Previous research has shown that the exact timing and position of the active dendrites can significantly affect neurons. While dendritic excitation can drive action potential spiking, dendritic inhibition serves as an opposing force to gate excitatory activities [34–36]. As described in [47], the self-adaptive multi-compartment (SAM) neuron model is composed of 2 dendritic cells and 1 soma cells. Figure 2.12a illustrates the morphology of a biological neuron, in which dendrites transmit both excitatory and inhibitory inputs from different paths, at the same time. The soma of the neuron has the spike adaptation mechanism, which can change the threshold depending on the firing mode of the neuron. It utilizes the spatial layout of different dendritic compartments to receive excitatory and inhibitory inputs. It also uses dendritic and somatic compartments to receive and send spikes, respectively. Compared with the traditional point-neuron model, the ability of the active dendrites in SAM can improve the ability to learn the low-resolution synapses [37].

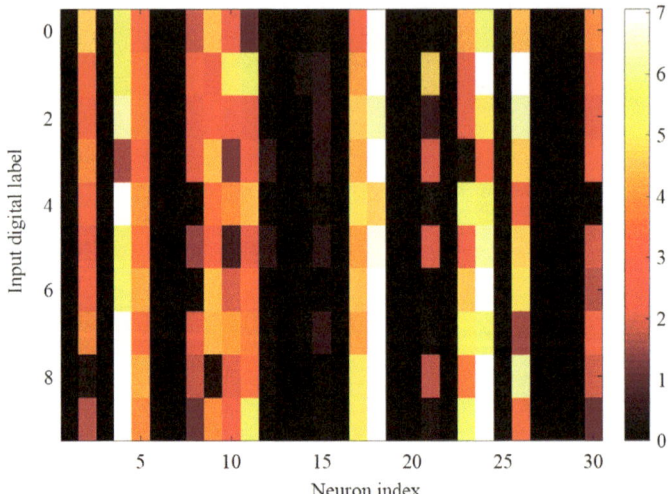

Fig. 2.10 Selectivity and tuning properties of 30 randomly selected hidden neurons in the SNN network with DEP algorithm. It is plotted by heat map with color called YlOrRd, whose color gradually changes across yellow, orange, and red [20]

The membrane potentials of soma and dendrite compartments evolve using the following formulas

$$
\begin{cases}
\tau \dot{V}_j(t) = -V_j(t) + RI_j(t) + g^{inh}\left(V_j^{den,i}(t) - V^{inh}\right) \\
\quad + g^{exc}\left(V_j^{den,e}(t) - V^{exc}\right) \\
\tau \dot{V}_j^{den,i}(t) = -V_j^{den,i}(t) + R^i I_j^i(t) \\
\tau \dot{V}_j^{den,e}(t) = -V_j^{den,e}(t) + R^e I_j^e(t)
\end{cases}
\tag{2.22}
$$

where τ is the membrane time constant and R is the membrane resistance of soma. R^e and R^i are the membrane resistance of the excitatory and inhibitory dendrites, respectively. The variables $V(t)$, $V^{den,i}(t)$, $V^{den,e}(t)$ are the membrane potentials of soma, inhibition and excitatory dendrite, respectively. The parameters g^{inh} and g^{exc} represent the synaptic conductance of inhibitory and excitatory dendrites, respectively. The parameters V^{exc} and V^{inh} indicate the reverse membrane potential of excitatory and inhibitory dendrites, respectively. Neuron index j indicates that the j_{th} neuron is to be updated, and when it is not in the refractory phase, it will give rise to a peak at time t.

The input current, $I_j(t)$, is defined as the weighted sum of spikes from external inputs and other neurons as

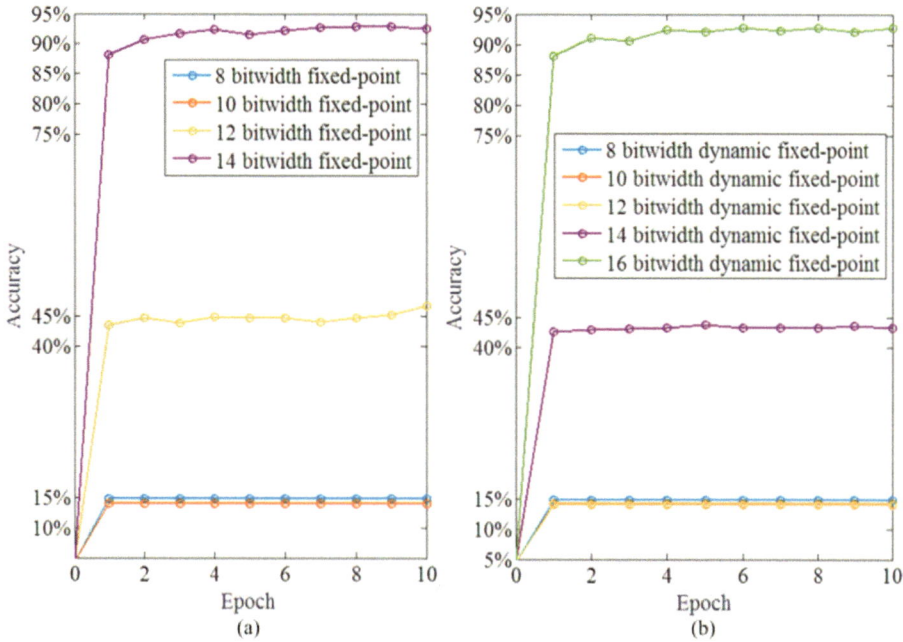

Fig. 2.11 Learning accuracy based on fixed-point and dynamic fixed-point representations. **a** Learning accuracy based on fixed-point representation with different bit width. **b** Learning accuracy based on dynamic fixed-point representation with different bit width [20]

$$
\begin{cases}
I_j(t) = \sum_{j=1}^{n} W_{ij}^{in} \alpha_i(t - d_{ij}^{input}) + \sum_{j=1}^{n} W_{ij}^{rec} \beta_i(t - d_{ij}^{rec}) \\[2mm]
I_j^i(t) = \sum_{j=1}^{n} W_{ij}^{iin} \alpha_i(t - d_{ij}^{iinput}) + \sum_{j=1}^{n} W_{ij}^{irec} \beta_i(t - d_{ij}^{irec}) \\[2mm]
I_j^e(t) = \sum_{j=1}^{n} W_{ij}^{ein} \alpha_i(t - d_{ij}^{einput}) + \sum_{j=1}^{n} W_{ij}^{erec} \beta_i(t - d_{ij}^{erec})
\end{cases}
\tag{2.23}
$$

where W_{ij}^{in}, W_{ij}^{iin}, and W_{ij}^{ein} represent the synaptic weights of soma, inhibitory dendrite and excitatory dendrite, respectively. W_{ij}^{rec}, W_{ij}^{erec}, and W_{ij}^{irec}, on the other hand, represent the recurrent synaptic weights of soma, excitatory dendrite and inhibitory dendrite, respectively. The constants d_{ij}^{input}, d_{ij}^{iinput}, d_{ij}^{einput}, d_{ij}^{rec}, d_{ij}^{irec}, and d_{ij}^{erec} represent the delays of input and recurrent synapses for soma, inhibitory dendrite and excitatory dendrite. The spike trains $\alpha_i(t)$ and $\beta_i(t)$ are modeled as sums of Dirac pulses, which represent the spike trains from input neurons and neurons with recurrent connections, respectively.

The SAM model is discretized with a time step $\Delta t = 1$ ms. The neural dynamics in discrete time can be, therefore, formulated as

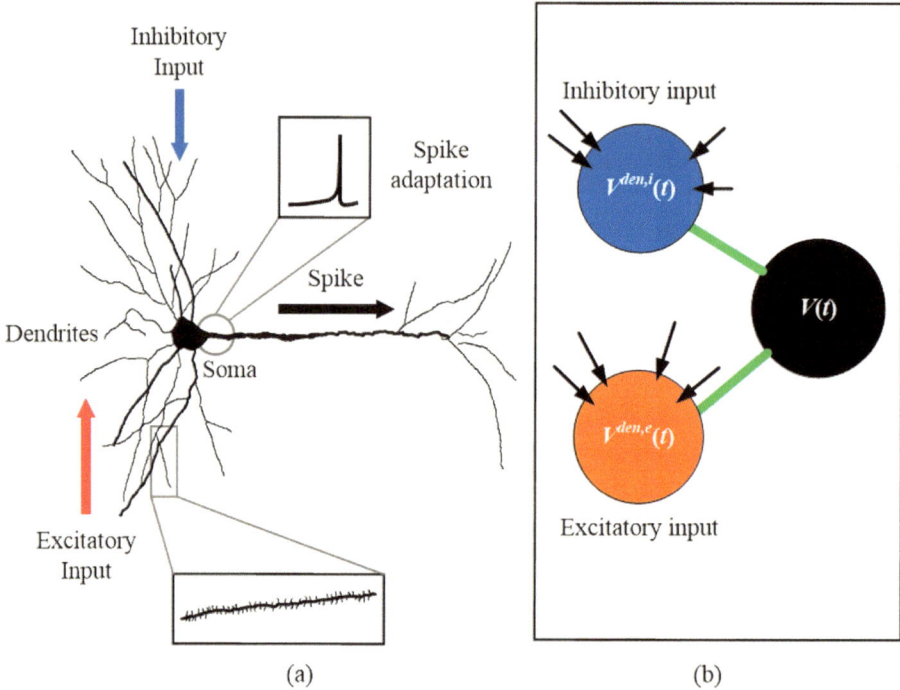

Fig. 2.12 Morphological architecture of a neuron with dendrites, which inspires the design of the SAM model. **a** Biological neuron with dendrites. **b** SAM neuron model [47]

$$
\begin{cases}
V_j(t + \Delta t) = \alpha V_j(t) + (1 - \alpha) R I_j(t) + (1 - \alpha) V_j^i(t) \\
\qquad\qquad + (1 - \alpha) V_j^e(t) - \Gamma_j(t) z_j(t) \Delta t \\
V_j^{den,e}(t + \Delta t) = \alpha V_j^{den,e}(t) + (1 - \alpha) R^e I_j^e(t) \\
V_j^{den,i}(t + \Delta t) = \alpha V_j^{den,i}(t) + (1 - \alpha) R^i I_j^i(t)
\end{cases}
\tag{2.24}
$$

where $\alpha = \exp(-\Delta t/\tau)$. Variable $z_j(t)$ represents the spike train of neuron j assuming values in $\{0, 1/\Delta t\}$. The dynamics of $\Gamma_j(t)$, representing the firing rate of neuron j, is changed with each spike, and is defined as

$$
\Gamma_j(t) = \tau_j^0 + \eta \cdot \tau_j(t)
\tag{2.25}
$$

where η represents a constant that scales the deviation $\tau_j(t)$ from the baseline τ_j^0. The variable $\tau_j(t)$ can be formulated as

$$
\tau_j(t + \Delta t) = \lambda_j \tau_j(t) + (1 - \lambda_j) z_j(t)
\tag{2.26}
$$

Table 2.3 Parameter settings of the SAM model [47]

Parameter	Value	Parameter	Value
R_m	1 Ω	R_m^e, R_m^i	1 Ω
τ	20 ms	V^{inh}, V^{exc}	0 mV
$d^{input}, d^{iinput}, d^{einput}$	5 ms	$d^{rec}, d^{irec}, d^{erec}$	5 ms
η	1.8	τ^0	0.01
τ_a	700 ms	g^{inh}, g^{exc}	1 nS

where $\lambda_j = \exp(-\Delta t/\tau_{a,j})$ and $\tau_{a,j}$ represents the adaptation time constant. The parameter values of the SAM model are listed in Table 2.3.

2.3.2 Self-adaptive Multi-compartment Neuron in SNN Architecture

In this section, SAM is tested in kinds of learning tasks. Figure 2.13 shows the SAM model of the network architecture. It consists of three layers, including input layer, hidden layer and output layer. The coding scheme of the input–output layer is chosen in accordance with the task to be executed. The blue solid and the green dotted lines indicate the feed-forward and side-inhibition synaptic connections, respectively. In the hidden layer, the dendrites and the soma of different neurons are attached by the lateral inhibition synapses, which are random and sparse. The dendrites in SAM model accept the neural information from the input layer, and the soma of the SAM model sends the peak to the output level.

In the SAM-based SNN architecture, the initial weights of the network are set according to a Gaussian distribution $W_{ij} \sim \frac{w_0}{\sqrt{n_{in}}} N(0, 1)$, where n_{in} represents the number of spiking neurons in the considered weight matrix. $N(0,1)$ represents the zero average unit variance Gaussian distribution, and $w_0 = \Delta t/R$ is a weight scaling factor that depends on the membrane resistance R and Δt. This scaling factor is used to initialize the network with a real firing rate required for effective training.

A deep rewiring algorithm is utilized since it can maintain the sign of each synapse during learning [38]. Thus, the sign is inherited from the initialization of the network weights. Therefore, it is necessary to initialize an effective weight for a given segment of inhibition and excitatory neurons. In order to do so, a sign $k_i = \{-1, 1\}$ is generated randomly for neuron i by sampling from a Bernoulli distribution. To avoid exploding gradients, the weights are scaled so that the maximum eigenvalue is less than 1. In this way, a big square matrix is produced, and the desired number of rows with equal probability is chosen. This dense matrix is then multiplied by a binary mask in order to generate a

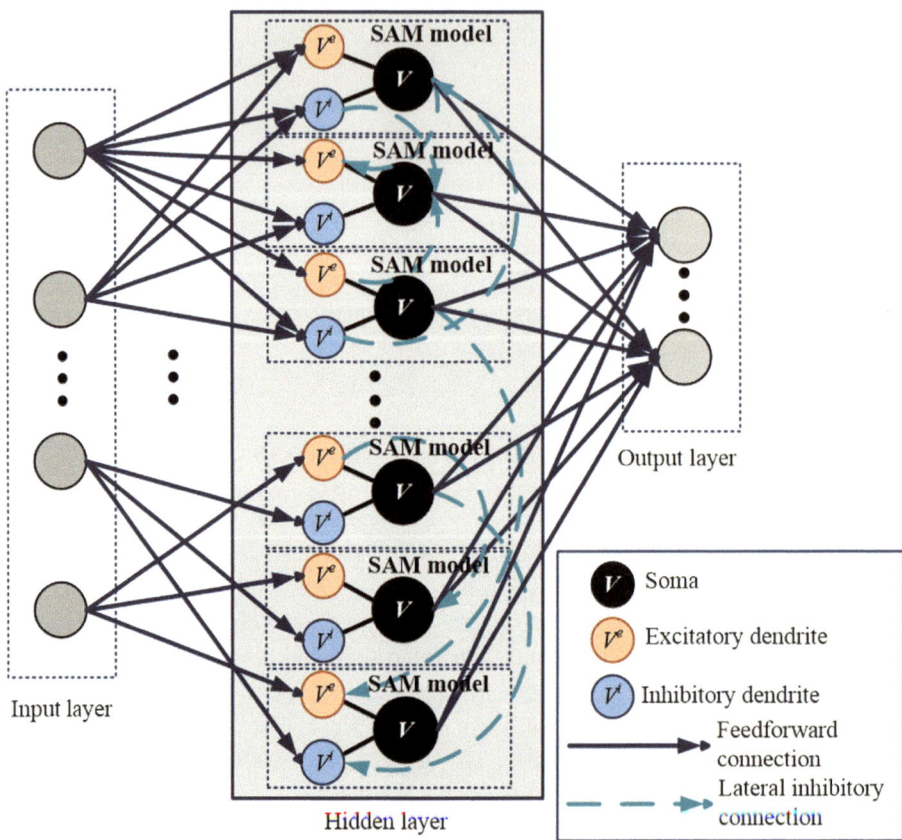

Fig. 2.13 Network architecture for learning and memory based on the SAM model. This architecture is comparable to a 2-layer network of point neurons. In the hidden layer, the dendrites and the soma of different neurons are randomly linked by the lateral inhibition synapse. The grey rings on the input and output levels indicate the input and output spiking neurons, which are not SAM neurons [47]

sparse matrix, as part of the deep rewiring algorithm, which allows for a dynamic disconnection of certain synapses and rewiring other synapses. The L_1-NORM regularized parameter is 0.01 and the temperature is 0 in the deep rewiring algorithm.

2.3.3   Learning with Self-adaptive Multi-compartment Model

In the traditional ANN model, the gradient of the loss function is calculated by back propagation. However, it is not possible to apply back propagation training in SNNs due to the non-differentiable nature of the spiking output. If the time is discretized, the gradient must be propagated either by successive time or by multiple time steps. A pseudo-derivative method is used for learning with a SAM model, which can be formulated as

$$\frac{dz_j(t)}{dv_j(t)} = k \max\{0, 1 - |v_j(t)|\} \tag{2.27}$$

where $k = 0.3$ (usually smaller than 1) is a constant that can suppress an increase in backpropagated errors through spikes by means of a pseudo derivative of amplitude to obtain steady performance. The variable $z_j(t)$ is the spike train of neuron j under the assumption that the values are in $\{0,1\}$. The variable $v_j(t)$ represents the normalized membrane potential, which is defined as

$$v_j(t) = \frac{V_j(t) - \Gamma_j(t)}{\Gamma_j(t)} \tag{2.28}$$

where Γ_j represents the firing rate of neuron j. For the purpose of providing the self-learning ability needed in reinforcement learning, this chapter uses the Proximal policy optimization (PPO) [39]. The implementation of this algorithm is simple, and it has the ability of self-learning. The clipped surrogate objective of PPO is defined as $O^{PPO}(\theta_{old}, \theta, t, k)$. The loss function with respect to θ is then formulated as

$$L(\theta) = -\frac{\sum_{k<K}\sum_{t<T} O^{PPO}(\theta_{old}, \theta, t, k)}{KT} + \mu_f \frac{1}{n}\sum_j \left\| \frac{\sum_{k,t} z_j(t,k) - \phi^{0*}}{KT} \right\|^2 \tag{2.29}$$

where ϕ^{0*} represents a target firing rate of 10 Hz and μ_f is a regularized hyperparameter. The variables t, k, θ indicate the simulation time step, the total number of epochs, and the current policy parameter [48]. For each training iteration, $K = 10$ episodes of $T = 2000$ time steps are generated with a fixed parameter θ_{old}, which is the vector of policy parameters before the update as described in [39]. The loss function $L(\theta)$ is minimized in each iteration using an ADAM optimizer [40].

In this section, the SAM-based SNN model is used to solve the problem of agent navigation task, as described in previous studies [41, 42]. The agent is required to learn how to find a target in a 2D region, and to then navigate to that destination from random locations in that region. This task is related to the well-known neuroscience paradigm of the Morris water maze task to study brain learning [43, 44]. In this task, at every time step, neurons in an input layer of a SAM-based network receive information about the present environment status, $s_e(t)$ and reward $r_w(t)$. The environment state $s_e(t)$ is represented by the coordinate of the agent's position. The location coordinates are coded by the input neurons in accordance with the Gauss Population Rate Code. Each neuron in the input layer is assigned a coordinate value with a firing rate of $r_{max} = \exp(-100(\xi_i - \xi)^2)$, in which ξ_i, ξ denotes a real coordinate and a preferred coordinate value, r_{max} is 500 Hz, and the instant reward $r(t)$ is coded by two sets of input neurons. Neurons in the first group spike synchronously when a positive reward is received, and the second group spike when the SAM model receives negative reward. The output of SAM network is expressed by 5 read neurons with membrane potential $V_i(t)$. The action vector $\zeta(t) = \left(\zeta_x(t), \zeta_y(t)\right)^T$, which is used to judge the motion of an agent in a navigational task, is computed from the Gauss distribution with the mean $u_x = \tanh(V_1(t))$ and $u_y = \tanh(V_2(t))$, and the variance $\phi_x = \sigma(V_3(t))$ and $\phi_y = \sigma(V_4(t))$. The final read neuron V_5 output is computed so as to predict a value function $u_\theta(t)$. It predicts the expected discounted sum of the future rewards $\Omega(t) = \sum_{t'>t} \gamma^{t'-t} \omega(t')$, where $\gamma = 0.99$ represents the discount factor and c represents the reward at time t'. Moreover, the SAM model is improved by adding the Gaussian noise with average 0 and standard deviation 0.03. According to the experiments, the addition of noise improves the performance of the model in navigation (Fig. 2.14).

2.3.4 Performance Evaluation

(1) Dynamical analysis of SAM model

In the first analysis, the changes of the neuronal membrane potential are investigated in both the soma and the excitatory/inhibitory dendrites, and examined the change of the threshold in response to the outside stimulus. When the external current is negative, the somatic membrane potential is suppressed, and the threshold of $\Gamma(t)$ does not vary. The amplitude and firing rate of neurons are raised with the increase of the input current, as is $\Gamma(t)$. Moreover, the amplitude of excitatory and inhibitory dendrites also increased with the increase of stimulating current.

In the second analysis of the SAM dynamics, the relation between the steady state spike rate or saturation threshold is investigated, as illustrated in Fig. 2.15. Figure 2.15a illustrates the variation of SAM soma firing rate in response to an external current that contains only excitatory, only inhibitory, and simultaneous excitatory and inhibitory inputs. The results indicate that inhibition input decreases the firing activity of the soma, whereas the increase in the external excitatory current up to 0.65 nA increases the firing rate until it is saturated at 135.5 Hz. The firing activity saturates at 0.77 nA when simultaneous inhibitory and excitatory currents are applied. The evolution of the dynamic threshold $\Gamma(t)$ is also illustrated in Fig. 2.15b with regard to the variation of the input current. Threshold $\Gamma(t)$ has a saturation value of 0.1585, and the inflection point with excitatory and simultaneous input occurs at 0.65 and 0.78 nA, respectively.

In Figs. 2.16 and 2.17, the influence of key parameters of SAM model on the saturation threshold and steady state spike frequency is discussed. These parameters include η, τ^0, τ_v and R. In this section, SAM's spiking behavior is investigated by changing two of the parameters and setting the rest to the default values shown in Table 2.2. The results show that the saturation threshold $\Gamma(t)$ can be improved by adding η, τ^0 and R. As illustrated in Fig. 2.16a, b and d. $\Gamma(t)$ reaches around 0.2 when $\tau^0 = 0.15$, $R = 1.25$,

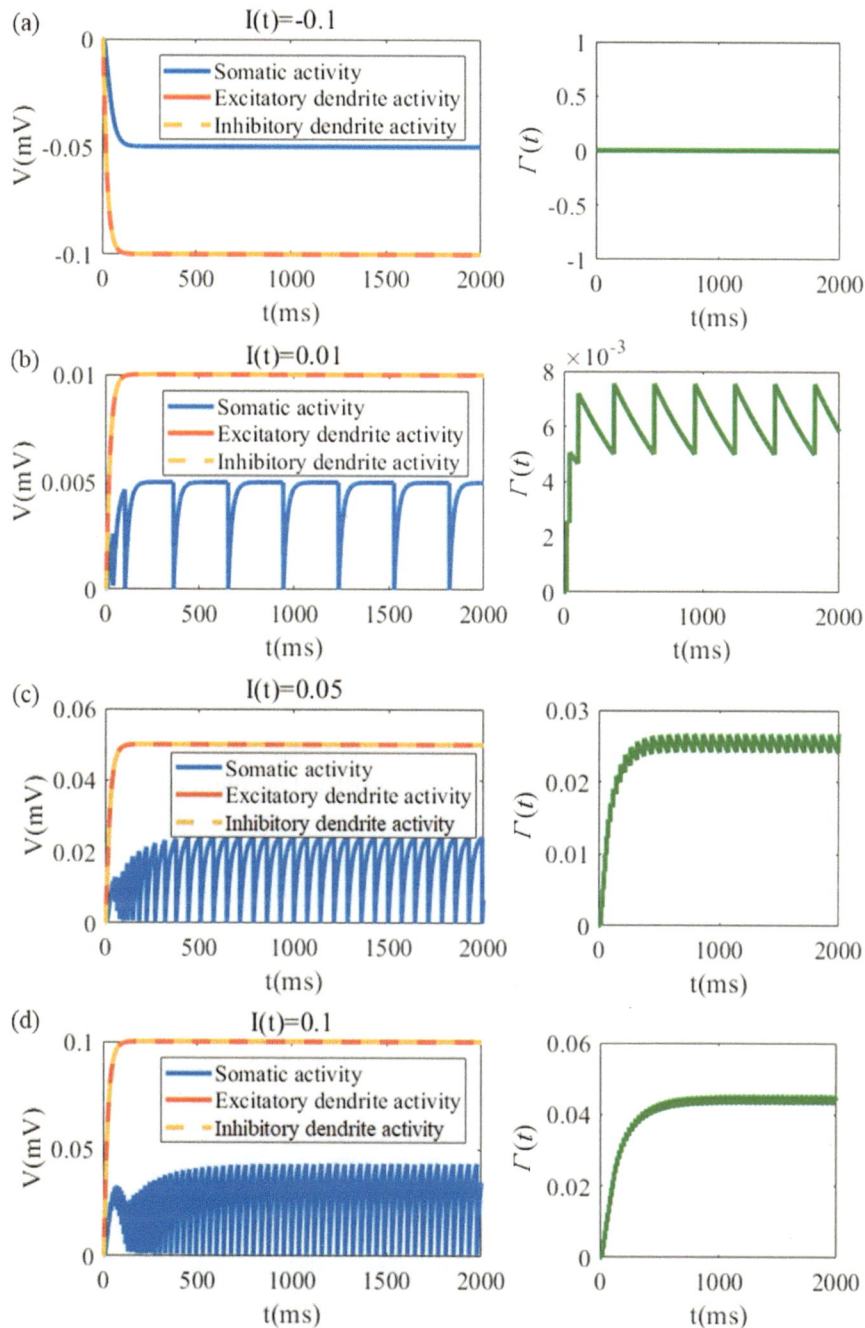

Fig. 2.14 SAM membrane potential (left plots) and dynamic threshold (right plots) behavior in response to negative and low positive external currents. **a** $I(t) = -0.1$. **b** $I(t) = 0.01$. **c** $I(t) = 0.05$. **d** $I(t) = 0.1$ [47]

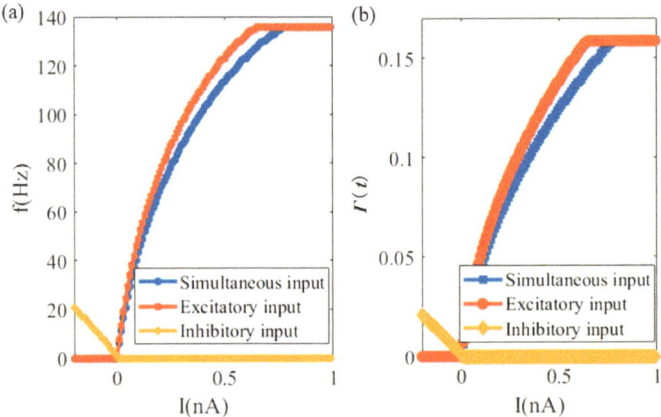

Fig. 2.15 Change in firing speed and threshold in response to a variation of outside input current. **a** Somatic firing rate. **b** Dynamic threshold $\Gamma(t)$ of the SAM [47]

or $\eta = 2.4$. The value of $\Gamma(t)$ is reduced to 0.05 when $\tau^0 = 0.02$, Rm = 0.55, or $\eta = 1$. However, as shown in Fig. 2.16c, e and f, the value of $\Gamma(t)$ is increased when τ is reduced. This indicates that the saturation threshold $\Gamma(t)$ is positively correlated with the critical parameters η, τ^0 and R. Moreover, the saturated threshold $\Gamma(t)$ has a negative relation with τ. Figure 2.16 shows the effect of the critical parameters on the steady state spiking frequency. As illustrated in Fig. 2.16a, c, and f, the SAM's spiking frequency is reduced by an increase in η, τ^0 and τ. On the other hand, as illustrated in Fig. 2.17b, d and e, the spike frequency will be increased in response to the rise of R, which indicates that the SAM spike frequency is positively correlated with R.

The results of the above studies on the dynamics of SAM give us a deeper insight into how SAM's internal parameters govern its dynamics.

(2) Supervised learning of the MNIST dataset using sequential spatiotemporal encoding

SNNs are believed to have the potential to replicate brain-like cognitive behavior due to their ability to encode spatial and temporal information. Conversely, most ANN models lack timing dynamics. In an experiment, a SAM-based SNN is fed an MNIST image pixel in a sequence of 784 time steps, one for each pixel. The input coding approach assigns a specific threshold value to each input neuron based on its corresponding grayscale value. The network's output is determined by the mean value of the output read in 56 time steps after the final input. The neural network is trained by minimizing the cross-entropy error of the label distributions and the Softmax of the averaged readout.

Table 2.4 shows that the SAM-based SNN model achieves a better average classification precision over MNIST than LSTM, RNN, and recurrent spiking neural network

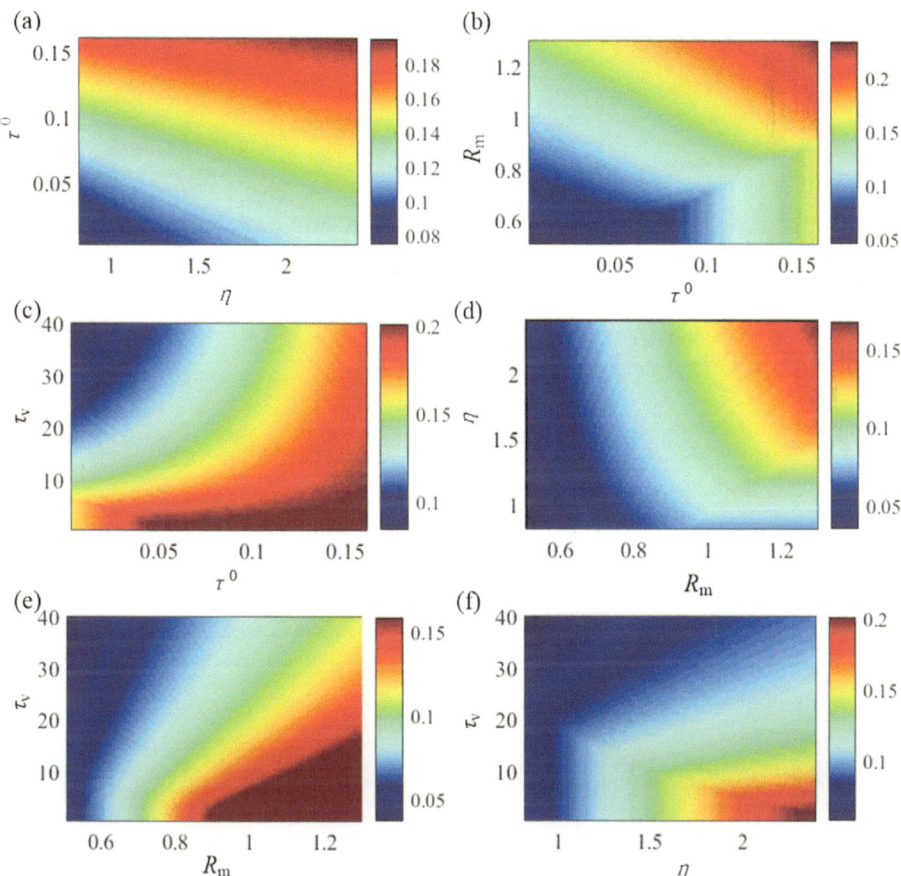

Fig. 2.16 Effects of critical parameters in the SAM model on the saturated threshold value, $\Gamma(t)$. The figures show the comprehensive effects of changes of **a** η and τ^0, **b** R and τ^0, **c** τ^0 and τ, **d** η and R, **e** τ and R, and **f** τ and η [47]

(RSNN) models [45, 46]. CNNs are not employed due to their lack of recurrent architecture and inability to interpret and retain working memory. Optimizing a sparse network connectivity architecture with 12% global connectivity yields optimal performance for SAM-based SNN models. This approach uses only some of the model's parameters compared to RNN or LSTM. Moreover, a longer input time of 2 ms has less of a detrimental effect on SAM learning compared to other models due to its working memory capability. The performance of the SAM-based SNN is then investigated in the classification of MNIST data with sequential space–time coding. In neuromorphic hardware, the SynOps is triggered only by active connections in the SNN model. Using the number of SynOps for a given precision demonstrates the efficiency of learning and potential energy consumption. Figure 2.18 shows that the SAM model requires fewer SynOps than RSNN

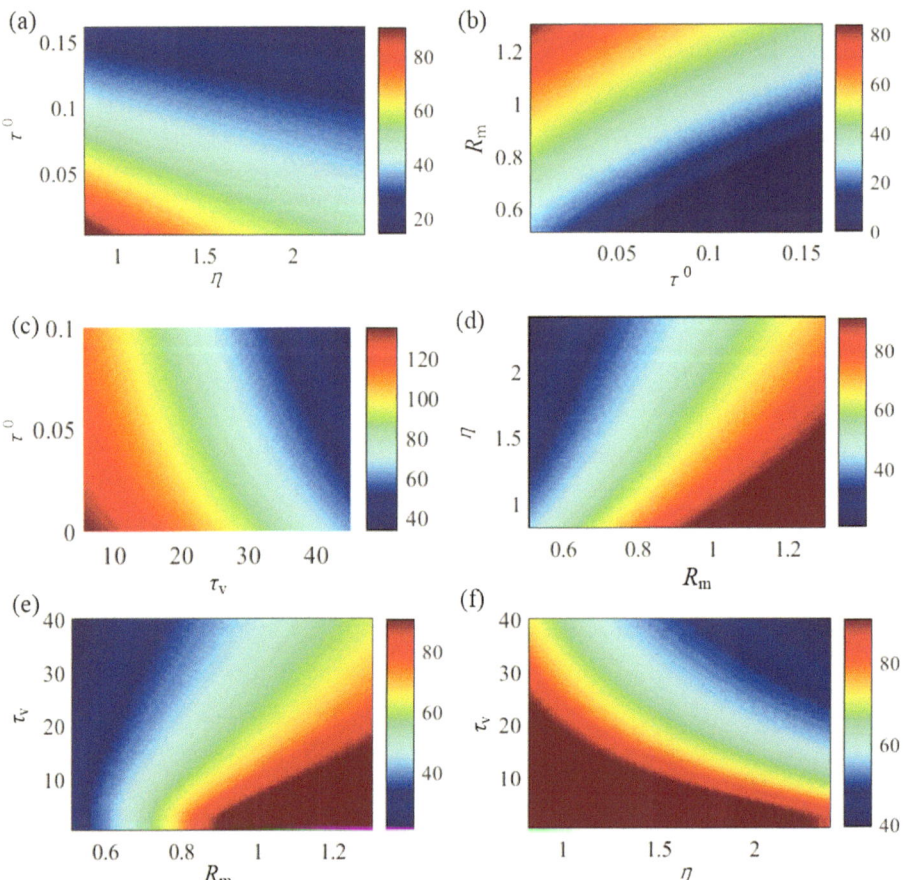

Fig. 2.17 Influence of key parameters in SAM model on steady firing rate. The figures indicate the comprehensive effects of **a** R and τ^0, **b** R and τ^0, **c** τ^0 and τ, **d** η and R, **e** τ and R, **f** τ and η [47]

and segregated dendritic learning (SDL) proposed by Guerguiev et al. [21]. The SAM model achieves the same accuracy as an SDL model with nearly 10 million fewer Syn-Ops and nearly 30% more precision than an RSNN model with the same SynOps count. The SAM model's temporal and spatial coding strategy results in fewer spikes, generating less power and higher precision than the rate-based coding strategy employed by the SDL model. While adaptation of sensitivity based on network precision in the error code compartment can significantly enhance the model's performance, this chapter focuses more on methodology than pursuing more accurate results.

In a separate experiment, the effectiveness of the SAM model is tested in distinguishing various temporal patterns. To achieve this goal, a five-class classification system and a template spike pattern for each category are generated. These templates were randomly

Table 2.4 Results of spatiotemporal pattern classification on sequential MNIST [47]

Model	Displayed time (ms)	Connectivity (%)	#Neurons	Dendrites (E/ I)	Mean accuracy (%)	Maximum (%)
LSTM	1	100	128	–	79.80	98.5
LSTM	2	100	128	–	48.20	98.0
RNN	1	100	128	–	71.30	89
RNN	2	100	128	–	30	67.9
RSNN	1	12	220	–	60.90	63.3
RSNN	2	12	220	–	34.6	51.8
SAM	**1**	**12**	**220**	**1.0/0.6**	**95.10**	**98.4**
SAM	**2**	**12**	**220**	**1.0/0.6**	**94.85**	**98.4**
SAM	1	50	220	1.0/0.6	94.60	97.7
SAM	2	50	220	1.0/0.6	94.35	97.7
SAM	1	80	220	1.0/0.6	94.10	99.2
SAM	2	80	220	1.0/0.6	94.25	98.4
SAM	1	12	220	0.6/0.1	94.10	98.4
SAM	2	12	220	0.6/0.1	93.35	97.7

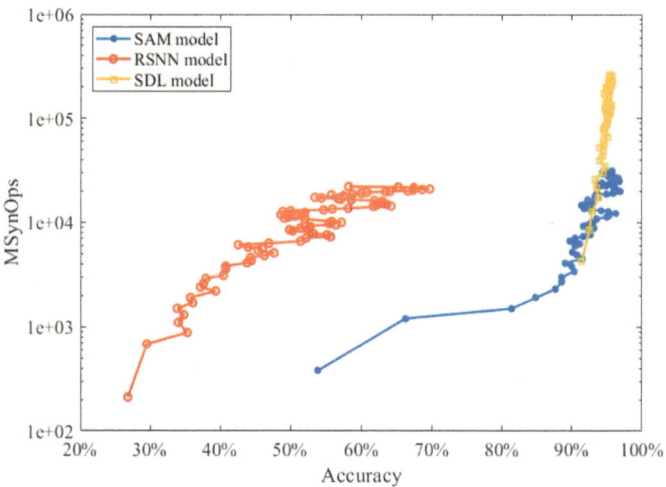

Fig. 2.18 Performance investigation of learning accuracy and power efficiency of SAM compared to state-of-the-art sequential encoding methods [47]

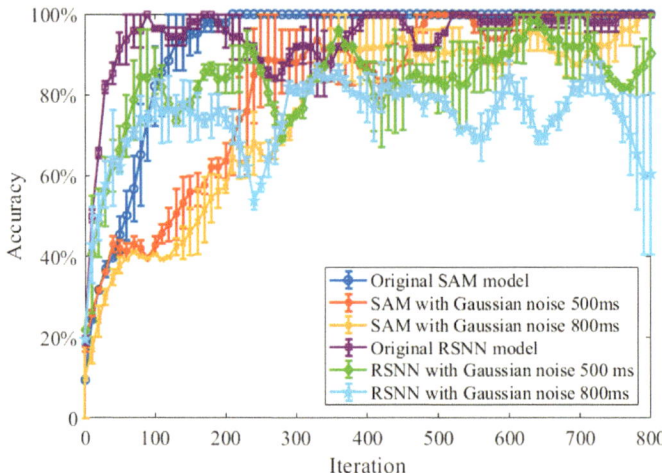

Fig. 2.19 Robustness of SAM and RSNN models to classify temporal spike patterns. The RSNN is compared, since it is built on an RSNN architecture, in which the spiking neurons are replaced by a SAM neuron model. The RSNN architecture is in reference [47]

generated and then fixed. Each spike pattern consisted of neurons firing a specific number of spikes over a period of time. To assess the impact of noise, a disturbance is added, in the form of a zero-average Gaussian noise with a standard deviation of $\sigma = 500$ ms or 800 ms, to each spike pattern of every class. As shown in Fig. 2.19, the SAM model outperformed the RSNN model in terms of robustness. This can be attributed to the SAM model's self-adapting mechanism, which enhances its learning ability and resilience. Additionally, the dendritic non-linearity of the SAM neural model enable it to learn better in the presence of noise.

Sparse coding is a notable feature of the SNN model, which is exemplified in Fig. 2.20, showing a grid diagram of a SAM-based SNN model used for spike pattern classification. The input patterns are encoded via population coding based on the firing probability of 80 input neurons. An additional input neuron is triggered once the spike pattern display is finished to initiate an output from the SAM network. When the input patterns are displayed in a sequence, the raster map of the neural activity in the hidden layer, consisting of 220 neurons, is presented in Fig. 2.20a. It's evident that the SAM-based SNN model uses sparsely active neurons to process neural information. Figure 2.20b shows the dynamic changes in the firing threshold of neurons in the hidden layer during the learning of the sequential spike pattern. The adaptive threshold is continuously adjusted and finally stabilizes at the saturation level.

Previously, the SNN model could identify background activity features using a spike-based approach [48]. In this section, the feature detection capabilities of the SAM model are evaluated. Eight images are generated, each representing a directional pattern, such as

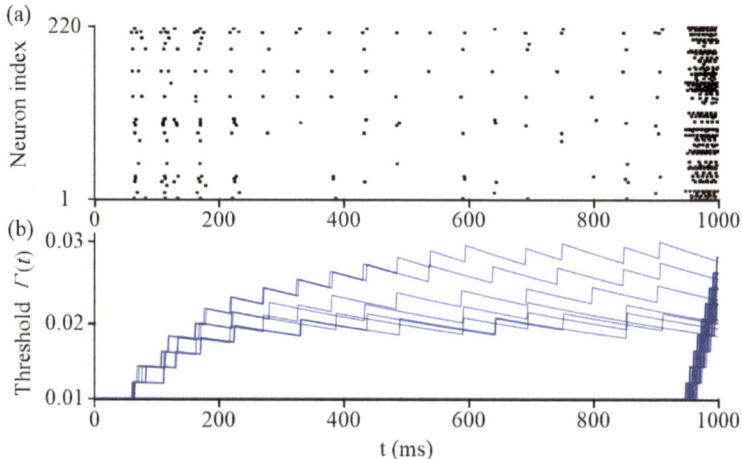

Fig. 2.20 Neural network dynamics in the temporal and spatial learning of spike patters. **a** Raster map of SAM neurons. **b** Adaptive threshold of SAM model evolution over time in the hidden layer [47]

0, 22.5, 45, 60, 75, 90, 112.5, 135, and 157.5°. Each image contains 729 (27 × 27) pixels, of which 10% are randomly selected to receive Gaussian noise, as shown in Fig. 2.21a. Figure 2.21b illustrates the learning performance of SAM and RSNN in detecting the features of each pattern. The results show that the spike-driven learning with memory capabilities of SAM enables successful detection of features in various patterns. Additionally, SAM is more resilient in identifying patterns contaminated by background noise and in situations where 40–80% of the peaks are randomly removed with certain probability. The performance of the SAM model outperforms the traditional RSNN model, as the tree non-linearity of SAM can extract more information from the input spike signal. Furthermore, the self-adaptation mechanism of the soma in the SAM model facilitates learning and storing spiking patterns, and it can adapt to different types of noise.

(3) SAM model for meta learning with working memory

In this section, a navigation task and a MNIST classification task are explored. These tasks require spike-based meta-learning ability, meaning that their performance depends on previous experience.

First, meta-learning ability of SAM for a navigation task is investigated. Figure 2.22 illustrates the SAM model's ability to manage navigation tasks flexibly. In a 2D arena simulation environment, a virtual agent is displayed as a point, and a SAM-based SNN model controls the agent's movements. The agent's starting and destination locations are randomly set with uniform probability. At each time step, the agent selects an action by producing a small velocity vector. When the agent reaches its destination, it receives a

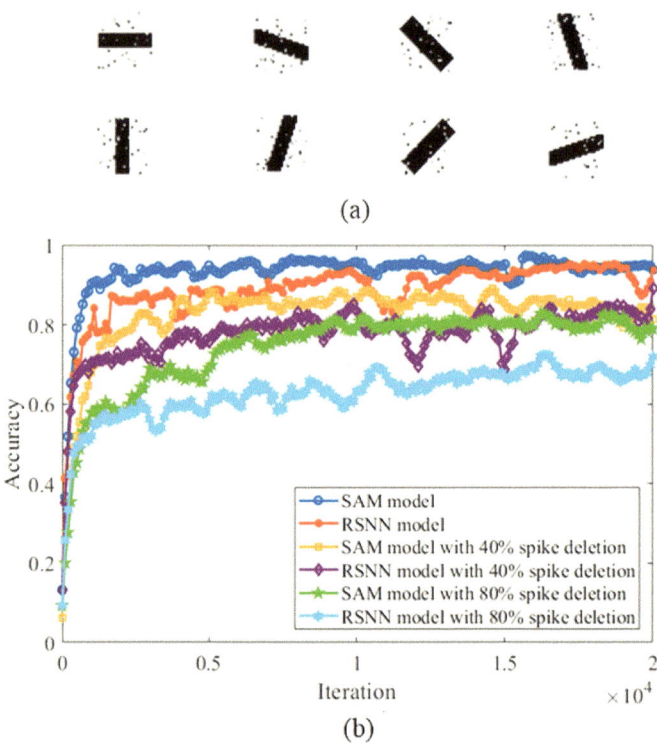

Fig. 2.21 Performance analysis in a spatiotemporal feature detection task. **a** Images in the feature detection task. **b** Performance comparison between SAM and RSNN models [47]

reward of 1. Figure 2.22a–c shows that SAM-based agents learn to navigate to the correct destination. Figure 2.22d shows the number of times an agent successfully arrived at its destination in each learning iteration. Each iteration contains a batch of 10 episodes, and the weights are iteratively updated across these episodes. The diagram shows that navigation performance decreases from 1000 to 2000 iterations, and reaches a maximum after 3000 iterations. This tendency is also illustrated by the navigational loss function $L(\theta)$, which reaches 0 and stays at 0 for most of the time after 300 iterations, as shown in Fig. 2.22e. Finally, the agent learns to take the shortest route to the destination, as shown in Fig. 2.22f. During training, there is a fluctuation in the 1800th iteration, but it returns to a steady state. These results demonstrate that SAM-based agents have meta-learning ability in navigation tasks due to their self-adapting mechanism characterized by tree nonlinearities.

Next, the behavior of SAM-based network is explored before, during, and after meta-learning in another navigation task. In this task, the start and end positions are randomly set to create five routes, as illustrated in Fig. 2.23. The objective is to maximize the

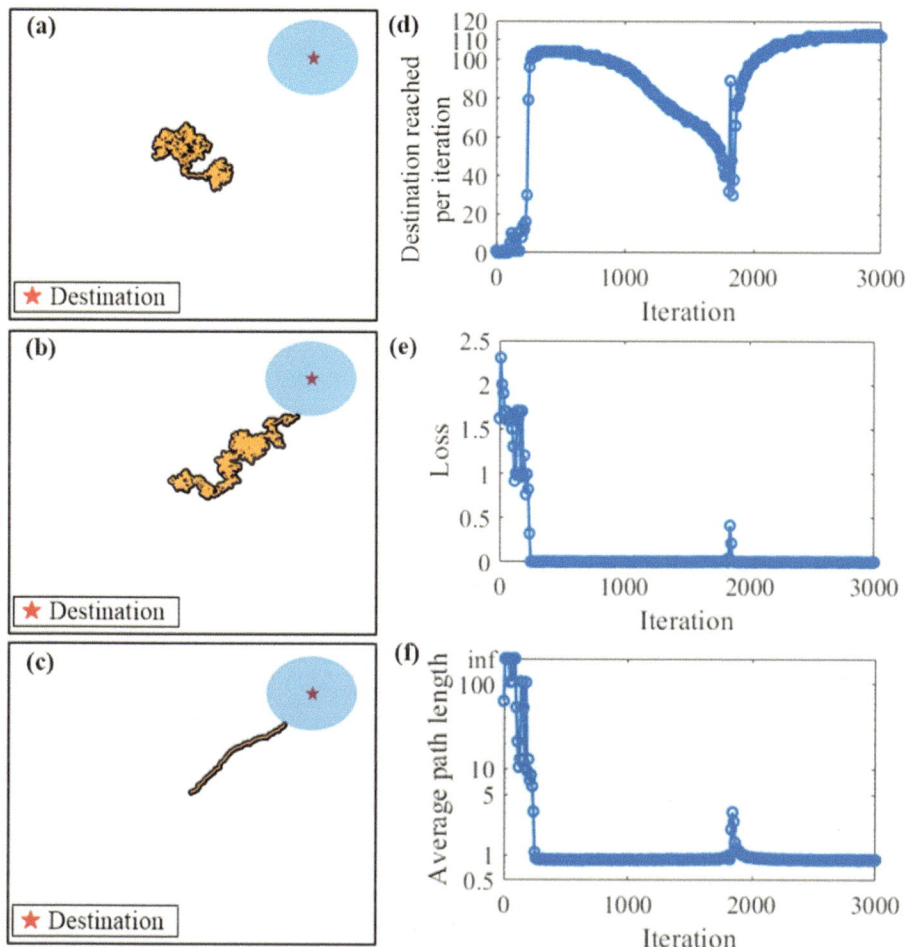

Fig. 2.22 Meta learning based on the SAM model in a navigation task. The orange points show the navigation path during the task. **a** Prior to the study. **b** During the course of study. **c** After the study. **d** Target reached in each iteration during the learning process. **e** During the learning process, the navigation loss function $L(\theta)$. **f** Average path length to the destination during learning [47]

number of destinations reached in each episode. In every episode, one of the best agents has to explore and remember the location of the target and use its prior knowledge to take the shortest route to the target. The first route is based on BPTT [49]. Figure 2.23 shows that an agent with a SAM model is capable of autonomous navigation because of its strong meta-learning capability.

This section also investigates the meta-learning ability of SAM for MNIST. To demonstrate this ability, the MNIST data set is split into two sections. The first section consists

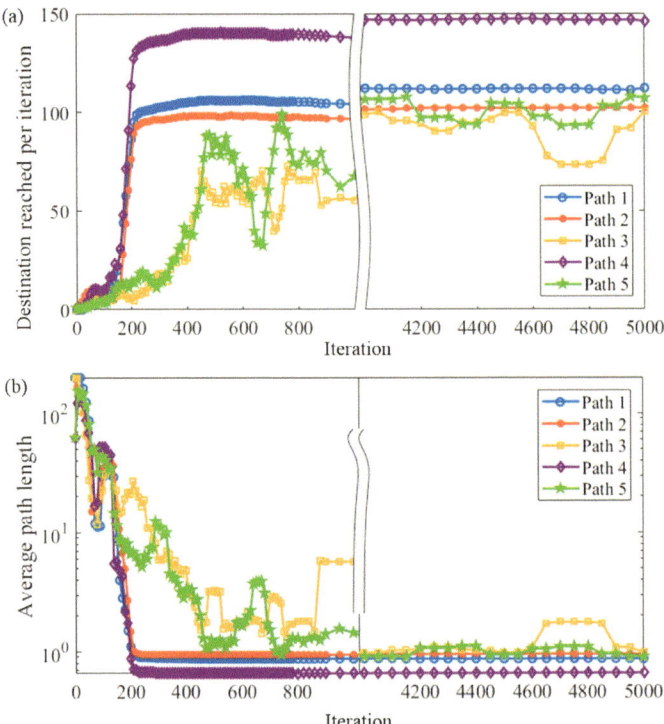

Fig. 2.23 Meta-learning ability of the SAM model in the five-path test. **a** Target reached in each iteration in five routes. **b** Average duration of the journey over the five routes [47]

of 30,000 images with 0–4 digits, and the second section contains 3000 images with 5–9 digits. In the first learning procedure, the SAM-based SNN model is trained with the MNIST dataset. The second portion of the dataset is used for the second learning procedure, where the SAM-based SNN model is further trained. Figure 2.24 shows that the second learning process performs better than the first one, indicating the SAM model's meta-learning capability. Additionally, the SAM-based SNN model outperforms the traditional RSNN model in both the first and second learning procedures. The RSNN model does not perform well in the first learning procedure. Furthermore, the SAM-based SNN model demonstrates a higher learning rate and faster convergence rate in the second learning procedure. These results indicate that the SAM-based SNN model has a powerful meta-learning capability in learning spatiotemporal patterns.

(4) Working memory for spatiotemporal learning

To further validate the working memory capacity of the SAM model, a store and recall task that is similar to the one described in [50] is studied in detailed. Specifically, it is a

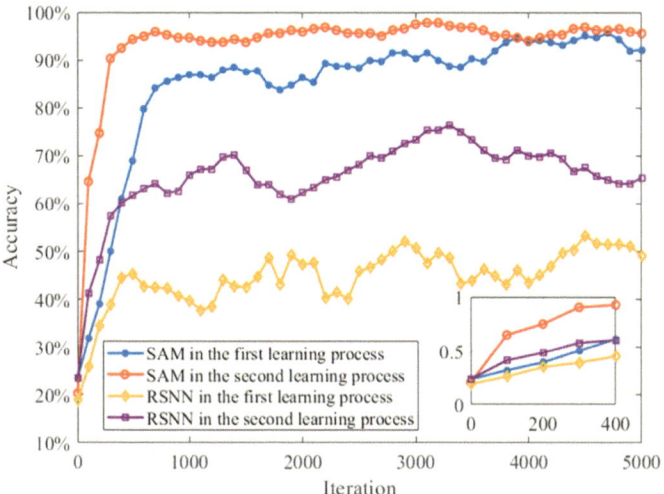

Fig. 2.24 Meta-learning capability of SAM model on sequential MNIST data set [47]

SAM-based SNN model that receives a series of frames represented by 10 rows of spikes over a specified time interval. Input #1 and input #2 use input neurons 1–10 and 11–20, respectively, for their spiking activities. For instance, the first frame in input #1 consists of 10 spike trains ranging from 200 to 350 s, as depicted in Fig. 2.25a. Additionally, the graph shows that the SAM-based SNN input neurons can utilize neurons 21–30 and 31–40 to receive random store and recall commands. A storage command directly focuses on a specific frame of an input stream, while the objective of the task is to recreate the frame upon receiving a recall command.

To perform the storage and recall operation, an SNN is trained with 20 SAM neurons in its hidden layer and 20 output sigmoidal neurons, similar to previous research [51]. Figure 2.25 illustrates a sample section of the test with the spiking activities of the SAM neurons at the beginning of training, indicating that the SAM-based SNN cannot perform the storage and recall function without training. Figure 2.25b shows the random spiking activity of the 20 SAM neurons before any learning, while the activation of the sigmoidal output neurons (Fig. 2.25c) and the threshold values of the 20 SAM neurons do not exhibit any particular behavior prior to learning. However, once training is complete, the SAM-based SNN is capable of performing the working memory task, as shown in Fig. 2.26. Furthermore, the dynamic thresholds are more consistent with each other when trained in the store and recall task. The kind of working memory exhibited by the SAM model refers to the activity-silent form of working memory in the human brain, which has been investigated by Wolff et al. [50].

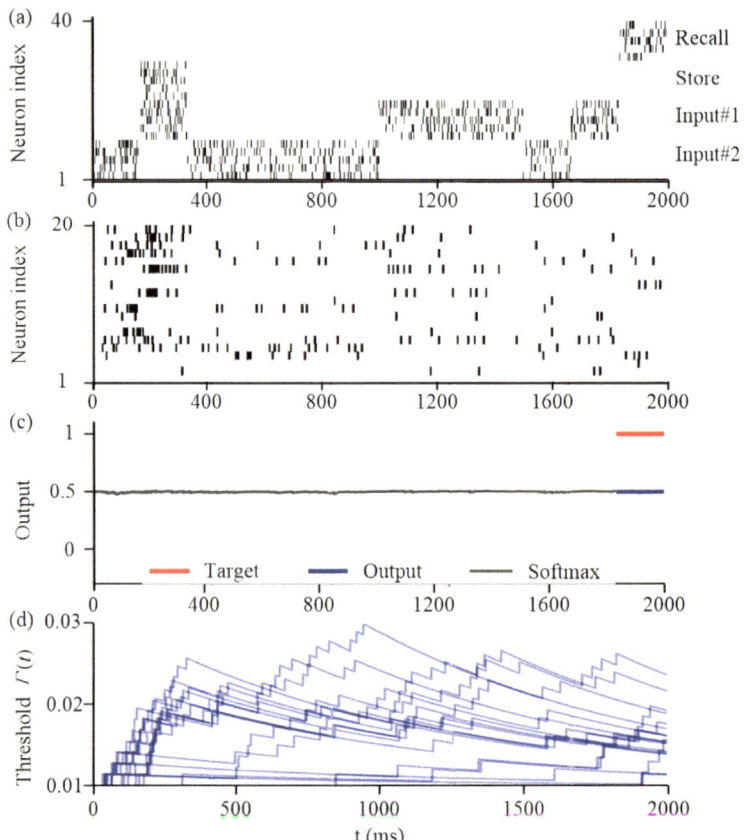

Fig. 2.25 Working memory capacity of SAM model at the start of training. **a** Spiking activity of the input neurons for the store and recall commands as well as the characteristic vectors # 1 and # 2. **b** SAM model spiking activity in suggested SNN architecture. **c** Traces of the sigmoidal reading neurons being activated. **d** Time evolution of the SAM model firing threshold $\Gamma_j(t)$ [47]

(5) Critical parameters of SAM in working memory

   In this section, the influence of key parameters on the SAM model's working memory performance are investigated, which is described in the previous section. As shown in Fig. 2.27, two values for β, the sum of Dirac pulses, which represent the spike trains from neurons with recurrent connections are considered as $\beta = 0.8$ and $\beta = 3$. This indicates that a high value of $\beta = 3$ will reduce the working memory capacity of the SAM. Moreover, high values of τ_v, the SAM membrane constant, will also decrease the working memory performance. Moreover, when the connectivity conductance of the

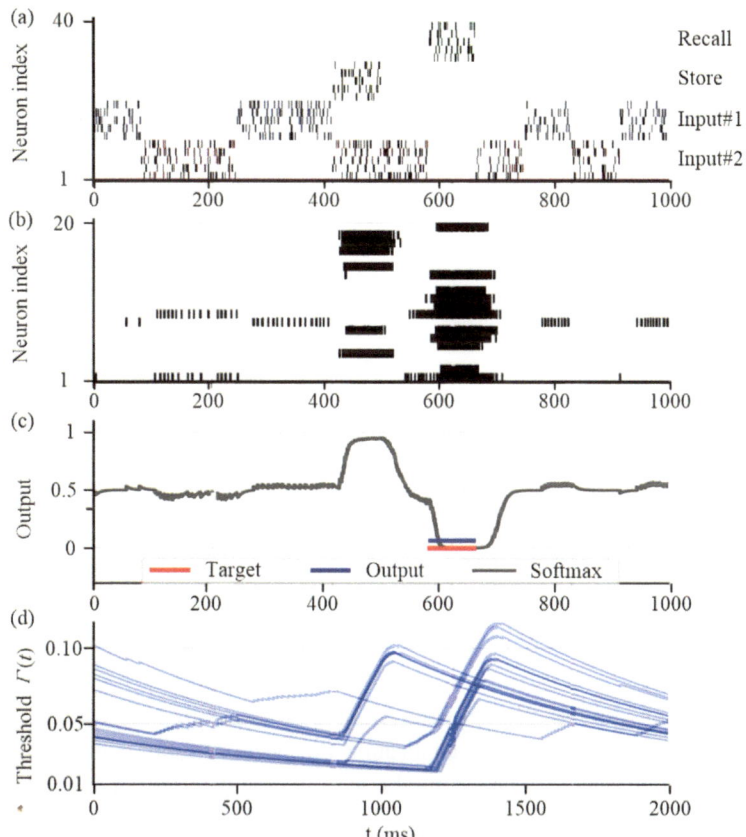

Fig. 2.26 Working memory capacity of SAM model at the start of training. **a** Spiking activity of the input neurons for the store and recall commands as well as the characteristic vectors # 1 and # 2. **b** SAM model spiking activity in suggested SNN architecture. **c** Traces of the reading neurons being activated. **d** Time evolution of the SAM model firing threshold $\Gamma_j(t)$ [47]

dendrites is reduced to $g_e = 0.6$ and $g_j = 0.1$, the learning convergence speed is lower than that of the initial values of $g_e = 1.0$ and $g_j = 0.6$. The SAM model, which has a lower beta value, and a higher membrane resistance of soma, R, will keep the memory property of the initial setup. Finally, Fig. 2.27 also demonstrates that the SAM model can improve the rate of learning convergence by using $\tau = 10$. Because the membrane constant τ is negatively correlated with the saturation threshold and the steady spiking frequency, it is shown that a higher stable spiking frequency and a saturated threshold value of the SAM model will further improve the learning convergence speed in working memory tasks. On the other hand, the higher the coupling strength of the dendrites, the slower the learning convergence will be.

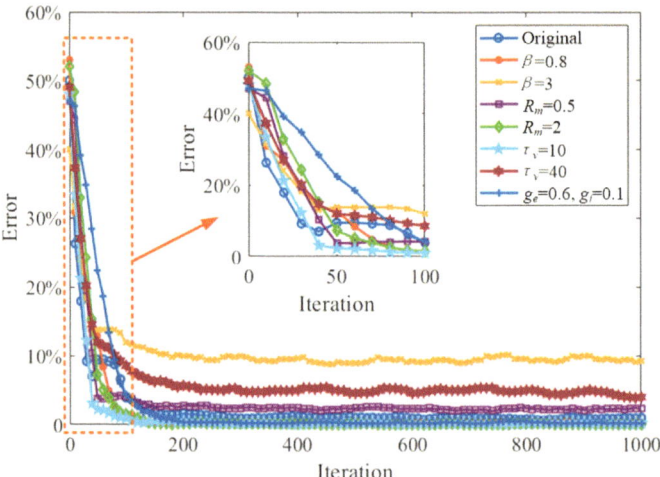

Fig. 2.27 Investigation on the effect of key parameters on the working memory capability of the SAM model used in store recall tasks [47]

2.4   Summary

This chapter mainly discusses spike-based learning models with multi-compartment models. Firstly, the basic concepts and application prospects of multi-compartment models and dendrites are introduced. The essential role of neuron dendrites in neural information processing is discussed. Secondly, the credit assignment mechanism based on separated dendrites and its application in SNN learning are explored. A biologically meaningful DEP algorithm [20] with dynamic fixed-point representation is presented. Finally, a multi-compartment spiking neuron model with self-adaptive characteristics is introduced in detail, and its application in SNN learning, especially meta-learning, is discussed. The learning performance is systematically analyzed. A novel neuron model, called SAM [47], is introduced, which is capable of learning and working memory. Through various experiments, SAM exhibits higher learning accuracy compared to state-of-the-art models in tasks such as supervised learning of the MNIST dataset using sequential spatiotemporal encoding, noisy spike pattern classification, sparse coding during pattern classification, and spatiotemporal feature detection. Additionally, SAM has been shown to be robust, efficient, and capable of meta-learning in the field of agent navigation and meta-learning of MNIST classification. SAM's working memory capabilities are explored for both temporal and spatial learning and have observed excellent performance. It has the potential to be used in various applications, including intelligent robots, IoT, and edge computing.

References

1. Koch C. Biophysics of computation: information processing in single neurons. Oxford University Press;1999.
2. Johnston D, Wu SM. Foundations of cellular neurophysiology. MIT Press;1995.
3. Petousakis KE, Apostolopoulou AA, Poirazi P. The impact of Hodgkin–Huxley models on dendritic research. J Physiol. 2022.
4. Harnett MT, Makara JK, Spruston N, et al. Synaptic amplification by dendritic spines enhances input cooperativity. Nature. 2012;491(7425):599–602.
5. Pettersen KH, Lindén H, Tetzlaff T, et al. Power laws from linear neuronal cable theory: power spectral densities of the soma potential, soma membrane current and single-neuron contribution to the EEG. PLoS Comput Biol. 2014;10(11):e1003928.
6. Mainen ZF, Sejnowski TJ. Influence of dendritic structure on firing pattern in model neocortical neurons. Nature. 1996;382(6589):363–6.
7. Wearne SL, Rodriguez A, Ehlenberger DB, et al. New techniques for imaging, digitization and analysis of three-dimensional neural morphology on multiple scales. Neuroscience. 2005;136(3):661–80.
8. Carnevale NT, Hines, ML. The NEURON book. Cambridge University Press;2006.
9. Bower JM, Beeman D. The book of GENESIS: exploring realistic neural models with the GEneral NEural SImulation System. Springer Science and Business Media;2012.
10. Ray S, Bhalla US, Wójcik DK. MOOSE: a parallel computational framework for multiscale systems biology. Front Neuroinformatics. 2018;12(88).
11. Stuart GJ, Spruston N. Dendritic integration: 60 years of progress. Nat Neurosci. 2015;18(12):1713–21.
12. Magee JC, Johnston D. A synaptic basis for dendritic integration in neocortical pyramidal neurons. Nature. 1997;385(6614):70–4.
13. Silver RA. Neuronal arithmetic. Nat Rev Neurosci. 2010;11(7):474–89.
14. Segev I, London M. Untangling dendrites with quantitative models. Science. 2000;290(5492):744–50.
15. Schneider-Mizell CM, Bodor AL, Collman F, et al. Structure and function of axo-axonic inhibition. Elife. 2021;10:e73783.
16. Lee DH, Zhang S, Fischer A, Bengio Y. Difference target propagation. In: Joint European conference on machine learning and knowledge discovery in databases;2015. p. 498–515.
17. Lillicrap TP, Cownden D, Tweed DB, Akerman CJ. Random synaptic feedback weights support error backpropagation for deep learning. Nat Commun. 2016;7(1):1–10.
18. Spratling MW. Cortical region interactions and the functional role of apical dendrites. Behav Cogn Neurosci Rev. 2002;1(3):219–28.
19. Larkum ME, Zhu JJ, Sakmann B. A new cellular mechanism for coupling inputs arriving at different cortical layers. Nature. 1999;398(6725):338–41.
20. Yang S, Gao T, Wang J, et al. Efficient spike-driven learning with dendritic event-based processing. Front Neurosci. 2021;15:601109.
21. Guerguiev J, Lillicrap TP, Richards BA. Towards deep learning with segregated dendrites. Elife. 2017;6: e22901.
22. Otsu N. A threshold selection method from gray-scale histogram. IEEE Trans Syst Man Cybern. 1978;8:62–66.
23. Courbariaux M, Bengio Y, David JP. Training deep neural networks with low precision multiplications. 2014. arXiv:1412.7024.

24. Bill J. Compensating inhomogeneities of neuromorphic VLSI devices via short-term synaptic plasticity. Front Comput Neurosci. 2010;4:129.
25. Sjöström PJ, Turrigiano GG, Nelson SB. Rate, timing, and cooperativity jointly determine cortical synaptic plasticity. Neuron. 2001;32(6):1149–64.
26. Bengio Y, LeCun Y. Scaling learning algorithms towards AI. Large-Scale Kernel Mach. 2007;34(5):1–41.
27. LeCun Y, Bengio Y, Hinton G. Deep learning. Nature. 2015;521(7553):436–44.
28. Mnih V, Kavukcuoglu K, Silver D. Human-level control through deep reinforcement learning. Nature. 2015;518(7540):529.
29. Van der Maaten L, Hinton G. Visualizing data using t-SNE. J Mach Learn Res. 2008;9(11):2579–605.
30. Bottou L, Cun YL. Large scale online learning. Adv Neural Inf Process Syst. 2004, 217–224.
31. Merolla PA, Arthur JV, Alvarez-Icaza R, Cassidy AS. A million spiking-neuron integrated circuit with a scalable communication network and interface. Science. 2014;345:668–673.
32. Qiao N, Corradi F. A reconfigurable on-line learning spiking neuromorphic processor comprising 256 neurons and 128K synapses. Front Neurosci. 2015;9:141.
33. Esser SK, Merolla PA, Arthur JV, et al. Convolutional networks for fast, energy efficient neuromorphic computing. PNAS. 2016;113:11441–6.
34. Ranganathan GN, Apostolides PF, Harnett MT, et al. Active dendritic integration and mixed neocortical network representations during an adaptive sensing behavior. Nat Neurosci. 2018;21(11):1583–90.
35. Poleg-Polsky A, Ding H, Diamond JS. Functional compartmentalization within starburst amacrine cell dendrites in the retina. Cell Rep. 2018;22(11):2898–908.
36. Grienberger C, Milstein AD, Bittner KC, et al. Inhibitory suppression of heterogeneously tuned excitation enhances spatial coding in CA1 place cells. Nat Neurosci. 2017;20(3):417–26.
37. Cazé RD, Stimberg M. Dendritic neurons can perform linearly separable computations with low resolution synaptic weights. F1000Research. 2020;9:1174.
38. Bellec G, Kappel D, Maass W, et al. Deep rewiring: training very sparse deep networks. 2017. arXiv:1711.05136.
39. Schulman J, Wolski F, Dhariwal P, et al. Proximal policy optimization algorithms. 2017. arXiv:1707.06347.
40. Kingma DP, Ba J. Adam: a method for stochastic optimization. 2014. arXiv:1412.6980.
41. Wang JX, Kurth-Nelson Z, Tirumala D, et al. Learning to reinforcement learn. 2016. arXiv:1611.05763.
42. Duan Y, Schulman J, Chen X, et al. Rl^2: fast reinforcement learning via slow reinforcement learning. 2016. arXiv:1611.02779.
43. Vasilaki E, Frémaux N, Urbanczik R, et al. Spike-based reinforcement learning in continuous state and action space: when policy gradient methods fail. PLoS Comput Biol. 2009;5(12):e1000586.
44. Morris R. Developments of a water-maze procedure for studying spatial learning in the rat. J Neurosci Methods. 1984;11(1):47–60.
45. Huh D, Sejnowski TJ. Gradient descent for spiking neural networks. Adv Neural Inf Process Syst. 2018;31.
46. Greff K, Srivastava RK, Koutník J, et al. LSTM: a search space odyssey. IEEE Trans Neural Netw Learn Syst. 2016;28(10):2222–32.
47. Yang S, Gao T, Wang J, et al. SAM: a unified self-adaptive multicompartmental spiking neuron model for learning with working memory. Front Neurosci. 2022;16:850945.
48. Masquelier T, Guyonneau R, Thorpe SJ. Spike timing dependent plasticity finds the start of repeating patterns in continuous spike trains. PLoS ONE. 2008;3(1):e1377.

49. Werbos PJ. Backpropagation through time: what it does and how to do it. Proc IEEE. 1990;78(10):1550–60.
50. Wolff MJ, Jochim J, Akyürek EG, et al. Dynamic hidden states underlying working-memory-guided behavior. Nat Neurosci. 2017;20(6):864–71.
51. Roy D, Chakraborty I, Roy K. Scaling deep spiking neural networks with binary stochastic activations. In: 2019 IEEE international conference on cognitive computing (ICCC). IEEE;2019. p. 50–58.

Spike-Based Learning with Information Theory

3

3.1 Information Theoretic Learning Principles

Information theory is a branch of mathematics that studies the quantification, storage, and communication of information [1]. It provides us with a set of tools to measure the amount of information in a signal, to compress it, and to transmit it over a noisy channel without loss [2]. On the other hand, statistical learning theory is concerned with the theoretical underpinnings of machine learning algorithms [3]. It provides us with a framework for understanding the performance of learning algorithms, the role of bias and variance in model selection, and the generalization properties of models.

Information theoretic learning (ITL) is a machine learning framework [4, 5] that blends ideas from information theory and statistical learning theory to develop models and algorithms that can effectively deal with complex and high-dimensional data while minimizing the risk of overfitting and sensitivity to noise. At the heart of ITL lies the concept of mutual information, which quantifies the information one random variable provides about another. In other word, mutual information measures the degree of dependency between two variables. In ITL's context, mutual information serves to identify pertinent features and reduce input space dimensionality. Essentially, mutual information guides the selection of the most important features for the learning algorithm while discarding less relevant ones. By reducing input space dimensionality, the learning algorithm can concentrate on the most crucial features, thereby enhancing the learning performance including generalization and robustness. For instance, consider a dataset containing numerous features that may be relevant to a particular prediction task. Mutual information helps pinpoint the most informative features for the task at hand, streamlining the learning algorithm's operation and improving its efficiency. Mutual information is a potent tool within ITL, facilitating the identification of relevant features and the reduction of input space dimensionality, ultimately resulting in improved learning performance.

S. Yang and B. Chen, *Neuromorphic Intelligence*, Synthesis Lectures on Engineering, Science, and Technology, https://doi.org/10.1007/978-3-031-57873-1_3

One of ITL's principal strengths is its versatility, as it can be applied to a wide spectrum of problems. For instance, in fields like image and speech recognition, ITL aids in identifying the most relevant features for recognizing and classifying images or sounds. In natural language processing, ITL can extract meaningful features from text data, enabling tasks like sentiment analysis or topic modeling. In bioinformatics, ITL can uncover patterns in biological data and predict the function of genes or proteins.

Below, some important learning principles of ITL are introduced in detail.

3.1.1 Maximum Correntropy Criterion

Information theoretic quantities establish a connection among information theory, non-parametric probability density function (PDF) estimators, and reproducing kernel Hilbert space (RKHS) in a simple and unconventional manner. Specifically, the correntropy, serving as a nonlinear local similarity measure in the kernel space between two arbitrary random variables, finds its origin in Renyi's entropy [6, 7]. Naturally, it becomes a robust cost for machine learning and signal processing [61–65]. Correntropy has found wide applications in signal processing and machine learning, which is defined by

$$CT_\sigma(X, Y) = E[k_\sigma(X, Y)] = \int\int k_\sigma(x, y) f_{XY}(x, y) dx dy \tag{3.1}$$

where X and Y represent two random variables, and $E[\cdot]$ represents the expected value with respect to the joint distribution of (X, Y). If the kernel function $k_\sigma(\cdot, \cdot)$ satisfies Mercer condition with bandwidth σ, and $f_{XY}(\cdot, \cdot)$ represents the joint PDF. In many practical situations, the joint PDF is usually unknown. With finite samples, correntropy can be obtained by a sample estimator as

$$C\hat{T}_\sigma(X, Y) \approx \frac{1}{N} \sum_{i=1}^{N} k_\sigma(x_i - y_i) \tag{3.2}$$

where $\{x(i), y(i)\}_{i=1}^{N}$ are N samples drawn from $f_{XY}(X, Y)$.

Correntropy can be employed to formulate robust cost functions in various applications within the realms of signal processing and machine learning. Let's consider a supervised learning scenario where the objective is to optimize a model, denoted as f. This model takes a random variable U as input and produces $Z = f(U)$ as the output, approximating a target variable D. Here, $f(\cdot)$ signifies an unknown mapping from the input to the output that requires learning. Within this learning task, a pivotal concern involves defining a loss function (or similarity measure) to compare Z with D. The minimum mean square error (MMSE) criterion has traditionally served as a cornerstone in supervised learning. Its goal is to minimize the cost of mean square error (MSE) $E[e^2]$, where $e = D - Z$ denotes the error variable. The integration of a linear feedforward model and mean square error results in a set of equations solvable through analytical methods. However, MMSE is optimal only under the assumption that the error variable adheres to a Gaussian distribution. Error distributions commonly tend to be skewed and with long tails, introducing challenges for MSE. Consequently, many optimal solutions may not be practical in real-world scenarios due to the criterion employed in the optimization. In response to the limitations of MSE, the literature has proposed various non-MSE optimization criteria. One of the current research focal points is the maximum correntropy criterion (MCC), demonstrating particular efficacy when the error distribution exhibits heavy tails.

In the context of MCC, the optimization (or training) of the model is carried out to maximize the correntropy between the target variable D and the output Z as follows:

$$f^* = \arg\max_{f \in F} C(D, Z)$$

$$= \arg\max_{f \in F} E[k_\sigma(D, Z)] \tag{3.3}$$

where f^* denotes the optimal model, and F stands for the hypothesis space. In a practical scenario, given a finite set of input-target samples $\{u(i), d(i)\}_i^N$, the model can be trained by maximizing a sample estimator of correntropy as follows:

$$f^* = \arg\max_{f \in F} \hat{C}(D, Z)$$

$$= \arg\max_{f \in F} \frac{1}{N} \sum_{i=1}^{N} k(e(i)) \tag{3.4}$$

where $e(i) = d(i) - z(i) = d(i) - f(u(i))$ represents the i-th error sample, and the input vector is denoted by $u(i) = [u_1(i), u_2(i), \cdots, u_M(i)]^T \in R^m$. It is important to note that $J_{MCC} = \frac{1}{N} \sum_{i=1}^{N} k(e(i))$ is neither concave nor convex but is a continuously differentiable function at any $e(i)$ ($i = 1, 2, ..., N$). The MCC algorithm has left an indelible mark across an extensive array of machine learning applications. Its prowess has been harnessed in fields such as signal processing, image processing, and speech recognition. Empirical evidence substantiates the MCC algorithm's superiority over traditional learning algorithms in terms of robustness and accuracy.

3.1.2 Minimum Error Entropy Criterion

Define the error between two random variables, X and Y, as $e = Y - X$. In ITL, entropy serves as a method to quantify the information contained in the error e. Utilizing Shannon's entropy, a well-known measure of entropy, it is defined as:

$$H(e) = - \int_{-\infty}^{\infty} p_e(x) \log p_e(x) dx$$

$$= E_e\big[-\log p_e(x)\big] \qquad (3.5)$$

where $p(\cdot)$ represents the probability distribution function of the error e, and $E_e[\cdot]$ denotes the expectation operator with respect to e. As an extension of Shannon's entropy, Renyi's entropy finds widespread use in ITL and is defined as follows:

$$H_\alpha(e) = \frac{1}{1-\alpha} \log \int_{-\infty}^{\infty} p_e^{\alpha}(x) dx$$

$$= \frac{1}{1-\alpha} \log I_\alpha(e) \qquad (3.6)$$

where α represents the entropy order, $\alpha > 0$, $\alpha \neq 1$, and $I_\alpha(e)$ is denoted as the order-α information potential as follows:

$$I_\alpha(e) = \int_{-\infty}^{\infty} p^{\alpha}(x) dx = E_e\big[p^{\alpha-1}(e)\big] \qquad (3.7)$$

where Renyi's entropy converges to Shannon's entropy as α approaches 1.

In practical applications, obtaining an analytical expression for error entropy is generally unfeasible, necessitating nonparametric estimation from samples. Since entropy is a functional of the PDF, a common approach is to substitute an estimate of the PDF directly into the sample mean approximation for the expectation. Kernel density estimation (KDE), also known as the Parzen windowing method, is a well-established and useful nonparametric technique. Using the KDE approach, the nonparametric estimate of the error's PDF is given by:

$$\hat{p}_e(x) = \frac{1}{N} \sum_{j=1}^{N} k_\sigma(x - e(j)) \qquad (3.8)$$

where $\{e(1), e(2), \ldots, e(N)\}$ are the error samples, and N is the sliding data length.

In the subsequent discussion, we focus on Renyi's entropy of order $\alpha = 2$. In this instance, minimizing $H(e)$ is equivalent to maximizing the quadratic information potential (QIP) I_2. Substituting the estimated PDF into the sample mean approximation, an empirical version of QIP can be expressed as:

$$\hat{I}_\alpha(e) = \frac{1}{N} \sum_{i=1}^{N} \hat{p}(e(i)) = \frac{1}{N^2} \sum_{i=1}^{N} \sum_{j=1}^{N} k_\sigma(e(i) - e(j)) \qquad (3.9)$$

In supervised learning under minimum error entropy (MEE), the objective is to bring the model output as close as possible to the target by minimizing the entropy of the error. Specifically, for $\alpha = 2$, the minimization of entropy $H_2(e)$ is tantamount to maximizing the information potential $I_2(e)$. Consequently, within the framework of MEE, the optimal model is determined by:

$$f^* = \arg\max_{f \in F} \hat{I}_2(e)$$

$$= \arg\max_{f \in F} \frac{1}{N^2} \sum_{i=1}^{N} \sum_{j=1}^{N} k_\sigma(e(i) - e(j)) \tag{3.10}$$

The QIP is computed through a double summation over all error samples, essentially constituting an adaptive cost function capable of adapting to the error distribution.

The MEE criterion [10–12] serves as a pivotal learning algorithm harnessed within both machine learning and signal processing domains, encompassing speech recognition, image processing, and control systems. In these applications, MEE consistently demonstrates superior robustness and accuracy when compared to traditional least-squares and mean-squared error algorithms.

3.1.3 Information Bottleneck

Information bottleneck (IB) criterion [17–19] is a theoretical framework and a learning algorithm that is used in machine learning and information theory. IB criterion seeks to find a compressed representation of the input data that preserves the relevant information required to predict the output variable. It has been shown to be particularly effective in applications where the input data has high dimensionality and the output variable is noisy or uncertain, and has been used in a wide range of machine learning applications, including clustering, classification, and regression. The IB algorithm has several variants, including the original IB algorithm proposed by Tishby et al. [4], the local IB (LIB) algorithm proposed by Chechik et al. [18], and the variational IB (VIB) algorithm proposed by Alemi et al. [19].

Rate distortion theory uses the information theoretic approach to explore the data compression problem, which is also called limited distortion source coding theory. The fundamental problem can be expressed as: under the condition of distortion measurement function $d(X, H)$, to seek the optimal coding strategy from source variable X to compressive variable H, where source variable X obeys the probability distribution $p(X)$. The coding strategy from X to H can be described by the conditional probability distribution $p(H|X)$. The function $R(D)$ requires that the strategy enables the minimum expected distortion value lower than the specified upper limit value D, and the mutual $I(X; H)$ between the obtained source variable X and the compressive variable H is minimum. The function can be formulated as

$$R(D) = \min_{\{p(H|X), E(d(x,t)\leq D)\}} I(X; H) \tag{3.11}$$

where $p(H|X) \geq 0$ represents the encoding strategy. The rate distortion theory function $R(D)$ is the minimum compressive information on all the conditional probability distribution $p(H|X)$ when meeting all the distortion limited situations.

The term $I(X; H)$ represents compressive mutual information, and can be obtained based on distribution $p(x) * p(H|X)$. The larger the probability that x is assigned to c, the larger the value of $p(H|X)$ reaches. When all the X is assigned to H, $I(X; H) = 0$ in this extreme condition. Otherwise, the lower the probability that X is assigned to H, the higher $I(X; H)$ is, which means the compression redundancy is. If the data X is not compressed, the value of $I(X; H)$ reaches the maximum value $H(X)$.

Distortion measurement function $d: X \times T \to R^+$, and $p(H|X)$ can describe the division of X. The expected value $E(d(X, H))$ should be lower than the specified upper limit value D, which can be formulated as

$$E(d(X, H)) = \sum_{x,c} p(X)p(H|X)d(X, H) \tag{3.12}$$

After introducing the Lagrange parameter β, it can be expressed as

$$F_{\min}[p(H|X)] = I(X; H) + \beta E(d(X, H)) \tag{3.13}$$

Then, by deriving $p(H|X)$ and making its reciprocal equal to 0, i.e., $\delta L / \delta p(H|X) = 0$, the minimum solution can be obtained as

$$p(H|X) = \frac{p(H)}{Z(X, \beta)} e^{-\beta t(X,H)} \tag{3.14}$$

where the probability normalization function $Z(X, \beta)$ obeys the normal distribution, and the optimal solution can be formulated as

$$Z(X, \beta) = \sum_{H} p(H)e^{-\beta d(X,H)} \tag{3.15}$$

which can be obtained by Blahut-Arimoto algorithm.

IB theory is developed based on the rate distortion theory. Due to the randomness of rate distortion theory in the selection of distortion measurement $d(x, h)$, the coding schemes obtained by different distortion measurements are quite different, and the final results of $R(D)$ are also different. Besides, the rate distortion function $R(D)$ depends too much on the selection of $d(x, h)$, while $d(x, h)$ has nothing to do with the source statistics. It is difficult to explain the rate distortion function, and it is also challenging to select an appropriate distortion measurement $d(x, h)$. IB theory can avoid the challenging problem in the selection of distortion measurement $d(x, h)$. It uses mutual information to describe the relevant information retained in the compression process without selecting the distortion measurement $d(x, h)$ in advance. The basic idea of IB theory is described as follows: suppose there are stochastic variables X and Y with certain correlation, where these two variables represent the source variable and relevant variable that describes the mode structure of X respectively. In the process of compressing the source variable X and obtaining its compressed representation T, the IB method tries to find an optimal coding scheme $p(h|x)$ about X, so that the bottleneck variable, i.e., compressed variable H, can save the most information amount about the relevant variable Y. Besides, X is compressed into H as much as possible to maximize the compression degree. Therefore, the distortion measurement $d(x, h)$ can be expressed based on IB theory as

$$d(x, h) = D_{KL}(p(y|x)||p(y|h)) \tag{3.16}$$

The three variables X, Y and H in the IB theory form IB Markov chain $H \rightarrow X \rightarrow Y$, i.e., $p(x, h, y) = p(x, h)p(y \mid x, h)$. Thus, the expected value of distortion measurement $E(d(x, h))$ can be described as

$$
\begin{aligned}
E(d(x, h)) =& E[D_{KL}(p(y|x)||p(y|h))] \\
=& \sum_{x \in X, h \in H, y \in Y} p(x, h) \sum_{y \in Y} p(y|x) \log \frac{p(y|x)}{p(y|h)} \\
=& \sum_{x \in X, h \in H, y \in Y} p(x, h, y) \log \frac{p(y|x)}{p(y|h)} \\
=& \left[\begin{array}{l} \sum_{x \in X, h \in H, y \in Y} p(x, h, y) \log \frac{p(x|y)}{p(x)} \\ - \sum_{x \in X, h \in H, y \in Y} p(x, h, y) \log \frac{p(y|h)}{p(y)} \end{array} \right] \\
=& I(X; Y) - I(H; Y) \tag{3.17}
\end{aligned}
$$

The rate distortion theory function $R(D)$ can be formulated based on IB theory as

$$R(D) = \min_{p(h|x):I(X;Y)-I(H;Y) \leq D} I(X; H) \tag{3.18}$$

Since $I(X; H)$ is a fixed value and can be ignored, it can be further formulated as follows

$$R(D) = \min_{p(h|x):I(H;Y \geq D)} I(X; H) \tag{3.19}$$

By introducing the Lagrange parameter β to solve the minimum value, the objective function of the IB theory can be described as

$$F_{\min}(p(h|x)) = I(H; X) - \beta I(H; Y) \qquad (3.20)$$

The IB approach can maintain a balanced situation between data compression and information retention by the Lagrange parameter β. As shown in Fig. 3.1, IB can control the balance between data compression and information retention by adjusting parameter β. When $\beta \to 0$, the IB approach focuses more on data compression. The mutual information is small, and the compression degree become larger. The information degree of H to Y is small. When $\beta \to \infty$, the IB method focuses more on information retention with less compression. The mutual information $I(X; H)$ takes the maximum value, i.e., $I(X; H) = H_{SE}(X)$, where H_{SE} represents the Shannon entropy. In addition, the following formulation is obtained by derivation as

$$p(h|x) = \frac{p(h)}{Z(x, \beta)} \exp(-\beta D_{KL}(p(y|x)\|p(y|h))) \qquad (3.21)$$

where $Z(x, \beta) = \sum_{h \in H} p(h) \exp(-\beta D_{KL}(p(y|x)\|p(y|h)))$ is the probability normalization function. From Markov chain $H \to X \to Y$ and related conceptual formulas, it can be formulated

$$\begin{cases} p(y|h) = \dfrac{1}{p(h)} \displaystyle\sum_{x \in X} p(y|x)p(h|x)p(x) \\[4mm] p(h) = \displaystyle\sum_{x \in X} p(h|x)p(x) \end{cases} \qquad (3.22)$$

which can be regarded as a formal solution of the objective function for IB theory.

Fig. 3.1 Relationship of hyper parameter β with data compression and information retention

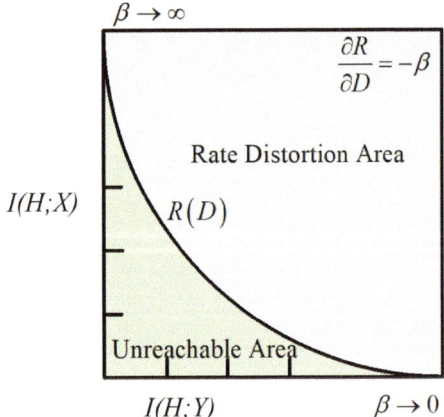

3.2   Robust Spike-Driven Few-Shot Learning

3.2.1   Ensemble Loss

A new objective function is introduced, which is a combination of different types of loss functions with trainable weights and it can find the best weights to generate a better applicable loss function. Another constraint is added to avoid having all weights with values close to zero. The ensemble loss function is expressed as

$$L = \sum_{i=1}^{K} \lambda_i L_i(y, \hat{y}), \sum_{i=1}^{K} \lambda_i = 1 \tag{3.23}$$

The optimization with N training samples can be expressed as

$$\underset{w,\lambda}{\text{minimize}} \sum_{i=1}^{N} \sum_{j=1}^{N} \lambda_j^2 L_j(y_i, \hat{y}_i)$$

$$s.t. \sum_{j=1}^{K} \lambda_j^2 = 1 \tag{3.24}$$

Based on the notion of the extended Lagrangian with this condition as the governing term, the modified objective function can be formulated as follows

$$\underset{w,\lambda}{\text{minimize}} \sum_{i=1}^{N} \sum_{j=1}^{N} \lambda_j^2 L_j(y_i, \hat{y}_i) + \eta_1 \left(\sum_{j=1}^{K} \lambda_j^2 - 1 \right) + \eta_2 \left(\sum_{j=1}^{K} \lambda_j^2 - 1 \right)^2 \tag{3.25}$$

The first and the second terms of the objective function induce the values of λ_i^2 approach to 0 but the third term satisfied $\sum_{j=1}^{K} \lambda_j^2 = 1$. The entire training process is described by Table 3.1.

The cross-entropy loss function, also known as log loss, is one of the most common back-propagation loss functions. The cross-entropy loss function increases with the deviation of the predicted probability from the real label, which can be expressed as

$$L_{ce}(\hat{y}_i, y_i) = - \sum_{i} y_i \log(\hat{y}_i) \tag{3.26}$$

In the model in this chapter, a label l^n for each image is used, which assumes a value of 1 only for images belonging to the same class of the image in the testing phase and assumes a value of zero otherwise. Then the formulation can be described as

$$E_C = \sum_{n=1}^{5} -l^n \log \sigma \left(y^{20+20 \cdot n} \right) - \left(1 - l^n \right) \log \left(1 - \sigma^{20+20 \cdot n} \right) \tag{3.27}$$

Table 3.1 Pseudo-code of the whole training process for the proposed method (From [52])

Input:

The training set T, parameters λ_i (Weights associated with each loss function), η_1, η_2 (Lagrangian weights), σ (Correntropy kernel bandwidth), and m (maximum number of iterations (epochs))

Base loss functions $\{L_j(X_i, y_i)\}_{j=1}^4$, $K=4$ (MMCC, Cross-entropy, MMSE based on firing rate, MMSE based on membrane potential)

Output:

Parameter W, λ_1, λ_2, λ_3, λ_4

1: Initiate Ensemble Loss Function using $\{L_j(X_i, y_i)\}_{j=1}^4$ and random λ_1, λ_2, λ_3, λ_4

2: Initialize parameters $W \sim N(0, \Sigma)$ and $t = 0$

3: **while** not converged **do**

4:      Select a mini-batch of training samples $\{X_i, y_i\}_{i=1}^N$ from training set T.

5:      Perform a forward path, calculate the loss and regularization term:

$$\sum_{i=1}^{N} \sum_{j=1}^{K=4} \lambda_j^2 L_j\left(y_i, \hat{y}_i\right) + \eta_1\left(\sum_{j=1}^{K=4} \lambda_j^2 - 1\right) + \eta_2\left(\sum_{j=1}^{K=4} \lambda_j^2 - 1\right)$$

6:      Perform a backward propagation by the BPTT algorithm

7:      Update W, λ_1, λ_2, λ_3, λ_4 by gradient descent algorithm.

8:      $t \leftarrow t+1$

return $\{W(t), \lambda_1(t), \lambda_2(t), \lambda_3(t), \lambda_4(t)\}$

As an extended version of correntropy, the mixture correntropy can be described as

$$CT_\sigma(X, Y) = E\left[\sum_{s=1}^{S} \lambda_s G_{\sigma s}(X, Y)\right] \tag{3.28}$$

where $\{G_{\sigma s}(., .)\}_{s=1}^S$ are S different Gaussian kernels depending on the kernel size σ_s. $\{\lambda_s\}_{s=1}^S$ are S mixture parameters that satisfy $0 \leq \lambda_s \leq 1$ and $\sum_{s=1}^S \lambda_s = 1$. The sample estimator of mixture correntropy can be expressed as

$$
\begin{aligned}
C\hat{T}_\sigma(X, Y) &= \frac{1}{N} \sum_{i=1}^{N} \left[\lambda G_{\sigma 1}\left(x_j, y_j\right) + (1-\lambda) G_{\sigma 2}\left(x_j, y_j\right)\right] \\
&= \frac{1}{N} \sum_{i=1}^{N} \left[\lambda \exp\left(-\frac{\|x_i - y_i\|^2}{2\sigma_1^2}\right) + (1-\lambda)\exp\left(-\frac{\|x_i - y_i\|^2}{2\sigma_2^2}\right)\right]
\end{aligned} \tag{3.29}
$$

An unknown parameter vector can be estimated through the maximization of the mixture correntropy, which quantifies the concordance between anticipated values and their estimated counterparts. To delve deeper into the properties of MMCC, refer to previous study [9] for comprehensive insights. The influence function of MMCC can be calculated as

$$\Psi(e) = \frac{\partial G_\sigma(e)}{\partial e} = -\frac{e}{\sigma^2} \exp\left(-\frac{e^2}{2\sigma^2}\right) \tag{3.30}$$

It is found that the MMSE influence function increases linearly with the magnitude of the estimation error, and the MMCC is restricted to the smaller error range. Due to the fact that the larger errors are generally caused by the outliers, MMCC can be used to mitigate the negative effects of outliers in robust learning.

The SNN model output only counts when all the images have been shown in full. In order to achieve a sparse firing mode, an extra term is added to regularize the spiking activity. Two kinds of regularizations are considered, namely, the firing rate regularizer and the voltage range regularizer. First, a term is added so that the mean firing rate f_j of all the neurons j approaches a predefined target firing rate f_{target}, which is defined as

$$\lambda_f E_{rate} = \lambda_f \sum_j \left(f_j - f_{\text{target}}\right)^2 \tag{3.31}$$

where f_j is computed as the average spike count, which is expressed as

$$f_j = \frac{1}{N_{batch} T} \sum_{n=1}^{N_{batch}} \sum_{t=1}^{T} z_j^{(n,t)} \tag{3.32}$$

where $z_j^{(n,t)}$ indicates the neural spikes in a particular batch with n, and T represents the total duration on a specific task. Moreover, the factor λ_f is a hyperparameter which measures the importance of the regularization of the firing rate.

Furthermore, in order to encourage the membrane potential to remain in a certain range, the membrane potential values are penalized, which is defined as

$$\hat{V}\left(v_j^{(n,t)}, A_{th\,j}^{(n,t)}\right) =$$

$$\frac{\lambda_v}{NT} \sum_{i=1}^{N} \sum_{t=1}^{T} \sum_j \left(\max\left(0, v_j^{(n,t)} - A_{th\,j}^{(n,t)}\right)^2 + \max\left(0, v_j^{(n,t)} - v_{th}\right)^2\right) \tag{3.33}$$

The parameters v_j^t and $a_{th\,j}^t$ denote the membrane potential and the adaptive firing threshold. The resulting threshold voltage $A_{th\,j}^{(t)}$. The factor λ_v is a hyperparameter which measures the importance of the membrane potential regularization.

3.2.2   Network Architecture

The HESFOL model includes a SFOL model with spiking neurons and an ensemble loss function. In Fig. 3.2, the learning approach is illustrated by the dotted box. The loss function includes MMCC, cross entropy loss function and MMSE. Suppose that in a multiclass dataset X, $x_i \in R^K$ represents the k-dimensional input. $y \in \{0, 1\}^C$ represents the one-hot encoding of the label. Figure 3.3 shows the HESFOL model, which is based on the ensemble loss function. In the reverse pass, the gradient of the loss function is returned to the network weights. Since the weights are set and adjusted to reduce the loss, the weights are updated in the reverse direction.

(1) Two-compartment spiking neuron model with adaptation mechanism

A two-compartment spiking neuron model is used for robust learning. Recent studies have shown that this approach can accelerate the convergence rate and reduce the number of spikes [20]. For this reason, a spiking neuron model with spiking dendrite is required. There are two variables in the soma compartment, which are the membrane potential v_j^t and the adaptive firing threshold $a_{th\,i}^t$. The resulting threshold voltage $A_{th\,i}(t)$ increase along with each output spike and decays back to the baseline threshold value v_{th} based on an adaptation time constant τ_a. Specifically, the soma compartment can be formulated

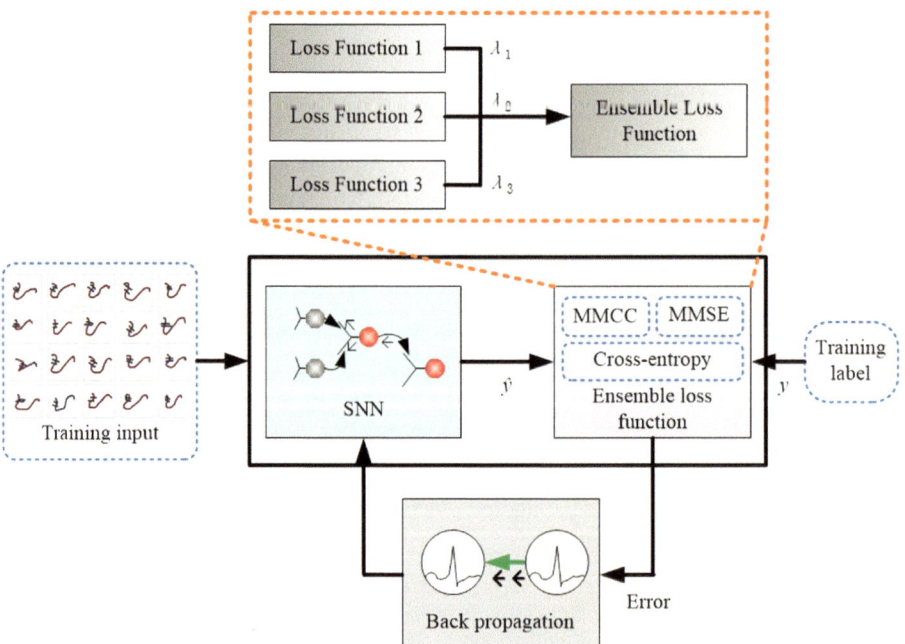

Fig. 3.2 Scheme of the HESFOL model [52]

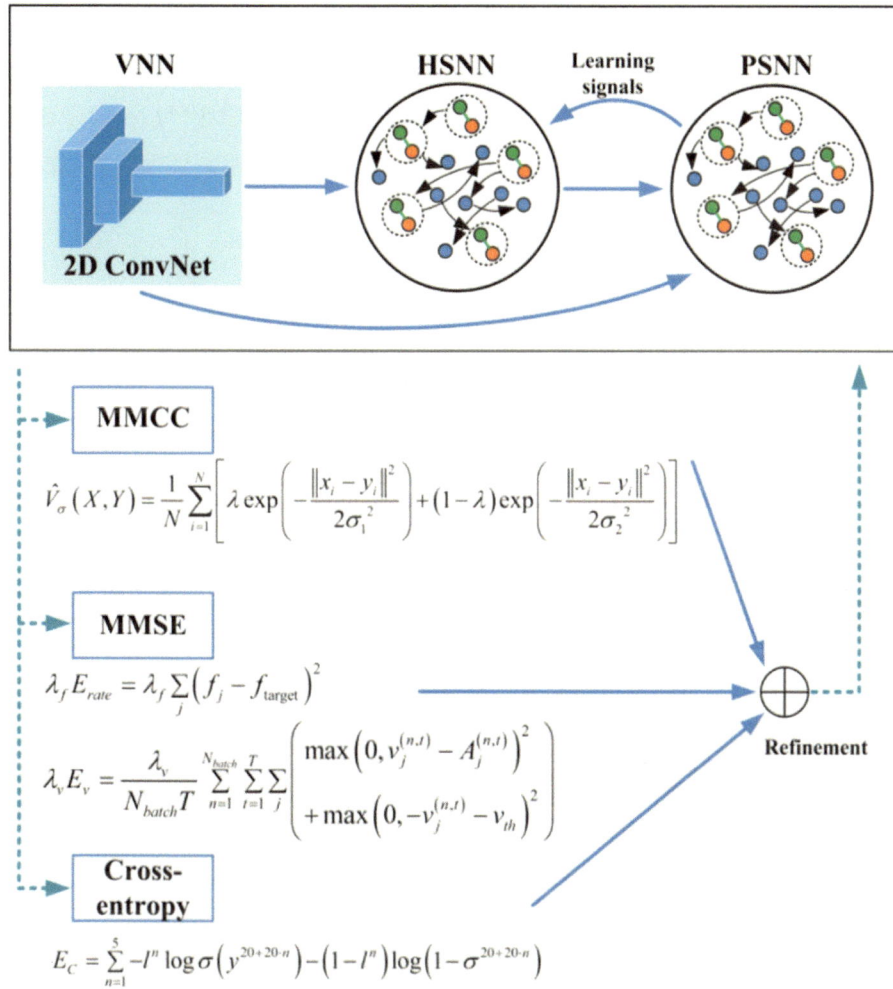

Fig. 3.3 Outline of HESFOL. For VNN, 2D convolution is used. Furthermore, there are two subnetworks, namely HSNN and PSNN. The learning signal is transferred to the HSNN from PSNN [52]

as

$$z_j(t) = H\left(v_j(t) - A_{th\,j}(t)\right) \tag{3.34}$$

$$A_{th\,j}(t) = \lambda a_{th\,j}(t) + v_{th} \tag{3.35}$$

$$a_{th\,j}(t) = \mu a_{th\,j}(t-1) + z_j(t-1) \tag{3.36}$$

where $\mu = e^{-\Delta t / \tau_a}$. The factor λ represents the impact of threshold adaptation. The general form of the spiking soma and dendrite models can be formulated as

$$\tau \frac{V_i(N+1) - V_i(N)}{\Delta T} = -V_i(N) + \frac{g_b}{g_l}\left(V_i^b(N) - V_i(N)\right)$$

$$+ \sum_{i \neq j} W_{ji}^{rec} z_i(N - D) \tag{3.37}$$

$$V_i^b(N) = \sum_{j=1}^{m} W_{ij} s_j^{input}(N) + b_i \tag{3.38}$$

where g_l and g_b are the representations of the leak conductance and the basal dendrite conductance, and ΔT is the integration step. The W_{ji}^{rec} is a representation of the synaptic weight between the neuron i to neuron j in the recurrent architecture, and D corresponds to the transmission delay. The parameter $\tau = C_m / g_l$ is a time constant, where C_m is the membrane capacitance. The variable z_i is a representation of the output spikes of the i_{th} spiking neuron. The variables V_i and V_i^b is a representation of the membrane potentials of soma and basal dendrite of the i_{th} neuron respectively. The term W_{ji} corresponds to the synaptic weights in the input layer, and the constant b_i is the bias term. The variable s^{input} can be calculated based on the following equation as

$$s_j^{input}(t) = \sum_k \kappa\left(t - t_{jk}^{input}\right) \tag{3.39}$$

where t_{jk}^{input} corresponds to the kth spiking time of the input neuron j, and the response kernel can be expressed as follows

$$\kappa(t) = \left(e^{-t/\tau_L} - e^{-t/\tau_s}\right)\Theta(t) \Big/ (\tau_L - \tau_s) \tag{3.40}$$

where τ_s and τ_L represent short and long time constant, and Θ represents the Heaviside step function.

(2) Spike-driven online learning model

In the HESFOL model, a regular LIF neuron model is employed, which is modeled according to the membrane potential $v_j(t)$ at time t. The membrane potential can integrate input current and fall to a resting potential according to its membrane time constant τ_m. Every time $v_j(t)$ reaches a threshold value, a spike is produced by the neuron as $z_j(t) = 1$. The regular spiking neuron model can be expressed as

$$z_j(t) = H\left(v_j(t) - A_{th\,j}(t)\right) \tag{3.41}$$

$$A_{th\,j}(t) = \lambda a_{th\,j}(t) + v_{th} \tag{3.42}$$

where W_{ji}^{rec} is the synaptic weight from neuron i to neuron j, and W_{ji}^{in} is the weight of input component $x_i(t)$ for neuron j. The coefficient λ is used to describe the decay speed of the membrane potential, and H_{ea} and d are the Heaviside step function and the transmission latency of recurrent spikes respectively. A refractory period t_{refrac} is used to set $z_j(t) = 0$ after a neural spike. The outputs from the HESFOL model are constructed by a weighted sum of low-pass filtered spikes, which can be expressed as

$$y_k(t) = (1 - v) \sum_{t' \leq t} \sum_j v(t - t') W_{kj}^{out} z_j(t') + b_k^{out} \tag{3.43}$$

where the W_{kj}^{out}, b_k^{out}, $v = e^{-\Delta t/\tau_{out}}$ and τ_{out} are the readout time constants.

In the HESFOL model, an associated eligibility trace is considered in every synapse. The eligibility trace $e_{ji}(t)$ is a representation of the effect of the weight W_{ji} on the spiking activity of the neuron j at time t, but it is necessary to consider the dependencies that do not include other neurons except i and j. Eligibility traces exist separately for input and recurrent synapses. The variable $h_j(t)$ is the hidden variables for a neuron j at time t. The dynamics of the eligibility trace is then defined as follows

$$e_{ji}(t) = \frac{\partial z_j(t)}{\partial h_j(t)} \cdot \varepsilon_{ji}(t) \tag{3.44}$$

$$\varepsilon_{ji}(t) = \frac{\partial h_j(t)}{\partial h_j(t-1)} \cdot \varepsilon_{ji}(t-1) + \frac{\partial h_j(t)}{\partial W_{ji}} \tag{3.45}$$

The eligibility vector $\varepsilon_{ji}(t)$ indicates that it is propagated in time as the HESFOL model is calculated. It is not possible to compute the term $\frac{\partial z_j(t)}{\partial h_j(t)}$ directly, since the relation of $z_j(t)$ and $h_j(t)$ includes the non-differentiable Heaviside function. Thus, the derivative in Eq. (4.25) can be replaced by a pseudo derivative which is described as

$$\Psi_j(t) = 0.3 \cdot \max\left(0, 1 - \left|\frac{v_{th} - v_j(t)}{v_{th}}\right|\right) \tag{3.46}$$

Using $h_j(t) = v_j(t)$, the eligible traces used in LIF dynamics can be expressed as

$$e_{ji}(t) = \psi_j(t) \cdot \overline{z}_i(t - d) \tag{3.47}$$

where $\overline{z}_i(t) = \sum_{t' \leq t} \alpha^{t-t'} z_i^{t'}$ is defined as the low-pass filtered presynaptic spiking activities of neuron i. In addition, the neuron's hidden variable, $h_j(t)$, also includes the trigger threshold $h_j(t) = (v_j(t), a_{th\,j}(t))$. The eligibility trace $e_{ji}(t)$ for an ALIF neuron model is defined as

$$e_{ji}(t) = \Psi_j(t)\left(\overline{z}_i(t - d) - \beta\varepsilon_{a,ji}(t)\right) \tag{3.48}$$

$$\varepsilon_{a,ji}(t) = \left(\rho - \beta \cdot \Psi_j(t-1)\right)\varepsilon_{a,ji}(t-1)$$
$$+ \Psi_j(t-1)\overline{z}_i(t-d-1) \qquad (3.49)$$

To make the HESFOL model more flexible, the derivative of the Heaviside function $\frac{\partial H_{ea}\left(v_j(t)-v_{th}\right)}{\partial v_j(t)}$ is substituted by the pseudo-derivative in the back pass.

$$\Psi_j(t) = 0.3 \cdot \max\left(0, 1 - \left|\frac{v_{th} - v_j(t)}{v_{th}}\right|\right) \qquad (3.50)$$

Besides, the derivative of the Heaviside function $\frac{\partial H_{ea}\left(v_j(t)-A_{th\,j}(t)\right)}{\partial v_j(t)}$ is replaced by the formulation as

$$\Psi(t) = 0.3 \cdot \max\left(0, 1 - \left|\frac{A_{th\,j}(t) - v_j(t)}{v_{th}}\right|\right) \qquad (3.51)$$

where the actual update of the initial synaptic weight W_{init} of the HESFOL model is implemented by applying Adam with a learning rate η_{rate}.

(3) Details of the HESFOL architecture

The HESFOL model, as shown in Fig. 3.3, comprises two components: the SFOL model and the ensemble loss. Inspired by the interplay between the hippocampus and prefrontal cortex in the human brain, the SFOL model consists of two modules: hippocampus-inspired SNN (HSNN) and PFC-inspired SNN (PSNN). The membrane potentials of neurons in the HSNN and PSNN modules are summed and integrated with external inputs. The readout of the HSNN module includes its weighted low pass filter spike sequences. If there is a family F of relevant learning tasks C, the HSNN module learns a particular task C from F using the learning signals provided by the PSNN module. The synaptic weight is updated every time the HSNN receives a new task C from family F. After the first stage of learning, the HSNN and PSNN parameters are fixed, and a new task C is selected from family F to assess HSNN learning performance. The SFOL model's coding module adopts a visual path processing mechanism, and a 2D ConvNet, inspired by the visual-pathway-inspired neural network (VNN), is used for input coding of images in a pixel array fashion. The 2D ConvNet, built on the nonspiking McCulloch-Pitts neural model, has three layers. The HSNN consists of 180 two-compartment LIF (TLIF) and 260 normal LIF neurons. In the first stage, the learning signal is transferred only from PSNN to HSNN. To optimize the outer loop in BPTT, the ensemble loss method is adopted, which includes MMCC, MMSE, and cross-entropy loss. Table 3.2 presents the hyperparameter values for the HESFOL model.

Table 3.2 Hyperparameters in HESFOL (From [52])

Hyperparameters	Description	Values
τ_m	Timing constant of membrane	15 ms
τ_{out}	Timing constant of readout neurons	10 ms
d	Synaptic transmission delay	1 ms
t_{refrac}	Refractory period duration	5 ms
f_{target}	Target firing rate	20 Hz
η_{out}	Learning rate of outer loop	2×10^{-3}
λ_f	Spike rate regularization	1.0
v_{th}	Threshold	1.0
λ_v	Voltage regularization	10^{-2}
t_{img}	Number of time steps per image	20 ms
τ_a	Adaptation timing constant	200 ms
η	Learning rate	1.915×10^{-3}
N_{HSNN}	Network size of HSNN	447
q_{ada}	Neuron fractions using adaptation	40.5%
β	Impact of threshold adaptation	0.4902
N_{batch}	Batch size for outer loop optimization	285
N_{PSNN}	Network size of the PSNN	239
τ_{LS}	Timing constant learning signals of readouts	10 ms
$f_{tarPSNN}$	Target firing rate for PSNN	20 Hz

3.2.3 Performance Evaluation

To begin, let's delve into the few-shot learning capabilities of the HESFOL model by applying spiking patterns subjected to non-Gaussian noise. In this section, a spatial–temporal spike pattern classification challenge encompassing firing rates spanning from 2 to 50 Hz is addressed. This spiking model effectively captures the temporal and spatial dynamics of a nerve group, where the firing rate and precise timing of spiking neurons encapsulate rich information regarding environmental input. To instantiate spike patterns for each category, non-Gaussian noise is incorporated into the corresponding template. This noise includes both Poisson noise and spiking deletion noise.

The construction of a few-shot learning dataset for the spiking model is undertaken, featuring 1000 classes, each with 25 samples. This is achieved by modeling 1000 spiking patterns based on a specific SNN and subsequently generating 25 spiking patterns per class by randomly modulating the neurons' firing rates. The classification task for few-shot learning in spiking patterns considers two types of non-Gaussian noise. In the initial method, Poisson noise with a standard deviation of σ_{noise} is introduced into the template,

Table 3.3 Test accuracy (%) of different ensemble parameter settings on Omniglot data set (From [5.2])

Groups	Loss	Values	Omniglot Accuracy (%)	Groups	Loss	Values	Omniglot Accuracy (%)
Group1	MMCC	0.1	90.6	Group5	MMCC	0.1	90.6
	Cross	0.9			Cross	0.9	
	Rate	0.5			Rate	0.5	
	Vol	0.5			Vol	0.5	
Group2	MMCC	0.1	90.6	**Group6**	**MMCC**	**0.2**	**93.1**
	Cross	1.3			**Cross**	**0.8**	
	Rate	0.3			**Rate**	**0.5**	
	Vol	0.3			**Vol**	**0.5**	
Group3	MMCC	0.1	92.2	Group7	MMCC	0.2	90.6
	Cross	1.0			Cross	1.3	
	Rate	0.45			Rate	0.25	
	Vol	0.45			Vol	0.25	
Group4	MMCC	0.1	91.4%	Group8	MMCC	0.2	89.8
	Cross	0.7			Cross	0.6	
	Rate	0.6			Rate	0.6	
	Vol	0.6			Vol	0.6	

giving rise to new noisy spatial–temporal pattern samples. In the second method, fresh noisy spiking patterns by incorporating random deletion noise into the template is generated. Here, each spike is randomly removed in accordance with the P_{del} probability. As depicted in Fig. 3.4, the HESFOL model demonstrates remarkable effectiveness across a diverse range of noisy environments. Notably, the loss function, comprising MMCC, MMSE, and cross-entropy loss, emerges as the most robust in terms of noise tolerance.

The HESFOL model undergoes further testing utilizing the Omniglot dataset. This dataset encompasses a total of 1623 classes and 32,460 images, which are categorized into 964 training classes and 659 test classes. The testing phase unfolds in two stages. In this scenario, a sequence of images is presented, with one image from the same class appearing in phase #2, as depicted in phase #1. The 2D CNN integrated into the model encompasses 15,488 neurons, featuring 16, 32, and 64 filters. These convolutional filters have a size of 3×3. To optimize the HESFOL model, both an average pooling layer and batch normalization layer are applied. Non-Gaussian noise in the form of SP noise to the Omniglot images by randomly flipping 15% of the images is added, thereby introducing a non-Gaussian element. The ensemble loss exhibits an upward trajectory with each iteration, as illustrated in Fig. 3.5. However, it is worth noting that the HESFOL

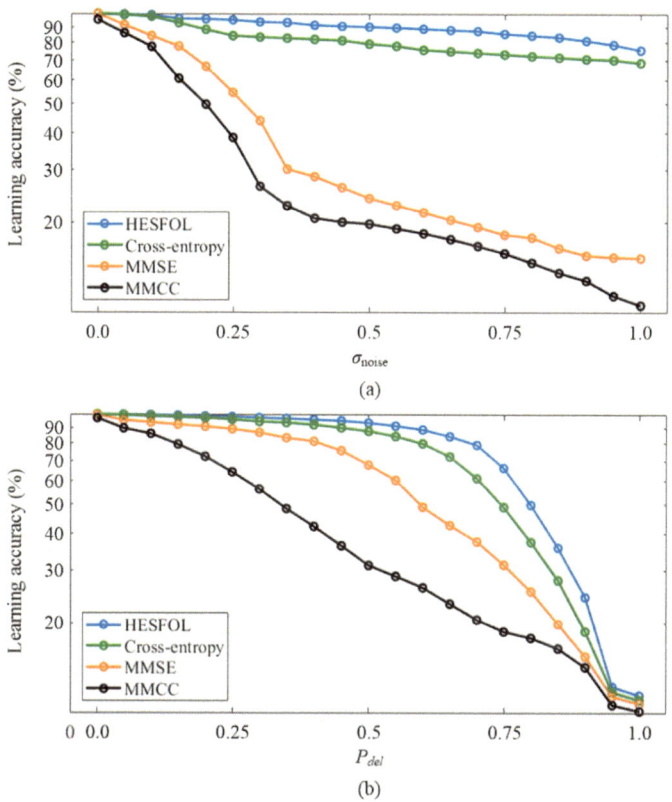

Fig. 3.4 Few-shot learning performance with non-Gaussian noise in spike signals for HESFOL and other models. **a** Few-shot learning accuracy with Poisson noise. **b** Few-shot learning accuracy with random deletion noise [52]

model's loss value rapidly diminishes, stabilizing at approximately 0.2 within the initial 1000 iterations.

The input signal contains values of 0 and 1, representing phase #1 and phase #2. The Omniglot dataset is displayed in a VNN with an array of 28×28 grayscale pixels. In phase #2, a single output determines if the displayed image belongs to the same class as in phase #1. The HESFOL model uses spike-based learning, and the PSNN receives spiking activity from the HSNN, along with phase ID input information. In the first stage, learning signals are sent only from PSNN to HSNN. Figure 3.6 illustrates the spiking activity of the HESFOL model in the Omniglot dataset, showing sparse spiking activities of the HSNN and PSNN subsystems in the few-shot learning task. To address the few-shot learning problem with non-Gaussian noise, the ensemble loss function combines MMCC, cross-entropy loss, and MMSE.

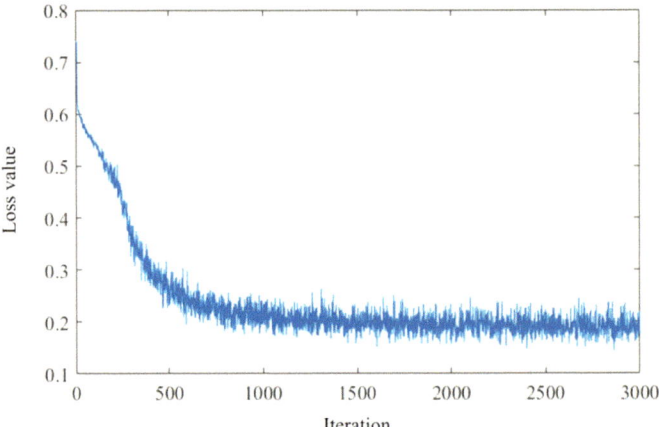

Fig. 3.5 Evolution of the loss value with the iteration [52]

The HESFOL model's ability for few-shot learning in robot control tasks is demonstrated. A 2-joint arm manipulator tracks a target trajectory in Euclidean coordinates (x, y), as shown in Fig. 3.7. The HESFOL model can learn to replicate a randomly generated target movement with the end-effector's actual movement in a motor control task. The learning task is divided into two parts, each with a training and a test phase. In the training phase, the PSNN provides the target motion in Euclidean coordinates, and the HSNN module receives the learning signal output by the PSNN. After the trial, the HSNN undergoes weight updating. In the testing phase, the HSNN reproduces the target motion without receiving the target trajectory. The HSNN input signal is the same for all trials, provided by a clock-like signal. The HSNN output signal is a motor command for the angular speed of the joints, $\Phi^t = (\Phi_1^t, \Phi_2^t)$. Figure 3.8 shows the trajectory generated by the HSNN as a solid line during both the training and testing phases. During training, the HSNN regenerates the target motion using a biorealistic sparse spiking activity when the PSNN transmits a learning signal to the HSNN. Figure 3.8 also illustrates the HESFOL model's learning signal and spiking activity. Figure 3.9 shows the mean squared error between the objective and actual motion in the testing phase. The HESFOL model with the ensemble loss function outperforms models with only one or fewer loss functions. The results demonstrate that HESFOL offers an effective neural mechanism for motor control and motor learning.

The contribution of each base loss function in the ensemble loss of the HESFOL model is investigated. The influence of ensemble parameters on few-shot learning using various datasets is evaluated, including spiking patterns and the Omniglot dataset. Overall, the cross-entropy loss function is the largest contributor to the HESFOL model, indicating that it has the highest weight.

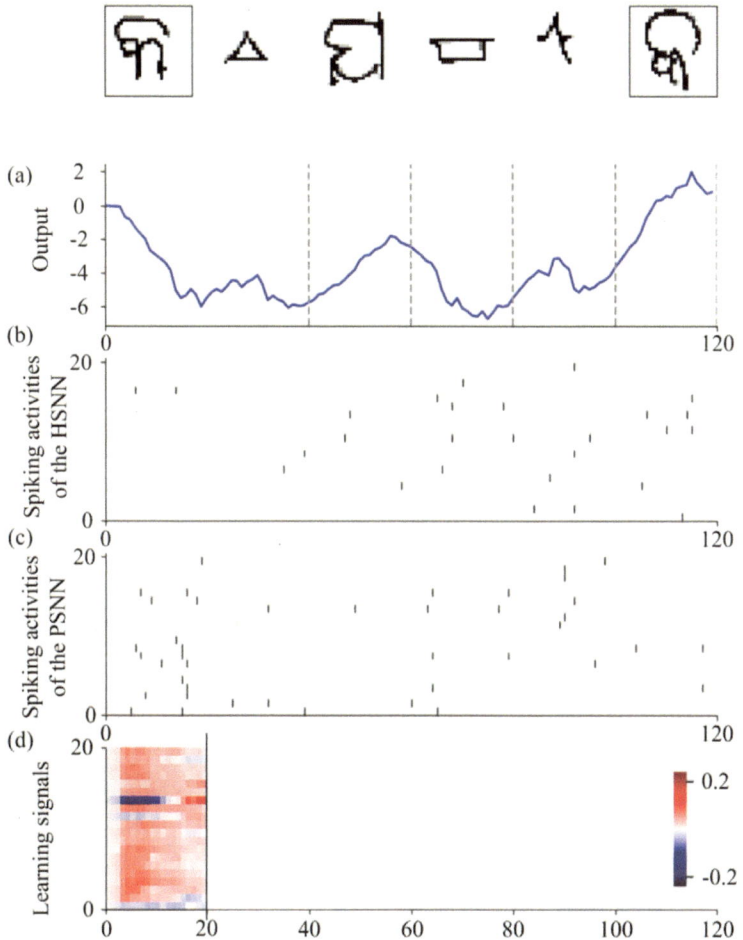

Fig. 3.6 HESFOL model was applied to the few-shot learning on Omniglot dataset. **a** Output of the read-out neuron **b** Spiking activities of the neurons in the HSNN module. **c** Spiking activities of the neurons in the PSNN module. **d** PSNN learning signals for HSNN neurons [52]

Fig. 3.7 Few-shot motor control of the end-effector of a 2-joint robotic arm [52]

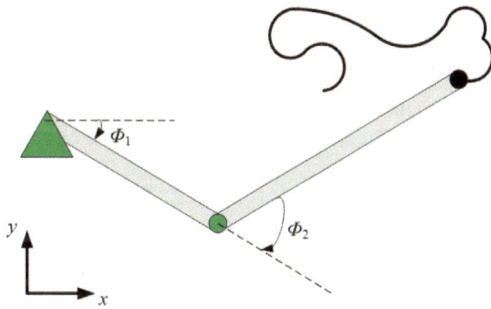

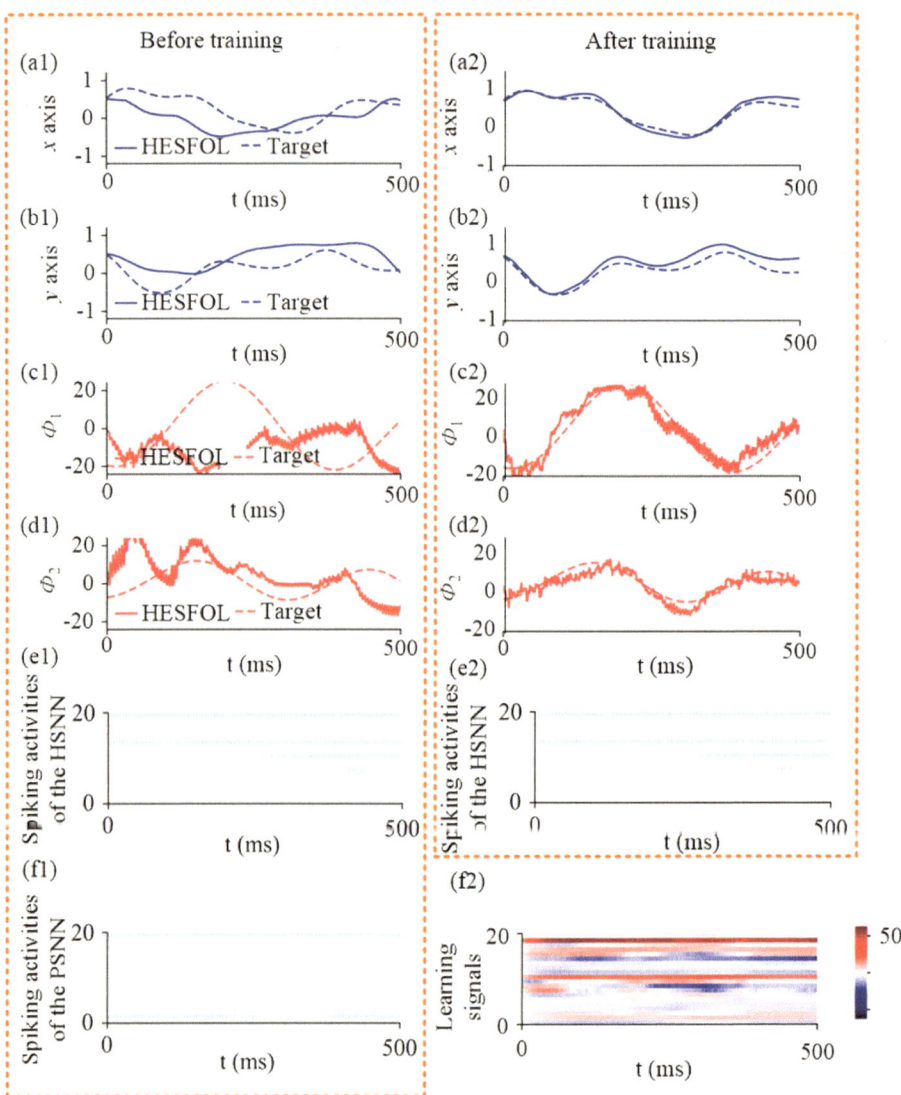

Fig. 3.8 Few-shot motor control performance of the HESFOL model, showing the ability to learn a new end-effector movement in just 500 ms. The figure illustrates the control performance and spiking activities before and after training, with (a1) showing the position on the x direction based on the HESFOL control before training and the target position on the x direction, and (a2) showing the same after training. Similarly, (b1) and (b2) show the position on the y direction before and after training, respectively. (c1) and (c2) display the motor command in the form of angular velocity of the joint and the target angular velocity on the x direction before and after training, respectively. (d1) and (d2) show the same for the y direction. The spiking activities of the HSNN are illustrated in (e1) before training and (e2) after training, while (f1) shows the spiking activities of the PSNN, and (f2) displays the learning signals produced by the PSNN for the HSNN. These results demonstrate the effectiveness of the HESFOL model for motor control and motor learning based on neural mechanisms [52]

Fig. 3.9 Control performance in terms of mean square error [52]

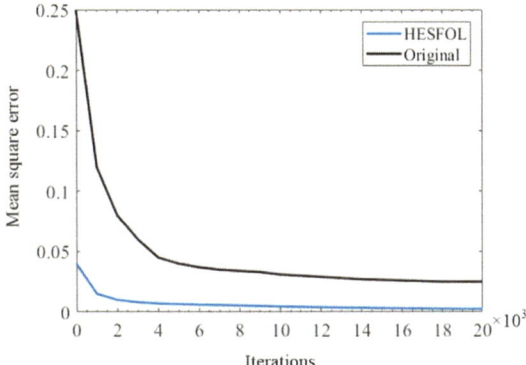

For the correntropy loss function, a weight of 0.1 is often appropriate for dealing with noise, particularly in the presence of outliers. As the SNN architecture relies on back-propagation of the gradient of the loss, it is susceptible to the problem of gradient vanishing. Thus, a loss function that emphasizes errors is preferable to an MMCC loss function. Consequently, the weight of the cross-entropy loss function is higher than the other base loss functions in the ensemble loss function of the HESFOL model.

To evaluate the few-shot learning performance of the HESFOL model directly, it is compared with other models, including ANNs and SNNs. Jiang et al. presented a new SNN model with an LSTM unit for few-shot learning, named MTSO (multi-timescale optimization) [21]. Since the HESFOL model does not use model augmentation to achieve optimal precision, it is fair to compare it with other non-augmented and unrefined models. The MTSO model without augmentation achieved high precision of 95.8%. MANN, presented by Santoro et al., achieved 82.8% accuracy on the Omniglot dataset [22]. Jiang et al. showed that the learning precision of CNN was only 92.1%, and the spiking CNN with L1 regularized method for sparsity achieved 92.8% in Omnilot [21]. Siamese Net achieved a precision of 96.7% with augmentation [23]. The HESFOL model achieved high precision of 93.1% on the Omniglot dataset with non-Gaussian noise, showing comparative performance in few-shot learning tasks. Although it is less accurate than Siamese Net, the HESFOL model is based on a spike-based model, so it has the advantages of low power consumption and high bio-credibility. Moreover, the HESFOL model is only 2.7% lower than MTSO, while being based on non-Gaussian noise data, unlike the pure dataset used by the MTSO model. This indicates that the HESFOL model can achieve high robust performance in few-shot learning without losing much accuracy.

In addition, the influence of critical parameters of the HESFOL model on few-shot learning performance is investigated. Three critical parameters are selected, including the timing constant of membrane τ_m, timing constant of readout neurons τ_{out}, and membrane potential threshold v_{th}. The Omniglot dataset is used to examine the HESFOL model. As illustrated in Fig. 3.10, the learning accuracy was improved by varying these parameters. Figure 3.10a revealed that the highest learning accuracy was achieved when $\tau_m = 15$ and

$\tau_{out} = 10$. In addition, Fig. 3.10b showed that $\tau_{out} = 10$ and $v_{th} = 1.0$ resulted in the highest learning accuracy. Furthermore, the optimal parameters for neural dynamics are established to achieve the classification tasks of few-shot learning. Results suggest that the SNN model with this set of parameter values has the highest long short-term memory performance and can store the prior experience for the current learning task. Since the HESFOL model is capable of achieving few-shot learning based on the meta-learning scheme, it also suggests that the model can learn to generalize well to new tasks with limited training data.

The loss function is a crucial component of a learning model, as it determines the impact of each sample on the model training. The value assigned to a sample by the loss function indicates its degree of participation in the training. If the loss function assigns a high value to an outlier sample, it can negatively affect the model parameters. To ensure robustness, learning machines must be designed to withstand the impact of outliers on the system's performance. For instance, a robust 0–1 loss function punishes all incorrectly classified samples with a value of 1.

The primary objective of any learning method is to classify unseen data effectively. Therefore, the classifier must be robust against data disturbance, and outliers must not

Fig. 3.10 Effects of the parameters on the learning performance over Omniglot data set [52]

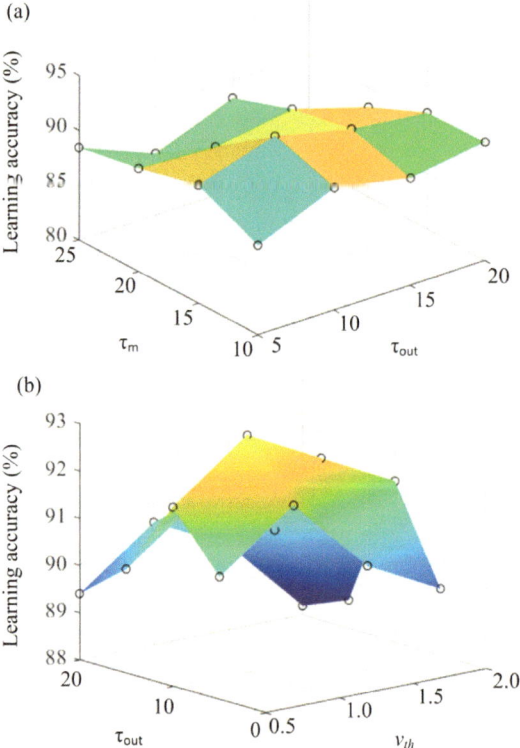

significantly affect the performance of the system. However, in a noisy environment, outliers can destroy the training data or test data. To handle noise effectively, one efficient approach is to use a robust loss function. A loss function can be deemed robust if there exists a constant k such that samples with $e_i > k$ are not given large values by the loss function, where e_i represents the error of the i_{th} sample.

Although a deep learning classifier may perform exceptionally well on the training data, it may not be able to generalize to unknown data. This phenomenon is referred to as over-fitting, where the classifier matches the training data too closely and lacks the ability to generalize. To address this problem, a better generalization method is to implement a more generic classifier through a loss function. MMCC is more sensitive to highly similar elements and less sensitive to those with significant differences. These features make MMCC particularly effective in reducing the impact of non-Gaussian noise on learning tasks, thereby inducing more robust spike-driven few-shot learning performance.

Additionally, the spiking dendrites in the HESFOL model have been shown to improve the robustness of few-shot learning, as demonstrated in previous studies [20]. This is because the non-linear calculation of the spiking dendrites can suppress input and transfer path noise, leading to increased learning efficiency. Moreover, the spiking dendrites can resolve the credit allocation problem and differentiate the information flow between the feedforward and recurrent paths, resulting in improved learning performance, including robustness.

Previous research has indicated that the minimum energy consumption of a synaptic operation on advanced neuromorphic systems is about 20 pJ [24, 25]. Using the Omniglot dataset, the HESFOL model would require approximately 60 spikes for the HSNN and 70 spikes for the PSNN classification task. Consequently, a single spike classification with the suggested HESFOL model would cost 2.6 pJ on this type of neuromorphic system, which is superior to current work (≈ 2 µJ) [26] and approximately 50,000 times more power efficient than existing GPU platforms [27]. Previous work has shown that the classification task using an improved DEP-based SNN model requires about 1011 SynOps to achieve the highest classification accuracy [20]. Compared with the most advanced SNN model, the HESFOL model can reduce the total induced spikes, i.e., power consumption, by 87.14%. Considering the causes of the low power consumption of the HESFOL model. First, the ensemble entropy theory is used, which can accelerate the learning speed and achieve maximal study precision, thus reducing the cost of power consumption during the learning process. Second, the few-shot learning method is used in the classification task, shortening the entire learning process and reducing power consumption. Finally, spiking dendrites are employed to reduce the number of spikes due to their nonlinear information processing capability. Consequently, the HESFOL model not only increases the learning precision and improves the learning robustness of SNN models but also reduces the power efficiency of SNN.

Previous work by Roy et al. provided an overview of current SNN training techniques for liquid state machines with a similar architecture to HESFOL [28]. Liquid

Table 3.4 Comparison with the representative LSM models with recurrent architecture (From [52])

Research	Application	Robustness	Few-shot learning
Soures and Kudithipudi, 2019	Video activity recognition	No	No
Wijesinghe et al., 2019	Image/speech recognition	No	No
Wang et al., 2020	Sitting posture recognition	No	No
Luo et al., 2018	Pattern classification	No	No
Al Zoubi et al., 2018	Emotion recognition	No	No
Panda and Roy, 2017	Visual recognition	Yes	No
HESFOL	Image classification	Yes	Yes

state machines (LSMs) are composed of non-structured, randomized recurrent networks coupled with a simple linear readout. As shown in Table 3.4, spiking dynamics have demonstrated remarkable performance on numerous sequential recognition tasks [29–31]. Soures and Kudithipudi proposed a deep LSM with STDP learning rule for video activity recognition [30]. Wijesinghe et al. presented an ensemble approach for LSM to enhance class discrimination, leading to better accuracy in speech and image recognition tasks when compared to a single large liquid [31]. Wang et al. presented a new LSM model for sitting posture recognition [32]. Luo et al. presented two different methods to improve LSM in real-time pattern classification from perspectives of spatial integration and temporal integration [33]. Zoubi et al. proposed LSM as an approach to automatically extract and predict features from raw electroencephalogram (EEG), with potential extension to a wider range of applications [34]. Despite the numerous approaches to the problem of sequential recognition, no approach has been able to overcome the few-shot learning problem. In this section, a new framework is proposed that achieves robust image classification and few-shot learning performance, which is superior to traditional LSM models based on recurrent architecture.

The ANN-SNN conversion requires fewer GPU computations for deep SNN training than supervised training with a surrogate gradient. Furthermore, it has been demonstrated to be one of the most effective approaches for large-scale datasets and networks. For example, Ding et al. suggested using a rate norm layer instead of the ReLU activation function in ANN training, which can be directly converted from ANN to SNN [35]. Zheng et al. also proposed a threshold-dependent batch normalization (tdBN) method based on emerging spatiotemporal back-propagation, enabling direct training of a very deep SNN and the efficient implementation of its inference on neuromorphic hardware [36]. However, none of these studies have solved few-shot learning problems, and learning robustness has not been a focus. In contrast, the HESFOL model presents a robust few-shot learning framework with an ITL approach, which is meaningful for combining machine learning approaches with brain-inspired SNN paradigms.

A key challenge is to provide effective training algorithms for SNN models to process complex datasets for more practical applications. Although shallow SNNs can be trained using the surrogate gradient descent method, they can only perform well on simple datasets like MNIST. In fact, the discrepancy between the forward spike activation function and backward surrogate gradient function during training limits the learning capability of deep SNNs. In recent years, SNNs have been successfully trained from scratch using a surrogate gradient descent method. For example, Kim and Panda introduced a technique known as batch normalization through time (BNTT) to train SNNs, which can dynamically vary parameters and have an implied influence as a dynamic threshold [37]. In addition, they also introduced a spike activation lift training method, which is essentially a threshold fine-tuning or initialization step ahead of actual training [38]. These two models can be used to train deep layer SNN models and can be tested on complex datasets like DVS-CIFAR10, CIFAR100, and Tiny ImageNet. They demonstrate high performance on SNN models that can be scaled up for more realistic applications. Due to the spiking dendrites in the HESFOL model, it has been shown that it is possible to resolve the credit allocation problem between feed-forward and feedback paths, making it significant for more complex tasks and actual situations. Earlier studies have proposed several possible approaches for SNNs to target complex tasks other than visual recognition. For example, Kim and Panda proposed an approach to analyze and explain the intrinsic spike behavior of SNNs to enable them to be widely used [39]. Kim et al. explored SNN applications beyond classification and presented semantic segmentation networks configured with spiking neurons [38]. Venkatesha et al. proposed a federated learning approach for SNNs that are decentralized and privacy-preserving [40]. In addition, Kim et al. presented PrivateSNN, which is designed to generate low-power SNNs based on the ANN model that has been pretrained without leaking sensitive information contained in a dataset [58]. These studies lead to the development of the HESFOL model in other areas, such as federated learning and privacy protection.

3.3   Robust Spike-Driven Meta-Learning

3.3.1   MeMEE Model

In this section, the SAM model is incorporated into a SNN framework, and test the accuracy of this new model on different types of learning tasks. The SNN model is illustrated in Fig. 3.11, which consists of 3 layers: input layer, hidden layer, and output layer. In Fig. 3.11, the solid blue lines indicate the feed-forward inhibition of the synaptic connections, and the red dotted lines indicate the inhibition of synaptic connections in the lateral direction. The dendrites and the soma of the hidden layer are connected by the lateral inhibition synapse, which is both random and sparse. The information is transmitted to the dendrites from the input layer, and the soma sends the spike signal to the output layer.

The scaling factor is significant that is used to initialize the SNN with a real firing rate required for efficient training.

A deep rewiring algorithm is used, which allows us to retain the signature of every synapse throughout the learning process [42]. Therefore, this sign is inherited from the initial weights of the network. Neurons are sampled from a Bernoulli distribution, producing

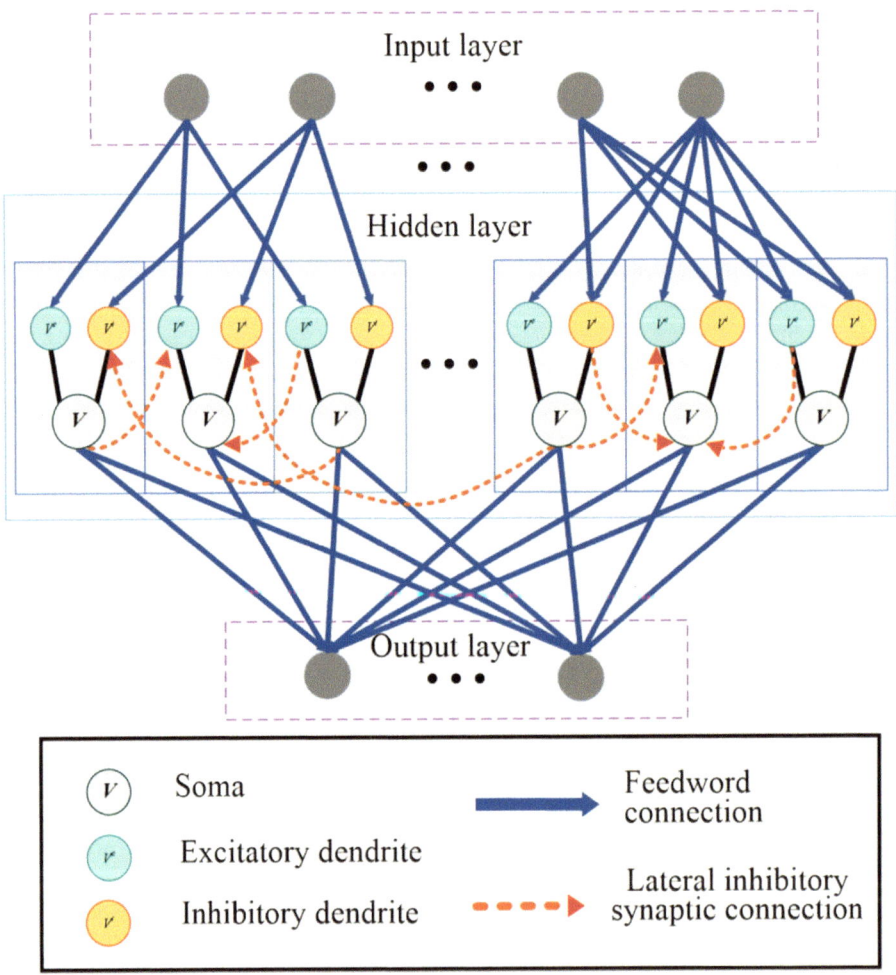

Fig. 3.11 Based on the SAM model, a network architecture for learning and memory integration. This model is similar to a 2-layer network of point neurons. The soma and dendrites of different neurons in the hidden layer were randomly attached to the lateral inhibition synapse. The grey rings on the input and output layers do not belong to SAM neurons, but stand for the incoming and outgoing ones. Input and output codes are defined for various tasks, as will be explained later in the experiment results [51]

the symbol sign $k_i \in \{-1, 1\}$ at random. Moreover, in order to avoid an explosion of gradient, the weights are scaled so that the maximum value is smaller than 1. A large square matrix is generated with the number of rows selected with uniform probability distribution. This square matrix is then multiplied by a binary mask to produce a sparse matrix, which is a part of the previously discussed depth rewiring algorithm. In this algorithm, the sparse connectivity level in the network can be kept by making dynamic disconnection of certain synapses at the same time as reconnecting others. The temperature parameter is set to 0 and the L1-NORM regularization parameter is 0.01.

In ordinary ANN models, the gradient of the loss function is calculated by backpropagation. In spite of this, it is not possible to directly apply the back-propagation training method to SNNs because of the non-differentiability of the spikes from SNN. Assuming that time is discrete, it is necessary to propagate the gradient through successive time or multiple time steps. In order to allow SNN models to learn during training, a pseudo derivative technique is employed as follows

$$\frac{dz_j(t)}{dv_j(t)} = k \max\{0, 1 - |v_j(t)|\} \tag{3.52}$$

where $k = 0.3$ (usually smaller than 1) is a constant value which can suppress an increase in backpropagated errors by means of a pseudo derivative of amplitude to obtain steady performance. The variable $z_j(t)$ is a spike train of neuron j, which is assumed to be in $\{0,1\}$. The variable $v_j(t)$ represents the normalized membrane potential, which is defined as follows:

$$v_j(t) = \frac{V_j(t) - \Gamma_j(t)}{\Gamma_j(t)} \tag{3.53}$$

where Γ_j represents the firing rate of neuron j. In order to provide the self-learning capability for reinforcement learning for the SAM model, the proximal policy optimization (PPO) algorithm is used [43]. The implementation of this algorithm is simple, and the model has the ability of self-learning. The clipped surrogate objective of this algorithm is defined as $O^{PPO}(\vartheta_{old}, \vartheta, t, k)$. Therefore, the loss function with respect to ϑ is formulated as:

$$L(\theta) = -\frac{\sum\limits_{k<K}\sum\limits_{t<T} O^{PPO}(\vartheta_{old}, \vartheta, t, k)}{KT} + \mu_f \frac{1}{n}\sum\limits_j \left\| \frac{\sum\limits_{k,t} z_j(t,k) - f^0}{KT} \right\|^2 \tag{3.54}$$

where f^0 stands for the target firing rate of 10 Hz, and μ_f stands for the regularized hyperparameter. Variables t and k denote the time step and the total number of epochs. This variable ϑ is the current policy parameter as defined in an earlier survey [44]. In each iteration of the training, a fixed parameter ϑ_{old} is used to generate $K = 10$ episodes of $T = 2000$ time steps, which is the vector of the policy parameters prior to the update, as shown in previous study [44]. Meanwhile, the ADAM optimizer [43] minimizes the loss function $L(\theta)$. MEE is a local optimal rule, but it is subject to the problem of shifting invariance. It can only locate the PDF error, but not the location of distribution. The function $G_{\sum 2}(\cdot)$ can be defined as the Gaussian kernel function with bandwidth σ.

$$G_{\sum 2}(x) = \frac{1}{\sqrt{2\pi}\sigma} \exp\left(-\frac{x^2}{2\sigma^2}\right) \tag{3.55}$$

To reduce the computation complexity, quantizing MEE (QMEE) is implemented. Thus, the information potential is expressed as

$$V^Q(e) = \frac{1}{N^2}\left(\sum_{i=1}^{N}\sum_{j=1}^{N} G_{\sum 2}\left(e_i - Q|e_j|\right)\right) = \frac{1}{N^2}\sum_{i=1}^{N}\sum_{j=1}^{M} \varphi_j G_{\sum 2}\left(e_i - c_j\right) \tag{3.56}$$

where $Q[\cdot]$ represents a quantization operator mapping each $\{e_i\}_{i=1}^{N}$ to one of $\{c_j\}_{j=1}^{M}$, resulting in a codebook $C = (c_1, c_2, c_3, ..., c_m)$. $\phi = (\varphi_1, \varphi_2, \varphi_3, ..., \varphi_m)$ is the number of samples that are quantized to the corresponding set $\{c_j\}_{j=1}^{M}$. Note that $\sum_{j=1}^{M} \varphi_j = N$. Theoretical evidence of the robustness is given in previous study [13].

The basic inner product is employed to measure the similarity, which is generalized from its vectors' application [16]. The inner product similarity of the continuous pdfs $f_X(x)$ and $g_X(x)$ can be expressed as

$$\langle f_X(x), g_X(x) \rangle = \int_X f_X(x) g_X(x) dx \tag{3.57}$$

The desired distribution $\rho_E(e)$, which is expressed in previous study [16] in detail, can be defined as follows:

$$\rho_E(e) = \begin{cases} \zeta_0, & e = 0 \\ \zeta_{-1}, & e = -1 \\ \zeta_1, & e = 1 \\ 0, & \text{otherwise} \end{cases} \tag{3.58}$$

where $\zeta_{d,i}(i = 0, -1, 1)$ denotes the corresponding density for each peak which is simplified into a Dirac-δ_D function.

The maximization of the similarity measure between the error PDF $f_E(e)$ and the expected distribution $\rho_E(e)$ can be expressed as

$$\max\langle f_E(e), \rho_E(e) \rangle$$
$$\Leftrightarrow \max \int_X f_E(e)\rho_E(e)dx \tag{3.59}$$
$$\Leftrightarrow \max \zeta_{d,0} f_E(0) + \zeta_{d,-1} f_E(-1) + \zeta_{d,1} f_E(1)$$

Furthermore, the model parameter can be expressed as

$$w^* = \arg\max \zeta_{d,0}\hat{f}_E(0) + \zeta_{d,-1}\hat{f}_E(-1)\zeta_{d,1}\hat{f}_E(1)$$

$$= \arg\max \left(\begin{array}{c} \zeta_{d,0}\dfrac{1}{N}\displaystyle\sum_{i=1}^{N} G_{\sum 2}(0 - e_i) \\[2ex] +\zeta_{d,-1}\dfrac{1}{N}\displaystyle\sum_{i=1}^{N} G_{\sum 2}(-1 - e_i) \\[2ex] \zeta_{d,1}\dfrac{1}{N}\displaystyle\sum_{i=1}^{N} G_{\sum 2}(1 - e_i) \end{array} \right)$$

$$= \arg\max \dfrac{1}{N^2}\sum_{i=1}^{N} \left(\begin{array}{c} N\zeta_{d,0}G_{\sum 2}(e_i) \\ +N\zeta_{d,-1}G_{\sum 2}(e_i + 1) \\ +N\zeta_{d,1}G_{\sum 2}(e_i - 1) \end{array} \right) \tag{3.60}$$

In fact, QMEE is able to converge the predicted errors $\{c_j\}_{j=1}^{M}$ to obtain a tight error distribution. In accordance with the method in previous study [16], QMEE is realized by using a predefined code book $C = (0, -1, 1)$, so that the error is limited to three locations, and the undesired double peak learning results are avoided. Thus, restricted minimum error entropy (RMEE) can be expressed as

$$E^R(e) = \dfrac{1}{N^2}\sum_{i=1}^{N} \left(\begin{array}{c} \varphi_0 G_{\sum 2}(e_i) \\ +\varphi_{-1}G_{\sum 2}(e_i + 1) \\ +\varphi_1 G_{\sum 2}(e_i - 1) \end{array} \right) \tag{3.61}$$

where $\phi = (\varphi_0, \varphi_{-1}, \varphi_1) = (N\zeta_{d,0}, N\zeta_{d,-1}, N\zeta_{d,1})$ which indicates the corresponding number for each quantization word $C = (0, -1, 1)$. The RMEE algorithm is used to maximize the internal product similarity of the error PDF $f_E(e)$ and the optimum 3-peak distribution $\rho_E(e)$. MEE is a particular form of QMEE in which the codebook is prefixed to $C = (0, -1, 1)$, and the learning error is convergent at the 3 positions.

The optimization problem is solved with half quadratic technology. In this section, a convex function $g(x) = -x\log(-x) + x$ is defined, and the information potential can be expressed as

$$E^R(e) = \sum_{i=1}^{N} \left(\begin{array}{l} \varphi_0 \left\{ u_i \dfrac{e_i^2}{2\sigma^2} - g(u_i) \right\} \\[2ex] +\varphi_{-1} \left\{ v_i \dfrac{(e_i+1)^2}{2\sigma^2} - g(v_i) \right\} \\[2ex] +\varphi_1 \left\{ s_i \dfrac{(e_i-1)^2}{2\sigma^2} - g(s_i) \right\} \end{array} \right) \triangleq J_{R1}(w, u_i, v_i, s_i) \tag{3.62}$$

In half-quadratic technique, it has the following relationship as

$$u_i^k = -\exp\left(-\frac{e_i^2}{2\sigma^2}\right) < 0$$

$$v_i^k = -\exp\left(-\frac{(e_i+1)^2}{2\sigma^2}\right) < 0$$

$$s_i^k = -\exp\left(-\frac{(e_i-1)^2}{2\sigma^2}\right) < 0$$

$$(i = 1, 2, ..., N). \tag{3.63}$$

By attaining the optimal (u_i^k, v_i^k, s_i^k) in the kth iteration, the information potential can be formulated as

$$E^R(e) = \sum_{i=1}^{N} \left(\begin{array}{l} \varphi_0 u_i (t_i - y_i)^2 \\[1ex] +\varphi_{-1} v_i (t_i + 1 - y_i)^2 \\[1ex] +\varphi_1 s_i (t_i - 1 - y_i)^2 \end{array} \right) \triangleq J_{R2}(w) \tag{3.64}$$

The $J_{R2}(w)$ by means of a gradient-based approach, since the objective function can be differentiated and continued. For instance, the gradient of $J_{R2}(w)$ can be expressed as

$$\frac{\partial}{\partial w} J_{R2}(w) = \sum_{i=1}^{N} \left(\begin{array}{l} \varphi_0 u_i \dfrac{\partial (t_i - y_i)^2}{\partial w} \\[2ex] +\varphi_{-1} v_i \dfrac{\partial (t_i + 1 - y_i)^2}{\partial w} \\[2ex] +\varphi_1 s_i \dfrac{\partial (t_i - 1 - y_i)^2}{\partial w} \end{array} \right)$$

$$= -2 \sum_{i=1}^{N} \left(\begin{array}{l} \varphi_0 u_i e_i \\[1ex] +\varphi_{-1} v_i (e_i + 1) \\[1ex] +\varphi_{-1} s_i (e_i - 1) \end{array} \right) x_i y_i (1 - y_i) \tag{3.65}$$

The detailed algorithm of the optimization and there convergence analysis for RMEE are given in previous study [16]. Since MEE has the property of shifting invariance, the

MEEC estimation result does not always approximate the real value. The combination of the RMEE criterion and the cross-entropy criterion is considered in order to obtain the global optimum solution. The cross-entropy loss function, also known as log loss, is one of the most popular back-propagation loss functions. Label l^n of every image is used, which is assumed to be only 1 for the image of the same class, and 0 if it is not. The cross-entropy formula can be expressed as:

$$J_2 = \sum_{n=1}^{5} -l^n \log \sigma \left(y^{20+20 \cdot n}\right) - \left(1 - l^n\right) \log\left(1 - \sigma^{20+20 \cdot n}\right) \tag{3.66}$$

The SNN model output is calculated only when all the images have been completely rendered. Therefore, for the novel criterion, the performance index can be formulated as

$$J_k(e) = \mu \left[\sum_{i=1}^{N} \left(\begin{array}{c} \varphi_0 u_i (t_i - y_i)^2 \\ +\varphi_{-1} v_i (t_i + 1 - y_i)^2 \\ +\varphi_1 s_i (t_i - 1 - y_i)^2 \end{array} \right) \right]$$

$$+ (1 - \mu) \left[\sum_{n=1}^{5} \left(\begin{array}{c} -l^n \log \sigma \left(y^{20+20 \cdot n}\right) \\ -\left(1 - l^n\right) \log\left(1 - \sigma^{20+20 \cdot n}\right) \end{array} \right) \right] \tag{3.67}$$

where μ represents a weighting constant.

3.3.2   Performance Evaluation

The problem of agent navigation is considered, which requires the network to have reinforcement learning capabilities. The agent must learn how to locate objects in a 2D region, and ultimately be able to navigate to locate objects in the region at random later on. This task is related to the famous Morris water maze task, which is designed to study brain learning [13]. In this task, a virtual agent is simulated as a point in a 2D simulation field, and the SNN model is used to control it. The location of the agent is randomly set with a uniform probability in the entire arena at the start of an episode. The agent produces a small velocity vector of the Euclidean norm and selects an action at each time step. It receives a reward value of "1" upon arriving at the destination.

The agent based on the SNN model is able to learn how to navigate towards the correct destination location after a meta-learning process. In this section, additional loss functions are introduced in order to ensure that the loss function aligns with the formula. Figure 3.12 shows the number of successful targets for each learning iteration. Each iteration is composed of a batch of 10 episodes, and the network weights are updated during the navigation task. The model is expected to explore until it reaches the target location and stores the position for each episode, and then it uses the prior knowledge

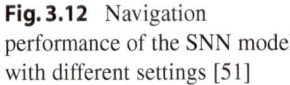
Fig. 3.12 Navigation
performance of the SNN model
with different settings [51]

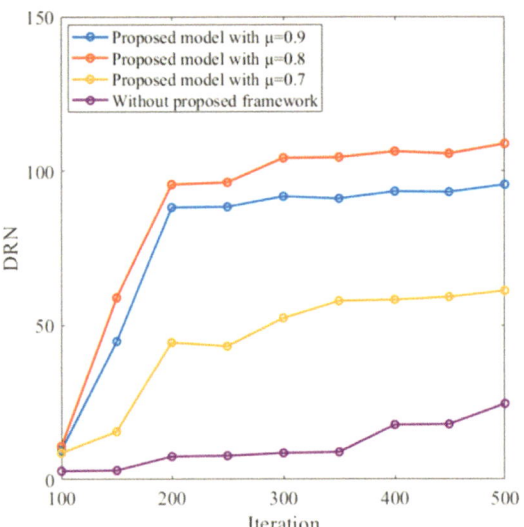

to search for the shortest path to the destination. These results demonstrate that the SNN model has the ability to meta-learn in autonomous navigation tasks.

To demonstrate the robustness performance, the problem of non-Gaussian noise is considered. The configuration of the store-recall task was described in detail in previous study [45]. In this task, the SNN model received a series of frames, represented by 10 spike trains over a given time. Inputs #1 and #2 were indicated by the spiking activity of the input neurons from #1 to #10 and from #11 to #20, respectively. As shown in Fig. 3.13, neurons #21 to #30 and #31 to #40 received the random-storage and recall commands, respectively. The storage command referred to a particular frame of the input data stream, which was reproduced upon receiving the recall command. Figure 3.13 depicts a test example with the spiking activities after working memory training, where the dynamic threshold varied with the learning process. The results demonstrate that the SNN model can display the function of working memory and accomplish the store-recall task. Moreover, it has been shown that the MeMEE model can perform meta-learning tasks with high accuracy. Since working memory is a vital feature and foundation for meta-learning, this suggests that the MeMEE model can exhibit meta-learning tasks based on its working memory mechanisms with a robust performance.

The SNN model's meta-learning ability is demonstrated in a transfer learning task based on the sequential MNIST (sMNIST) dataset. The sMNIST dataset is split into two sections: the first section contains 30,000 images for the numbers 0, 1, 2, 3, and 4, while the second section contains 30,000 images for the numbers 5, 6, 7, 8, and 9. In the first phase, the SNN model is trained on the first part and then use the second part for training. In the second stage, 10% salt & pepper noise is added as non-Gaussian noise to evaluate the system's performance. Figure 3.14 illustrates MeMEE's performance and compares

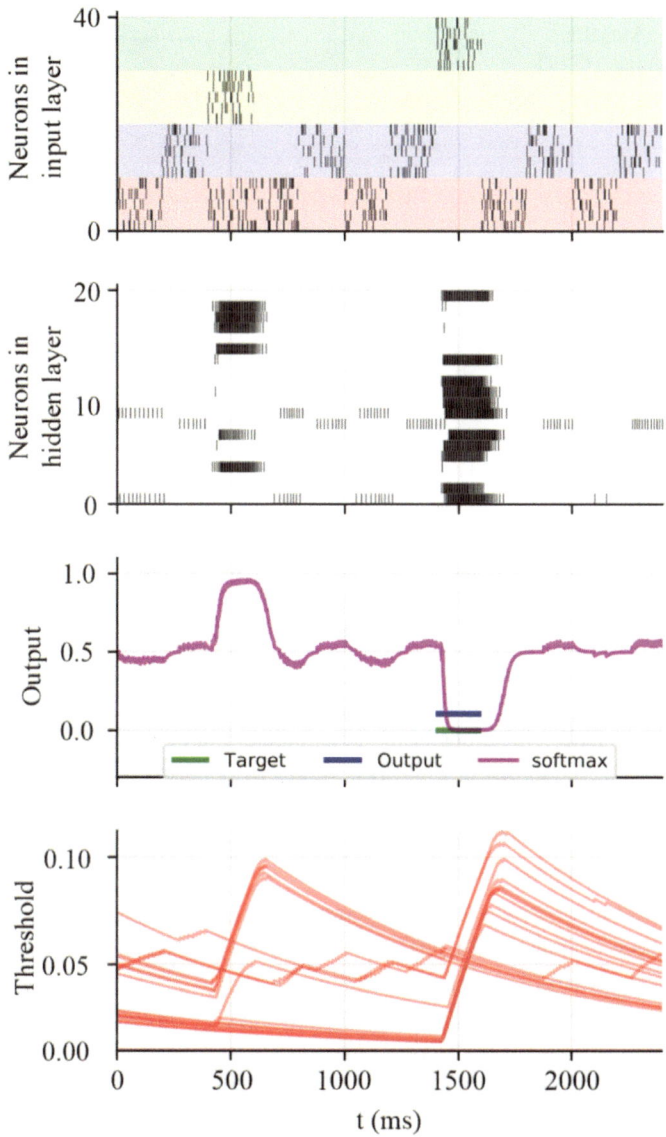

Fig. 3.13 Working memory capability of the SNN model after training [51]

it with other equivalent models such as RSNN and traditional LIF-based SNN models that do not have RMEE standard. The results show that the SNN model outperforms the others for three reasons. Firstly, this model has the ability of meta-learning, allowing it to explain transfer learning ability, and its transfer learning performance is superior to the RSNN model, considering accuracy and convergence speed. Secondly, using the RMEE

Fig. 3.14 Meta learning capability of the MeMEE model on sequential MNIST data set [51]

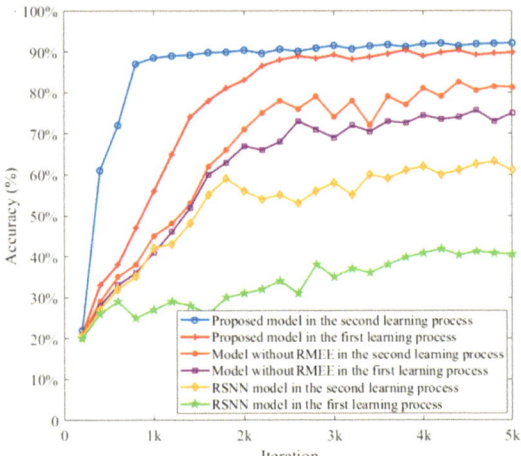

criterion as a loss function makes the model more robust to non-Gaussian noise than models without RMEE. The results indicate that MeMEE based on RMEE is more robust in learning sequential spatial–temporal patterns.

In this chapter, the impact of different loss functions are investigated on the MeMEE model by evaluating its performance on the sMNIST dataset. To demonstrate the model's robustness, salt & pepper noise are added to the dataset at various levels ranging from 3.19% to 19.13%. Different values of the μ parameter are considered, which is set in the range of 0.3 to 1.0. As shown in Fig. 3.15, values of μ between 0.7 and 0.9 can lead to higher learning accuracy for sequential visual recognition. The RMEE criterion improves the robustness of the MeMEE model compared to when $\mu = 1$, as the latter achieves a precision of only 83.6% with 3.19% non-Gaussian noise. The results demonstrate that RMEE enhances the model's learning precision under salt & pepper noise, which suggests that it can be used to improve the robustness of the MeMEE model in learning sequential spatial–temporal patterns.

3.4 Information Bottleneck for Spike-Based Learning

3.4.1 SIBoLS Model

When considering to apply the IB theory in BPTT training algorithm for deep SNNs, a critical problem is how to represent the hidden information of deep spiking convolutional network. In addition, another critical problem is how to approximate the information entropy to get the optimal loss function. Therefore, these two difficulties should be first considered and solved. The spike output tensor in each layer can be represented as a tensor with dimensions $\{g_b, \Omega, \dagger, \varkappa, \flat\}$, where $g_b, \Omega, \dagger, \varkappa$, and $\flat$ represent the batch size,

Fig. 3.15 Effects of loss parameters on the learning performance of sequential classification [51]

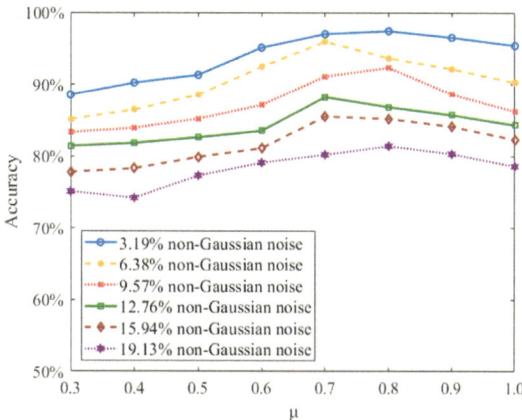

channel number, pattern height, pattern width and time step. In the spike-based information bottleneck with learnable state (SIBoLS) model, only the dimension $\{\dagger, \varkappa, \flat\}$ of the spike output tensor in each layer is considered to simplify the explanation. In the SIBoLS framework, both the input information and the output labels are considered as the approximated representation of information. Nevertheless, the spike output tensor in each layer is a sparse zero–one matrix formed by $\{\dagger, \varkappa, \flat\}$, which cannot demonstrate the corresponding meaning of information representation. When $\{\dagger, \varkappa, \flat\} = 1$, a spike is generated at position $\{\dagger, \varkappa\}$ in time $\flat$. However, multiple spikes can be generated at different time steps during training a single image, which induces multi-spike contradictions for the time-domain coding strategy of the spike output tensor.

In addition, mutual information in $\{\dagger, \varkappa, \flat\}$ is required to be calculated based on the input data, spike output tensor, and output labels. Conventional research has applied IB to deep neural networks by approximating mutual information with adding noise and computing a bound. Nevertheless, a tighter bound may not be perfect as investigated by Tschannen et al. [46]. Therefore, in the SIBoLS framework, the mutual information is approximated by Hilbert–Schmidt independence criterion, which provides a robust computation with a time complexity $O(m^2)$, where m represents the number of data points. Hilbert–Schmidt independence criterion is the Hilbert–Schmidt norm of the cross-covariance operator between the distributions in RKHS, which can be described as

$$
\begin{aligned}
L(P_{XY}, \Xi, \Upsilon) &= \|C_{XY}\|^2 \\
&= E_{XYX'Y'}\big[k_X(X, X')k_{Y'}(Y, Y')\big] \\
&\quad + E_{XX'}\big[k_X(X, X')\big]E_{Y'}\big[k_Y(Y, Y')\big] \\
&\quad - 2E_{XY}\big[E_{X'}\big[k_X(X, X')\big]E_{Y'}\big[k_Y(Y, Y')\big]\big]
\end{aligned}
\tag{3.68}
$$

where k_X and k_Y represent kernel functions. $[\Xi, \Upsilon]$ represents the Hilbert space, and E_{XY} represents the expectation over E_{XY} and Y. Suppose $D := \{(x_1, y_1), ..., (x_m, y_m)\}$

contain m i.i.d. samples in P_{XY}, where $x_i \in R^{d_x}$ and $y_i \in R^{d_y}$. Then it can be formulated as

$$L(P_{XY}, \Xi, \Upsilon) = (m-1)^{-2}\text{tr}(K_X H K_Y H) \tag{3.69}$$

where $K_X \in R^{m \times m}$ and $K_Y \in R^{m \times m}$ have entries $K_{X\,i,j} = k(x_i, x_j)$ and $K_{Y\,i,j} = k(y_i, y_j)$, and $H \in R^{m \times m}$ represents the centering matrix $H = I_m - \frac{1}{m}1_m 1_m^T$. The kernel distance between features is calculated as follows:

$$k(x, y) = \exp\left(-\frac{1}{2\sigma^2}\|x - y\|^2\right) \tag{3.70}$$

Nevertheless, since most of the elements in the sparse matrix of spike output tensor are 0, the mutual information is a very low value. It results in a small loss function value, which is not proper for the effective training of SNNs. Therefore, the SIBoLS considers the mechanism transfer from conventional ANN to SNN. The sparse spike tensor output is not suitable for information representation in a multi-layer SNN. It induces the loss of information representation characterized based on membrane potential. Therefore, for the SIBoLS framework, a learnable hidden state related to the membrane potential evolution is constructed for the dynamic representation of neural information. Especially, the previous SAM model in this network architecture is applied [47], which has dynamic adaptive neural threshold $\Gamma_j(t)$. It is varied along with spiking activities, which can be formulated as

$$\Gamma_j(t) = \tau_j^0 + \eta \cdot \tau_j(t) \tag{3.71}$$

where η denotes a scaling parameter from the baseline τ_j^0 to deviation $\tau_j(t)$. The variable $\tau_j(t)$ can be expressed as

$$\tau_j(t + \Delta t) = \lambda_j \tau_j(t) + (1 - \lambda_j)z_j(t) \tag{3.72}$$

where $\lambda_j = \exp(-\Delta t / \tau_{a,j})$ and $\tau_{a,j}$ denotes the adaptation time constant. By introducing the learnable adaptive threshold in SAM model, the propose SIBoLS framework can be applied to achieve higher spike-based learning performance. The objective function can be formulated as

$$L_i = I(X; V_i) - \beta I(V_i; Y) \tag{3.73}$$

where represents the membrane potential of spiking neurons in the ith layer. A combination of the cross-entropy loss and the feature-correlation loss function is considered as

$$L_i = CE\left(Y, D^T V_i\right) - \beta \|C(V_i) - C(Y)\|_F^2 \tag{3.74}$$

where D^T represents the decoder matrix for decoding local layer features V_i to $\tilde{Y}$. $C(X)$ represents the correlation matrix, where an element is formulated as

$$c_{ij} = c_{ji} = \frac{\tilde{x}_i^T \tilde{x}_j}{\|\tilde{x}_i\|_2 \|\tilde{x}_j\|_2} \tag{3.75}$$

The hyper parameter β represents a weight parameter for the loss function.

Suppose a deep SNN with VGG architecture contains L hidden layers $T_i(\cdot) : \mathbf{R}^{d_{i-1}} \to \mathbf{R}^{d_i}$, which leads to hidden representations $\mathbf{Z}_i \in \mathbf{R}^{g_b \times d_i}$, where $i \in \{1, ..., L\}$ and g_b represents batch size. By implementing the IB principle, the original learning objective is modified and formulated as

$$\mathbf{Z}_i^* = \arg\min_{\mathbf{Z}_i} \text{SIBoLS}(\mathbf{Z}_i, \mathbf{X}) - \beta \cdot \text{SIBoLS}(\mathbf{Z}_i, \mathbf{Y}) \tag{3.76}$$

where $\mathbf{X} \in \mathbf{R}^{g_b \times d_x}$ and $\mathbf{Y} \in \mathbf{R}^{g_b \times d_y}$ represent the input and the label respectively. $i \in \{0, ..., L\}$ and L represent the hidden layer number. d_x and d_y denote the dimensionalities of the input and output variables. d_y represent the classes number for the classification task. The hyper parameter determines the balance of SIBoLS objectives. The SIBoLS of each term can be

$$\text{SIBoLS}(\mathbf{Z}_i, \mathbf{X}) = \text{tr}\left(\tilde{\mathbf{K}}_{Z_i} \tilde{\mathbf{K}}_X\right) \tag{3.77}$$

$$\text{SIBoLS}(\mathbf{Z}_i, \mathbf{Y}) = \text{tr}\left(\tilde{\mathbf{K}}_{Z_i} \tilde{\mathbf{K}}_Y\right) \tag{3.78}$$

where $\tilde{\mathbf{K}}_X = \overline{\mathbf{K}}_X \left(\overline{\mathbf{K}}_X + \varpi m \mathbf{I}_m\right)^{-1}$ and $\tilde{\mathbf{K}}_Y = \overline{\mathbf{K}}_Y \left(\overline{\mathbf{K}}_Y + \varpi m \mathbf{I}_m\right)^{-1}$. $\overline{\mathbf{K}}_X$ and $\overline{\mathbf{K}}_Y$ represent centered kernel matrices, and ϖ stands for a small constant. The equation reveals that the optimal hidden representation can obtain a balance between independence from unnecessary input details and output dependence. A single output layer $O(\cdot) : \mathbf{R}^{d_L} \to \mathbf{R}^{d_y}$ is used based on a Softmax function for classification. The SIBoLS architecture is shown in Fig. 3.16.

Deep networks are critical to recognize intricate input patterns to effectively learn hierarchical representations. SIBoLS in a deep convolutional SNN architecture is first considered, containing an input layer, several hidden layers, and a classification layer. The hidden layers are formed by multiple convolutional and pooling layers, denoting the processing procedure as a feature extractor, and convolutional layers contain trainable synaptic weights. Average pooling technique is used with fixed parameters, and an additional thresholding method is employed to generate output spikes. Specifically, VGG architecture is employed, which employed small convolutional kernels uniformly within the network architecture [48]. Effective stacking of convolutional layers can be realized based on small kernels, and the number of parameters in deep networks can be minimized consequently. Therefore, spiking VGG architecture is used to build deep SNN model with

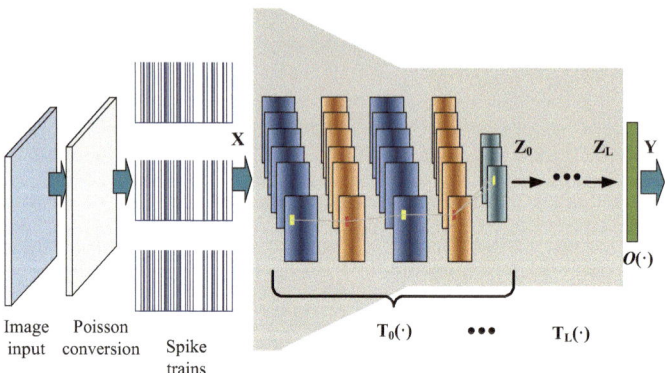

Fig. 3.16 The SIBoLS framework for deep SNN with VGG architecture. Hidden representations at the last layer are used for final classification [60]

small convolutional kernel. The membrane potentials of spiking neurons can be considered as the information representation to calculate local loss. The BPTT algorithm is used in the training process of deep SNN, which can be formulated based on chain rule as

$$\Delta W_l = \sum_t \frac{\partial L}{\partial W_l^t} = \begin{cases} \sum_t \frac{\partial L}{\partial O_l^t} \frac{\partial O_l^t}{\partial V_l^t} \frac{\partial V_l^t}{\partial W_l^t} \\ \sum_t \frac{\partial L}{\partial V_l^T} \frac{\partial V_l^T}{\partial W_l^t} \end{cases} \tag{3.79}$$

where O_l and V_l represent output spikes and membrane potential at the lth layer respectively. For the output layer, the derivative of the loss function to the membrane potential v_i^T at final time step T as follows

$$\frac{\partial L}{\partial v_i^T} = \frac{e^{v_i^T}}{\sum_{k=1}^{C} e^{v_k^T}} - y_i \tag{3.80}$$

Spiking neurons in hidden layers output spikes when the membrane potential v_i^t is larger than the firing threshold, resulting in the non-differentiability problem. Thus, an approximated gradient is used with the expression as

$$\frac{\partial o_i^t}{\partial v_i^t} = \alpha \max\left\{0, 1 - \left|\frac{v_i^t - v_{th}}{v_{th}}\right|\right\} \tag{3.81}$$

where α represents a damping factor for gradients in back propagation. Finally, the network weight parameters in the lth layer based on the gradient value as

$$W_l = W_l - \eta \Delta W_l \tag{3.82}$$

In the SIBoLS framework, a learnable parameter υ is introduced. Then the membrane potential of spiking neurons in the hidden layers can be formulated as follows:

$$V_i^t = \zeta V_i^{t-1} + \upsilon_i^t \left(\frac{\sum_j w_{ij} S_j^t - \mu_i^t}{\sqrt{\left(\sigma_i^t\right)^2 + \tau}} \right) \tag{3.83}$$

where the calculation of the mean μ_i^t and variance σ_i^t for each time t is shown in Table 3.5. The variable υ represents a learnable hidden state to update the membrane potential. An exponential average is applied to compute the mean value $\overline{\mu}_i^t$ and variance value $\overline{\sigma}_i^t$ for each time step t, which normalizes the test data at inference. The gradient of V_i versus υ_i can be formulated as

$$\frac{\partial V_i}{\partial \upsilon_i} = \frac{\partial}{\partial \upsilon_i}(\upsilon_i - V_i)H(V_i - \upsilon_i)$$

$$= (\upsilon_i - V_i)\delta_D(V_i - \upsilon_i) + 1 \approx \eta_1 S_i(k) + \eta_2 \tag{3.84}$$

where $H(x)$ and $\delta_D(x)$ represent the Heaviside step and the Dirac delta functions separately. The Dirac function is obtained dependent on hyperparameters η_1 and η_2.

Suppose that $I_i^k = \sum_j w_{ij} S_j^k$ is an input signal for the training of deep SNN, then the gradient value through lower layers can be calculated as follows:

Table 3.5 SNN layer for SIBoLS framework (From [60])

Input: mini-batch (B) at time step $t\left(x_{\{1,\ldots,m\}}^t\right)$, learnable parameter $\left(\gamma^t\right)$, update factor (α)

Output: $\left\{y^t = SNN_{\gamma^t}\left(x^t\right)\right\}$

1: $\mu^t \leftarrow \frac{1}{m}\sum_{b=1}^m x_b^t$

2: $\left(\sigma^t\right)^2 \leftarrow \frac{1}{m}\sum_{b=1}^m \left(x_b^t - \mu^t\right)^2$

3: $\hat{x}^t = \frac{x^t - \mu^t}{\sqrt{\left(\sigma^t\right)^2 + \varepsilon}}$

4: $y^t \leftarrow \gamma^t \hat{x}^t \equiv SNN_{\gamma^t}\left(x^t\right)$

5: % Exponential moving average

6: $\overline{\mu}^t \leftarrow (1 - \alpha)\overline{\mu}^t + \alpha\overline{\mu}^t$

7: $\overline{\sigma}^t \leftarrow (1 - \alpha)\overline{\sigma}^t + \alpha\overline{\sigma}^t$

Table 3.6 The proposed SIBoLS training algorithm (From [60])

Input: input pixel *(X)*; label *(Y)*; timestep *(T)*

Output: updated network weights

1: **for** $i \leftarrow 1$ to *max iterations* **do**

2:   get a mini batch X

3:   **for** $t \leftarrow 1$ to T **do**

4:     $O \leftarrow Poisson\ Generator(X)$

5:     **for** layer $l \leftarrow 1$ to *L-1* **do**

6:       $\left(S_l^t, U_l^t\right) \leftarrow \left(\zeta, U_l^{t-1}, SNN_{v^i}\left(W_t, S_{l-1}^{t-1}\right)\right)$

7:     **end for**

8:     $U_l^t \leftarrow \left(U_l^{t-1}, SNN_{v^i}\left(W_t, S_{l-1}^{t-1}\right)\right)$

9:   **end for**

10:   $L \leftarrow \sum_{L-2}^{L} SIBoLS\left(U_l^T, Y\right)$

11: **end for**

$$\frac{\partial L}{\partial I_b^t} = \frac{1}{g_b\sqrt{(\sigma^t)^2 + \tau_{index}}} \begin{pmatrix} g_b\dfrac{\partial L}{\partial \hat{I}_b^t} - \displaystyle\sum_{l=1}^{g_b}\dfrac{\partial L}{\partial \hat{I}_l^t} \\ -\hat{I}_b^t\displaystyle\sum_{l=1}^{g_b}\dfrac{\partial L}{\partial \hat{I}_l^t}\hat{I}_l^t \end{pmatrix} \tag{3.85}$$

where g_b and τ_{index} represent the batch size and batch index. A learnable parameter v is used to find an optimum for efficient inference as follows:

$$v^t = v^t - p\Delta v^t \tag{3.86}$$

$$\Delta v^t = \frac{\partial L}{\partial v^t} = \frac{\partial L}{\partial V^t}\frac{\partial V^t}{\partial v^t} = \sum_{l=1}^{g}\frac{\partial L}{\partial V_l^t}\hat{I}_l^t \tag{3.87}$$

The binary $S(t)$ is not used in the SIBoLS framework, which can solve non-differentiable problem of the binary spiking trains. In addition, the membrane potential is considered as an information representation, which can be employed as a kind of state register to avoid using another time-dimension tensor value for realizing real-time online learning. It can save more memory resource. Since the membrane potential with stable limited cycle can be calculated in real time, the time steps can be significantly cut down and the learning efficiency can be improved dramatically.

Table 3.6 presents the SIBoLS training algorithm for deep SNN. The weighted sum of the input binary spiking activities is added to the spiking neurons to update membrane potential. For last three layers, the input is computed without leaky constant, so that each Softmax layer is used to calculate three probability distributions. The sum of cross entropy loss is then obtained with these probability distributions and approximated weight gradients, which is the key point of SIBoLS. A normalization layer is used to calculate the statistical parameters (i.e., μ_t and σ_t) and the global average for the mean and variance values is calculated. During the inference process, the time step in the early exit process is determined based on the value of υ in SIBoLS. The normalization of test input based on global $\overline{\mu}_t, \overline{\sigma}_t$ is then classified by SIBoLS. The model is a VGG architecture with BPTT and spike neural network. Taking VGG9 as an example, it consists of seven convolutional layers, three pooling layers, and a fully connected layer. Spiking LIF model is employed following each layer, resulting in output spikes. At the same time, BPTT is used before each LIF neuron receiving input information. Finally, information bottleneck criterion is used for optimization. For VGG11 or VGG16, there are only difference in the number and sequence of layers

3.4.2 Performance Evaluation

(1) Learning robustness

In this section, learning performance in comprehensive experiments are carried out on public classification data sets. The deep SNN architecture with spike-based learning strategy are first trained based on IB framework and reported based on large-scale data sets. A detailed comparison is presented with the previous SNN training approaches. The SIBoLS is evaluated to explore its effectiveness and efficiency from different perspective.

For neuromorphic systems, a critical problem is the sensitivity to variations in operating environments. For the robustness evaluation on deep SNN, the SIBoLS framework on three static data sets is first evaluated. 60,000 images are contained in CIFAR-10, including 50,000 and 10,000 for training and testing separately with 10 categories. RGB color with 32×32 pixels are employed for all images. The spike generator produces a Poisson spike train for each pixel of the input image with frequency proportional to the pixel intensity. Rate coding is used because of the reliable performance in different tasks. In this scheme, each pixel value is compared to a random number between minimum and maximum possible pixel intensity at each time step. If the random number is larger than the pixel intensity, the Poisson spike generator produces a spike. In contrast, no spike is generated.

The learning robustness of SIBoLS on synaptic noise is first investigated, which is induced by non-ideality on neuromorphic hardware. Gaussian noise is added on the synaptic weight update results during training, which can be described as

$$w_{noise} = \Delta w + N\left(0, \sigma_{SN}^2\right) \tag{3.88}$$

where w_{noise} denotes the noisy weight update value, and Δw denotes the weight change based on SIBoLS. The variable σ_{SN} represents the standard variance of the Gaussian noise on synapses. The robustness of SIBoLS is compared to the model without IB (WIB) theory, with the variance σ_{SN} ranging from 0.1, 0.4, 0.7 to 1.0. The learning accuracy increases along with the training epoch increasing as shown in Fig. 3.17. The SIBoLS model with $\sigma_{SN} = 0.1$ and the WIB model in all conditions are getting into a stable accuracy at the 110th training epoch. The learning accuracy of SIBoLS is superior to the WIB model with all the σ_{SN} levels of synaptic noise. The SIBoLS model with $\sigma_{SN} = 0.4$, 0.7, and 1.0 get into a stable accuracy level after the 150th training epoch. It also suggests that the accuracy improvement based on SIBoLS under the conditions $\sigma_{SN} = 0.1, 0.4, 0.7$, and 1.0 are 6.1%, 8.4%, 8.6% and 10.7% respectively. Therefore, SIBoLS can improve the learning robustness of deep SNN under the synapse noise on neuromorphic hardware significantly.

The learning robustness of SIBoLS with background noise on neuronal membrane potential is further explored. The training process is contaminated by background noise based on Gaussian white noise with mean value of 0 and variance σ_{MP}. Figure 3.18 shows the robustness performance of SIBoLS with $\sigma_{MP} = 0.1, 0.2$, and 0.3 respectively, and compare the performance with the WIB model. It reveals that the accuracy of SIBoLS significantly outperforms the WIB model at all the levels of variance σ_{MP}. The improvement values of the learning accuracy compared to WIB are 6.69%, 6.88%, and 7.63% respectively, which means that the SIBoLS can enhance the learning robustness of deep SNN with the membrane potential noise induced by background noise on neuromorphic system.

Fig. 3.17 Learning robustness of SIBoLS with synaptic noise. The variance σ_{SN} is selected as 0.1, 0.4, 0.7, and 1.0 respectively [60]

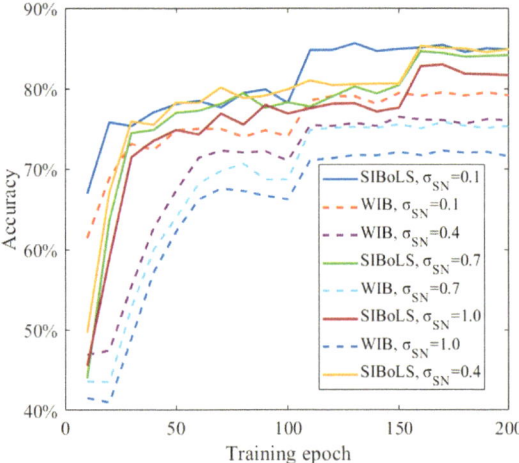

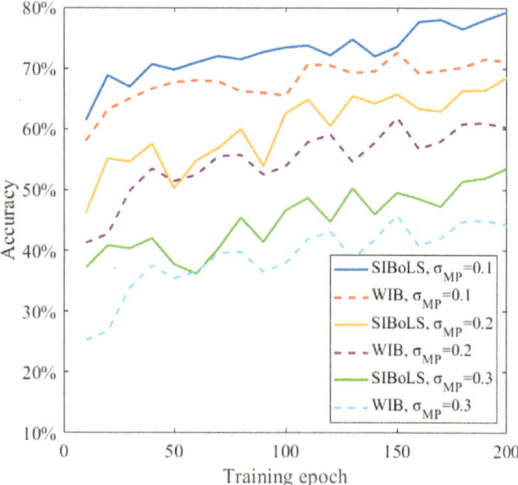

Fig. 3.18 Learning robustness of SIBoLS with membrane potential noise. The variance σ_{MP} is selected as 0.1, 0.2, and 0.3 respectively [60]

The effect of membrane constant and learning parameter on the learning accuracy of SIBoLS is explored in Fig. 3.19. Two important parameters are considered, which are the learning rate and the membrane constant ζ respectively. It reveals that the best values of learning rate and the membrane constant are around 0.1 and 1.0 respectively. It also reveals that the performance will be reduced significantly as the membrane constant increasing over 1.0. Since the membrane constant has critical impact on the spiking rate of LIF model, it also suggests that the spiking rate under 1.0 of the membrane constants can induce the best learning performance for SIBoLS.

In addition, the SIBoLS's robustness is tested with the ablation experiment. Five types of noise are considered, including Gaussian, salt-and-pepper (SP), uniform, Gamma, and Rayleigh noise. The uniform noise is expressed as

$$z = \alpha_{uni} + (\beta_{uni} - \alpha_{uni})u(0, 1) \qquad (3.89)$$

where $u(0, 1)$ is distributed between zero and one uniformly. β_{uni} is defined as the noise density in the following section. The Rayleigh noise can be formulated as

$$z = \alpha_{ray} + \beta_{ray}\sqrt{-\ln[1 - u(0, 1)]} \qquad (3.90)$$

where α_{ray} and β_{ray} represent the critical parameters of the Rayleigh noise. The parameter β_{ray} is defined as the noise density. Gamma noise is represented by n exponentially distributed noise, which is described by

$$z = z_1 + z_2 + \cdots + z_n \qquad (3.91)$$

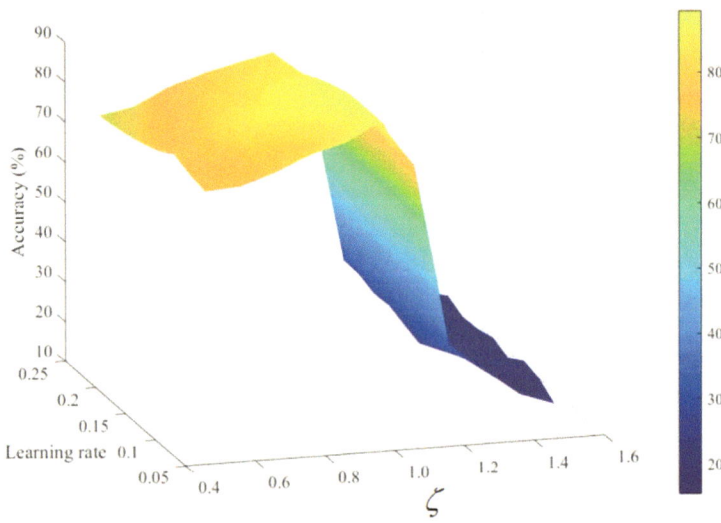

Fig. 3.19 Impact of critical parameters on the learning accuracy of SIBoLS on CIFAR-10 [60]

where $z_i = -\frac{1}{\alpha}\ln[1 - u(0, 1)]$. When $n = 1$, the Gamma noise can be considered as the exponential noise. For this noise, $1/\alpha$ is defined as the noise density. The SP noise is generated by sparsely mixed white and black color that is produced by sharp disturbance.

As shown in Fig. 3.20, the robustness of SIBoLS is superior to the model without IB framework on SP noise. In addition, three values of time steps are also investigated, including 5 ms, 10 ms, and 25 ms. The noise density is defined by the ratio of image pixel number contaminated by SP noise to the total pixel number. It reveals that SIBoLS with 25 ms has better performance than 5 ms under the SP noise conditions. The SIBoLS with all the three time step values achieve better robustness compared to the condition without IB learning rule in the ablation experiment. Therefore, the SIBoLS framework provides a robust approach for spike-based machine learning.

Figure 3.21 shows the robustness of SIBoLS on Gaussian noise. The noise density means the standard deviation of the Gaussian noise. The Gaussian noise can be expressed as

$$f(x) = \frac{1}{\sqrt{2\pi}\sigma} e^{-\frac{(x-\mu)^2}{2\sigma^2}}. \tag{3.92}$$

TS represents time step, which includes 5, 10, and 25 ms respectively. WIB represents the SNN model without IB in the ablation condition. It reveals that SIBoLS with $TS = 25$ ms achieves the highest accuracy compared to the other situations. The robustness of SIBoLS with uniform, Gamma, and Rayleigh noise are demonstrated in Fig. 3.22, Fig. 3.23, and Fig. 3.24 respectively. The uniform noise can be from all these figures, it reveals that SIBoLS can promote the robustness of the deep SNN. The robustness

Fig. 3.20 Effects of the SP noise on average learning accuracy of SIBoLS [60]

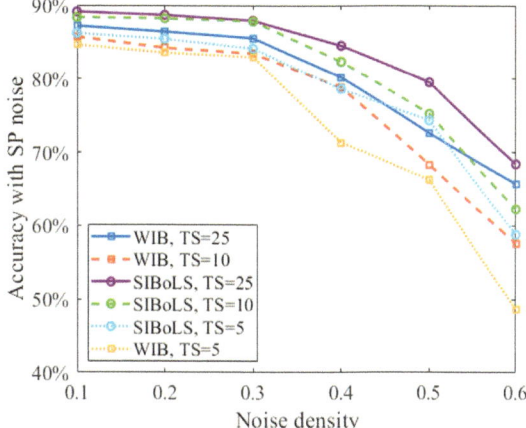

Fig. 3.21 Effects of the Gaussian noise on average learning accuracy of SIBoLS [60]

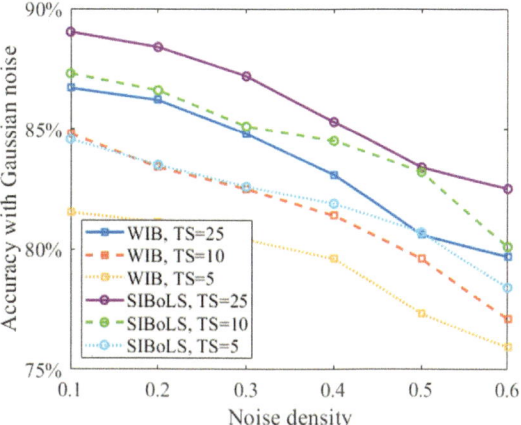

improvement of SIBoLS is significantly meaningful and effective in real-time applications and noise interfered situations.

The learning robustness with different types of noise based on SIBoLS is compared with other ITL-based methods for training SNNs [51, 52]. ITL-based methods are considered for the fair comparison with the same network architecture and neuronal parameters and structures, because these studies focus on the improvement of objective function of BPTT learning in SNNs. Other approaches, such as the methods based on intrinsic plasticity or lateral inhibition to improve the learning robustness, are not considered to be compared. It is because these models and techniques can also be applied in the SIBoLS framework to further improve the learning robustness. A more complicated CIFAR100 data set is used for robustness test, which contains 100 categories of images. Five types of noise are considered, including SP noise, Gaussian noise, uniform noise, Gamma noise,

Fig. 3.22 Effects of the
uniform noise on average
learning accuracy of SIBoLS
[60]

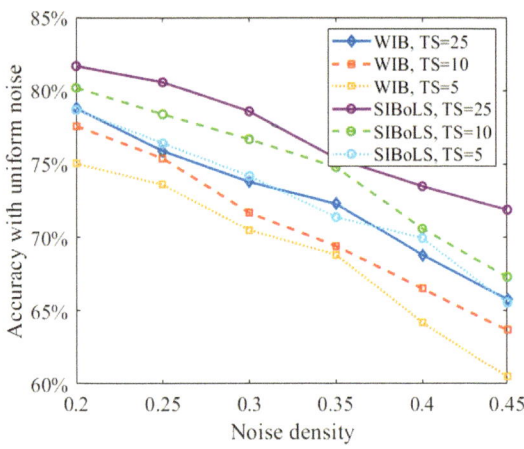

Fig. 3.23 Effects of the
Gamma noise on average
learning accuracy of SIBoLS
[60]

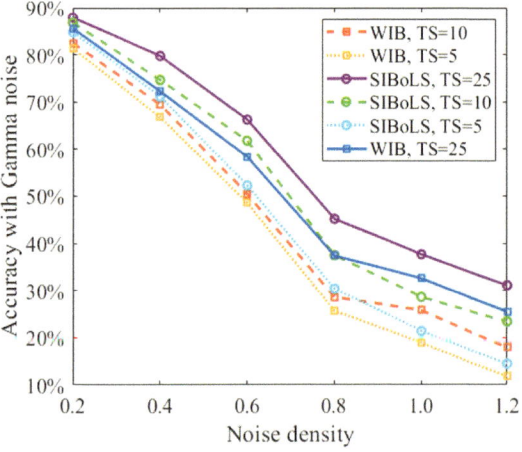

Fig. 3.24 Effects of the
Rayleigh noise on average
learning accuracy of SIBoLS
[60]

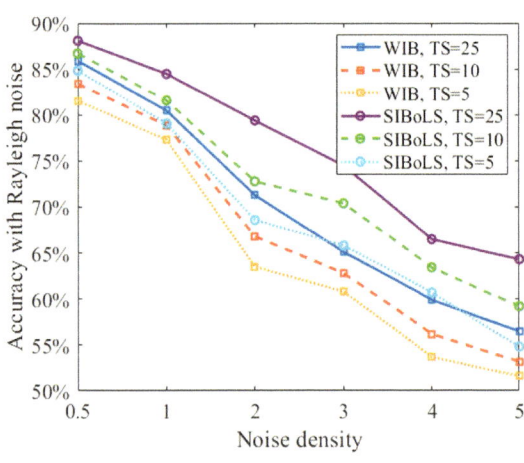

and Rayleigh noise. As demonstrated in Table 3.7, the SIBoLS outperform the others on both the pure data and noisy one. Along with the noise density increasing, SIBoLS demonstrates more significant robustness improvement in these situations compared to methods [51, 52]. It also suggests that SIBoLS is the state-of-the-art ITL method for BPTT in deep SNNs to improve the learning robustness in different situations of noise.

DVS-CIFAR10 is a discrete event-stream data set based on event-driven camera. For neuromorphic DVS-CIFAR10 data set, ASC method is not suitable because ANNs cannot realize the dynamical spiking activities. To pre-process the data set, the size of 128×128 images is downsampled to 42×42. The total number of time steps in the original time frame data into 20 intervals, and the spikes are accumulated within each time interval. Figure 3.25 shows the detailed network architecture with SIBoLS for DVS-CIFAR10 data set, and the network architecture is based on the previous work [53]. AP and FC represent average pooling and fully connected configuration respectively. It suggests that the SIBoLS can improve the learning robustness when dealing with event-based data in neuromorphic applications.

Table 3.7 Comparison with current ITL-based SNN learning methods on different levels of noise

Density of SP noise	0	0.1	0.2	0.3	0.4
Method in [51] (%)	61.8	60.4	59.3	58.1	55.7
Method in [52] (%)	62.4	61.2	60.5	59.2	57.1
SIBoLS (ours) (%)	**66.1**	**64.8**	**63.7**	**62.4**	**60.2**
Density of Gaussian noise	0	0.1	0.2	0.3	0.4
Method in [51] (%)	61.8	60.7	59.6	58.6	56.4
Method in [52] (%)	62.4	61.1	60.3	58.9	57.2
SIBoLS (ours) (%)	**66.1**	**65.8**	**64.6**	**63.1**	**61.7**
Density of uniform noise	0	0.1	0.2	0.3	0.4
Method in [51] (%)	61.8	57.6	51.8	46.3	40.2
Method in [52] (%)	62.4	58.8	53.5	48.2	42.7
SIBoLS (ours) (%)	**66.1**	**63.5**	**59.6**	**55.3**	**50.2**
Density of Gamma noise	0	0.2	0.4	0.6	0.8
Method in [51] (%)	61.8	49.7	37.6	19.8	10.0
Method in [52] (%)	62.4	52.1	40.6	24.8	10.0
SIBoLS (ours) (%)	**66.1**	**57.2**	**44.3**	**32.4**	**18.6**
Density of Rayleigh noise	0	1	2	3	4
Method in [51] (%)	61.8	53.2	43.5	32.4	20.8
Method in [52] (%)	62.4	55.8	46.1	36.7	26.2
SIBoLS (ours) (%)	**66.1**	**59.6**	**51.3**	**42.7**	**33.8**

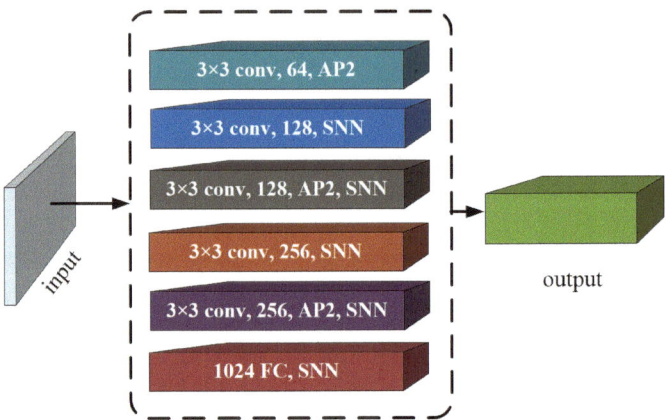

Fig. 3.25 Network architecture with SIBoLS for DVS-CIFAR10 data set [60]

The robustness improvement based on SIBoLS is explored in comparison with WIB. Figure 3.26 shows the experimental results, while normalizing the noise intensity to [0,1]. It reveals that the robustness improvement can be enhanced along with the normalized noise intensity increasing. It demonstrates that the improvement on synaptic noise is higher than membrane potential (MP) noise. In addition, the robustness improvement on uniform noise is larger than other types of noise for input images. It suggests that the SIBoLS can not only improve the robustness for static image, but also make sense when dealing with event-based data set for neuromorphic processors.

(2) Energy efficiency

To demonstrate the low-power advantages of SIBoLS, the energy consumption of SIBoLS is further evaluated on digital neuromorphic architecture. In fact, the total energy consumption is dependent on floating point operations (FLOPs) number. The FLOPs for the r_{th} convolutional layer in the model can be calculated as:

$$FLOPs(r) = n_{KS}^2 \times n_{OFM}^2 \times n_{IC} \times n_{OC} \tag{3.93}$$

where n_{KS} represents the kernel size, and n_{OFM} represents the output feature map size. The variables n_{IC} and n_{OC} denote the input and output channel numbers. For the r_{th} linear layer in the model, the FLOPs can be computed as:

$$FLOPs(r) = n_{IC} \times n_{OC} \tag{3.94}$$

Siring rate at layer r is defined to calculate the average spiking rate as follows:

$$s_R(r) = \frac{n_{SPN}}{n_{NEU}} \tag{3.95}$$

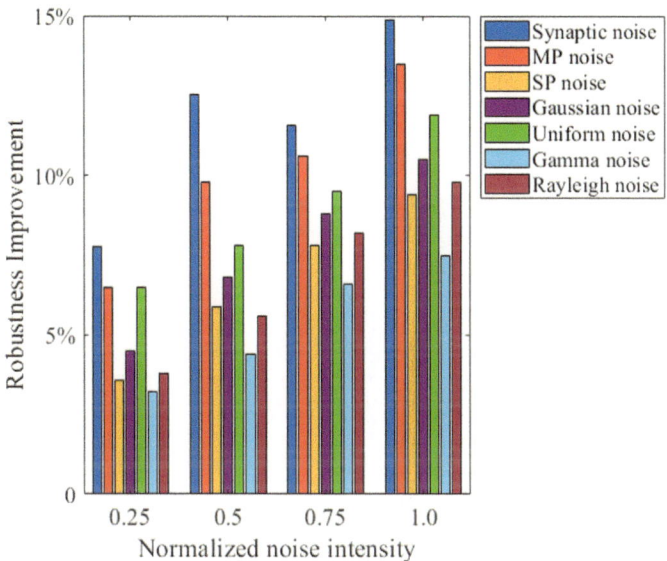

Fig. 3.26 Robustness improvement of average accuracy on CIFAR-100 data set using SIBoLS and WIB methods with different types of noise along with the normalized noise intensity increasing [60]

where n_{SPN} and n_{NEU} represent the neural spikes of layer r over all time steps and neuron number at layer r respectively. The $s_R(r)$ for every layer as the spike trains through several time steps is compared to other approaches in Fig. 3.27, including ASC, surrogate back propagation (SBP), SIBoLS without IB (SWIB), and SIBoLS. It reveals that the SIBoLS can significantly enhance the spike sparsity across the second to the seventh convolutional layers in comparison with other approaches. The Conv#1 and full connected layer induce slightly more spiking rate, but significantly lower than the other ASC and SBP approaches.

In addition, the $FLOP_{SNN}$ of SNN model can be obtained by multiplying the firing rate $s_R(r)$ with FLOPs as follows:

$$FLOP_{SNN}(r) = FLOPs(r) \times s_R(r) \tag{3.96}$$

Since SNN is event-driven with binary spike processing based on 0 and 1, the MAC operation can be simplified to just accumulating (AC) operation. Total energy consumption of ANNs and SNNs can be calculated as follows

$$E_{ANN} = \sum_r FLOPs(r) \times E_{MAC} \tag{3.97}$$

$$E_{SNN} = \sum_r FLOP_{SNN}(r) \times E_{AC} \tag{3.98}$$

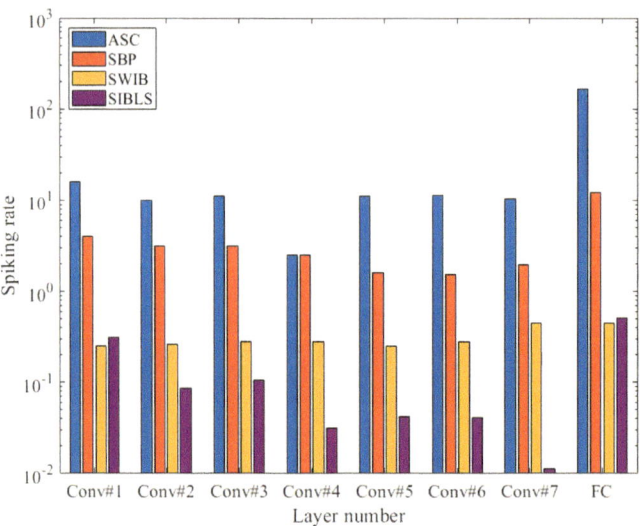

Fig. 3.27 Spiking rate of SIBoLS for each layer [60]

As demonstrated in Table 3.5, the energy consumption of MAC and adder are 4.6pJ and 0.9pJ on TrueNorth based on 45 nm complementary metal–oxide–semiconductor (CMOS) technology respectively. Therefore, the energy cost of MAC is 5.1 times larger than adder. Since SIBoLS employs spike-based coding and learning schemes, it can significantly cut down the energy consumption compared to ANNs.

According to the above equations and Table 3.8, the energy consumption in each layer is calculated and obtained. As shown in Fig. 3.28, the power consumption of SIBoLS is compared with ASC, SBP, and SWIB. It shows that the power consumption using SIBoLS is significantly lower than ASC and SBP methods in all the layers. In Conv#1 and FC layers, the power consumption of SIBoLS is slightly higher than SWIB. However, in all the six layers from Conv#2 to Conv#7 layers, the power consumption of SIBoLS is all lower than SWIB with a larger amount. Therefore, it reveals that the SIBoLS framework can not only cut down the power consumption of ASC method, but also further improve the energy efficiency of the conventional SNN learning methods.

Table 3.8 Energy consumption of 32-bit floating-point operations (From [60])	Operations	Energy (pJ)
	32-bit floating-point multiplication	3.7
	32-bit floating-point adder	0.9
	32-bit floating-point MAC	4.6
	32-bit floating-point AC	0.9

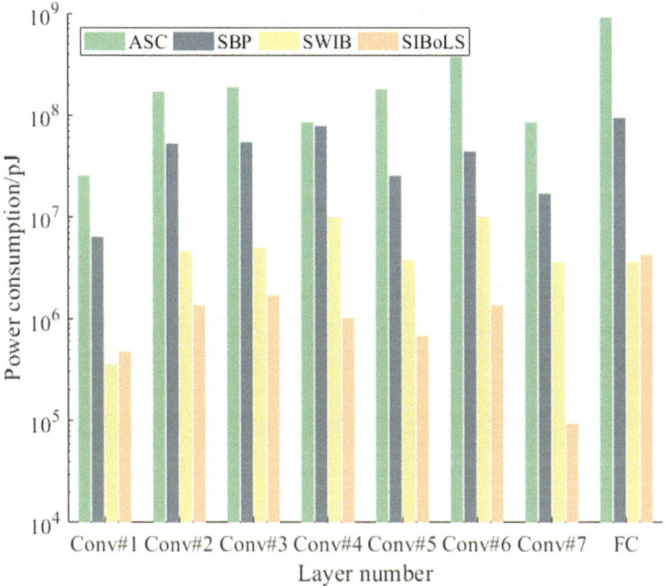

Fig. 3.28 Comparison of energy consumption between ASC, SBP, SWIB, and SIBoLS [60]

3.5   Nonlinear Information Bottleneck for Robust Spiking Neural Network

3.5.1   SNIB Model

Suppose that detailed recordings are made for meteorological data, X, by a remote weather station, and the data are encoded and transmitted to a central server, H, for the prediction of weather condition Y. The information sent from remote weather station to central server should be compressed if the channel capability between them is low. In order to predict weather condition Y with high $I(Y; H)$ in future days optimally, a compressed representation of X with a low channel capacity should be found, i.e., $I(X; H)$ by minimizing the IB objective. Therefore, there is a critical trade-off between final prediction accuracy and channel capability.

Most of studies do not obtain optimal bottleneck variables H based on the settlement of constrained optimization problem, but use the maximum Lagrangian relaxation as follows

$$L_{IB}(H) \doteq I(Y; H) - \beta I(X; H) \tag{3.99}$$

where $\beta \in [0, 1]$ represents a Lagrange multiplier to guarantee the constraint $I(X; H) \leq R$. It can determine the balance between compression and prediction. Maximal compression of X is obtained if $\beta \to 1$, and $I(X; H) = I(Y; H) = 0$ with the optimal S is obtained if $\beta = 1$. Prediction of Y will be favored by IB if $\beta \to 0$, and no penalty on $I(X; H)$ and $I(Y; H) = I(X; Y)$ is satisfied with the optimal H if $\beta = 0$.

In recent years, there have been some IB approaches for continuous non-Gaussian random variables [56], such as variational IB (VIB) [57]. These methods employ variational upper bound as follows

$$
\begin{aligned}
I_\theta(X; H) &= D_{KL}(P_\theta(H|X)\|R(H)) - D_{KL}(P_\theta(H)\|R(H)) \\
&\leq D_{KL}(P_\theta(H|X)\|R(H))
\end{aligned}
\tag{3.100}
$$

where R represents surrogate marginal distribution over H, which can be standard multi-variable normal or log-uniform distribution. The variational lower bound for $L_{IB}(H)$ can be formulated as

$$
\begin{aligned}
L_{IB}(H) &\geq \mathrm{E}_{Q_B(Y,H)}\big[\log P_\theta(Y|H)\big] \\
&\quad - \beta D_{KL}(P_\theta(H|X)\|R(H)) + \mathrm{const}
\end{aligned}
\tag{3.101}
$$

where the encoding map $P_\theta(H|X)$ can be a deterministic function contaminated by Gaussian noise or multiplicative.

In order to understand the information processing process in the network, the information processing process is considered as two sub-tasks of encoding and decoding, so the information can be regarded as a Markov chain, that is, from the input information X, to the encoded information H, and finally converted to the output classification result Y. The intermediate presentation H is represented by the membrane potential in SNIB. Let $P_\alpha(H|X)$ represents the conditional probability, where α indicates parameter vector. The mutual information $I(X; H)$ can be calculated based on the joint distribution. Besides, the mutual information can be formulated based on the joint distribution as

$$
Q_\alpha(Y, H) \doteq \int P_\alpha(H|X)Q(X, Y)dX
\tag{3.102}
$$

$$
Q_\alpha(X, H) = P_\alpha(H|X)Q(X)
\tag{3.103}
$$

Considering IB loss, the maximum $L_{IB}(H)$ is expected to obtain, which can be written as

$$
L_{IB}(\alpha) = I_\alpha(Y; H) - \beta I_\alpha(X; H)
\tag{3.104}
$$

According to the definition of mutual information, the mutual information between the output result Y and the encoded information H can be expressed as follows

$$
I_\alpha(Y; H) = H(Q(Y)) - L_{CE}(Q_\alpha(Y|H))
\tag{3.105}
$$

The non-negativity of Kullback–Leibler (KL) divergence results is formulated as

$$I_\alpha(Y; H) = H_{SE}(Q(Y)) - H_{SE}(Q_\alpha(Y|H))$$
$$\geq H_{SE}(Q(Y)) - H_{SE}(Q_\alpha(Y|H))$$
$$- D_{KL}\big(Q_\alpha(Y|H)||P_\varphi(Y|H)\big)$$
$$= H_{SE}(Q(Y)) + E_{Q_\alpha(Y,H)}\big[\log P_\varphi(Y|H)\big]$$
(3.106)

where $H_{SE}(\cdot)$ represents Shannon entropy, and the second term can be replaced with the following equation as

$$E_{Q_\alpha(Y,H)}\big[\log P_\phi(Y|H)\big] = - D_{KL}\big(Q_\alpha(Y|H)||P_\varphi(Y|H)\big)$$
$$- H_{SE}(Q_\alpha(Y|H))$$
(3.107)

where $D_{KL}(\cdot||\cdot)$ represents KL divergence. The KL divergence is a measure of how different two probability distributions are from each other. It is a non-symmetric measure, which means that the KL divergence from distribution P to distribution Q is not necessarily the same as the KL divergence from distribution Q to distribution P. The KL divergence is often used in information theory, statistics, and machine learning to quantify the amount of information lost when one distribution is used to approximate another. In simple terms, the KL divergence tells us how much additional information is needed to encode data from one distribution. It is defined as the expected logarithmic difference between the probabilities of the two distributions, where the expectation is taken over the probability distribution of the first distribution. The approximation of the joint distribution of X and Y is obtained based on empirical distribution in training data set as follows

$$Q(X, Y) \approx \frac{1}{N} \sum_i \delta(X_i, X)\delta(Y_i, Y)$$
(3.108)

where the encoding map is defined as the sum of deterministic function $f_\theta(X)$ along with Gaussian noise as follows

$$H = f_\alpha(X) + N\left(f_\alpha(X), \sum_\alpha(X)\right)$$
(3.109)

the bottleneck variable H can be viewed as a mixture distribution. The non-parametric upper bound on mutual information is used, which is formulated as follows

$$I_\alpha(X; H) \leq \hat{I}_\alpha(X; H)$$
$$\doteq -\frac{1}{N} \sum_i \log \frac{1}{N} \sum_j e^{-D_{kl}[N(f_\alpha(X_i), \Sigma_\alpha(X_i))||N(f_\alpha(X_j), \Sigma_\alpha(X_j))]}$$
(3.110)

The mutual information is bounded based on the pairwise KL divergence. The KL divergence can be formulated as

$$D_{KL}[N(\mu', \Sigma')||N(\mu, \Sigma)]$$
$$= \frac{1}{2}\left[\ln\frac{\det\Sigma}{\det\Sigma'} + (\mu' - \mu)\Sigma^{-1}(\mu' - \mu) + tr(\Sigma^{-1}\Sigma') - d\right] \qquad (3.111)$$

Then the tractable lower bound can be expressed with IB Lagrangian hyper-parameter as follows

$$l_{IB}(\alpha) = I_\alpha(Y; H) - \beta I_\alpha(X; H)$$
$$\geq E_{Q_\alpha(Y|H)}\left[\log P_\varphi(Y|H)\right] - \beta \hat{I}_\alpha(X; H) \qquad (3.112)$$

where the additive constant $H_{SE}(Q(Y))$ is ignored. It can be defined as the nonlinear IB objective function. However, for a one-to-one mapping relationship, that is, one input information corresponds to only one classification output, Y is a deterministic function of X. A critical approach to optimize the IB Lagrangian is to optimize the nonlinear IB Lagrangian formation, which can define lower bound as follows

$$l_{SNIB}(\alpha) = I_\alpha(Y; H) - \beta[I_\alpha(X; H)]^n$$
$$\geq E_{Q_\alpha(Y,H)}\left[\log P_\varphi(Y|H)\right] - \beta\left[\hat{I}_\alpha(X; H)\right]^n \qquad (3.113)$$

where the compression term $\hat{I}_\alpha(X; H)$ can be obtained from data, and the expectation term can be estimated as follows

$$E_{Q_\theta(Y,H)}\left[\log P_\phi(Y|H)\right] \approx \frac{1}{N}\sum_i \log P_\phi(Y_i|H_i) \qquad (3.114)$$

where H_i denotes samples from $P_\theta(H|X_i)$. Furthermore, squared IB Lagrangian is introduced on the basis of focusing on optimizing the bounds of the IB Lagrangian equation, which is more effective for improving network performance.

In the formula of nonlinear information bottleneck, two quantities are required to be calculated, namely the encoding mapping $P_\alpha(H|X)$ between the input information X and the encoding information H, and the decoding mapping P_φ $(Y|H)$ between the encoding information H and the output information Y, thereby optimizing the encoded information by maximizing the nonlinear information bottleneck objective function.

The encoded information H consists of two parts. One part is the deterministic function $f_\alpha(X)$ determined by the parameter α in the neural network, and the other part is the zero-centered Gaussian noise with covariance $\sum_\alpha (X)$, the sum of these two parts is considered as encoded information, and use this information to characterize the bottleneck layer. Since noise is uncertain, the neural network after adding noise is random. The variance of Gaussian noise is specified using the homoscedastic noise model, and the size of the set noise variance can be changed by changing the size of σ^2. Therefore, the mutual information between X and H can be rewritten as

$$\sum_\alpha (x) = \sigma^2 I \qquad (3.115)$$

$$I_{\alpha\,max}(X;H) = -\frac{1}{N}\sum_i \log\frac{1}{N}\sum_j e^{-\frac{1}{2\sigma^2}[f_\alpha(x_i)\|f_\alpha(x_j)]} \tag{3.116}$$

The parameter σ^2 is a trainable parameter in the α parameter set and needs to be set reasonably. The Gaussian component of the mixture distribution of the encoded information H will have a large difference in standard deviation. The initial value $\sigma^2 = 1$ is considered. In the SNIB model, the optimized objective can be expressed as

$$L_{SNIB}(\alpha) \doteq E_{Q_\alpha(Y,H)}\left[\log P_\varphi(Y|H)\right]$$
$$- \beta\left[D_{KL}\left(P_{H|X}(H|X)\|R(H)\right)\right]^n \tag{3.117}$$

where $R(H)$ is considered as the standard Gaussian $N(0,1)$. The index n denotes the order of KL divergence, and $n = 2, 3, 4$ represents the SIB, CIB and QIB respectively.

3.5.2 HOSIB Model

The surrogate gradient method is a type of supervised method that derives a transfer function to formalize the accumulated effect of spikes from the event-based update of neuronal membrane potential activities. BPTT scheme can be applied by SNNs for efficient training due to the similarity between RNN and SNN. Figure 3.29 shows the training process of SNNs with BPTT in the high-order spike-based information bottleneck (HOSIB) model. Neural states in SNNs are updated at each time step with spatiotemporal information within the whole time window in the forward process. After the time window ended, the backward process begins with calculated loss function. BPTT enables SNNs to achieve high performance in classification tasks and scaled to deep layers to deal with complicated problems. As shown in Fig. 3.30, we construct a spike-based information-bottleneck representation to explore the intrinsic information if input patterns to improve the generalization and robustness ability of the BPTT training in different SNN architectures.

Suppose Y is the deterministic function of X, it is difficult to optimize IB Lagrange to approximate IB curve by different β values. Therefore, this study uses the second-order form to realize the optimization of IB Lagrange. In this study, the HOSIB framework is first presented by introducing second-order IB (SOIB) principle for spike-based machine learning. The mathematical formulation of the HOSIB with SOIB principle can be expressed as follows

$$L_{SO}(H) = I^2(X;H) + \beta p^2(y|h), \tag{3.118}$$

where $\beta > 0$ represents a Lagrange hyper parameter as a balance between prediction and compression. The variables x, y, h represent the input, target output and bottleneck

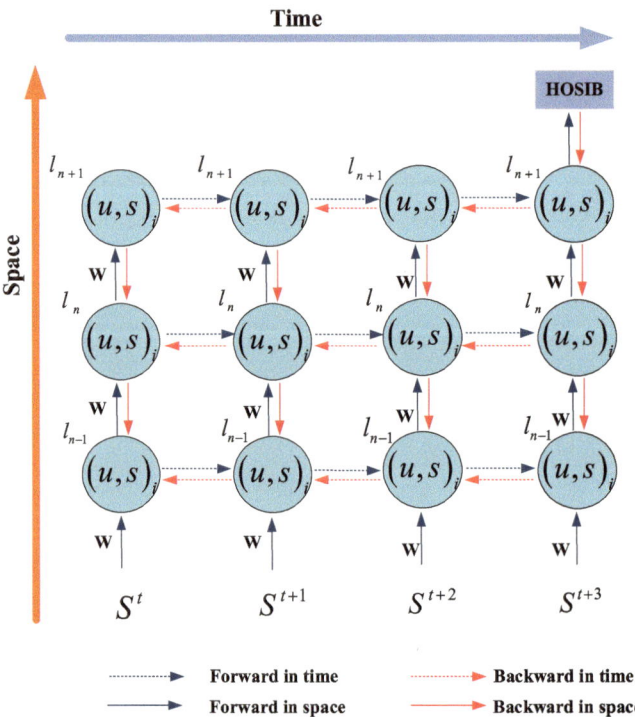

Fig. 3.29 BPTT training scheme for SNNs unfolded in time. A spiking neuron layer is represented by one node where l and i represent the layer index and neuron index respectively. The variable U and s denote the neuronal membrane potential and dynamical omitted spikes separately. Synaptic weights between hidden layers are represented by W, and the loss function is represented by L

variables, separately. Conditional entropy $p(y|t)$ can be formulated as

$$P(y|h)=p(y) - I(y;h). \tag{3.119}$$

The entropy $p(y)$ is merely based on the distribution of y, and is considered fixed or with a relatively slight disturbance for the whole data set $Ð$ or a training data batch. Therefore, the second-order optimization function L_{SO} stands for a concave function of the IB principle for local or global minimum.

The optimal parameter set ω is defined, which can obtain the best learning performance. Besides, suppose $\hat{Y}$ denotes the representation predicted by a parameter set φ, and $\hat{y}$ denotes a random sample of $\hat{Y}$. Therefore, it can be expressed as

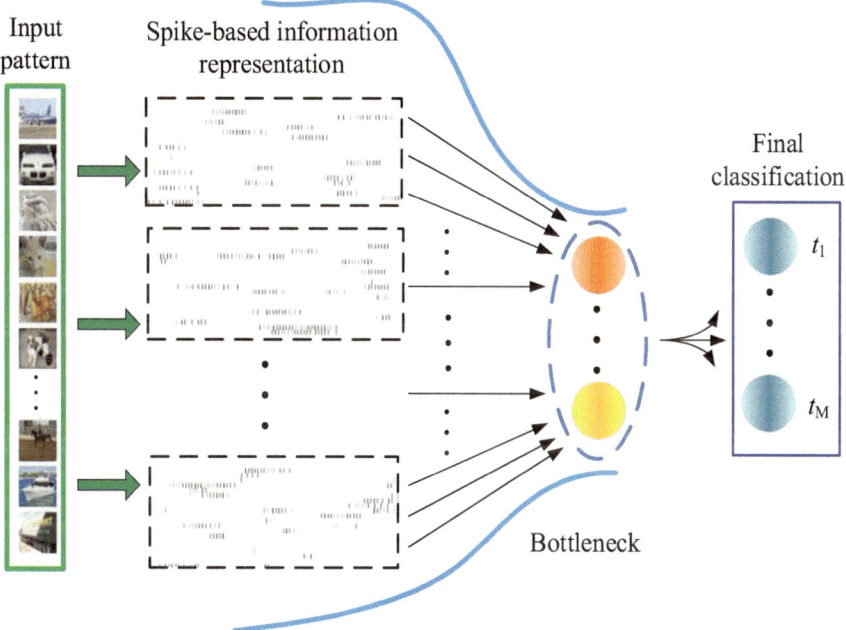

Input pattern

Spike-based information representation

Final classification

t_1

t_M

Bottleneck

Fig. 3.30 Illustration of the proposed HOSIB framework for surrogate gradient learning of SNNs

$$H\left(P_{Y|H}^{\varphi}(y|h)\right) \leq H\left(P_{Y|H}^{\varphi}(y|h)\right)$$
$$+ D_{KL}\left(P_{Y|H}^{\omega}(y|h) || P_{Y|T}^{\varphi}(y|h)\right) \tag{3.120}$$
$$= -E_{P_{Y|H}^{\omega}(y|h)}\left[\log P_{Y|H}^{\varphi}(y|h)\right],$$

where $E(\cdot)$ represents the expectation. The KL divergence is a nonnegative value, so the inequality holds. We transform the entropy of conditional PDF $P_{Y|H}^{(\Psi)}(y|h)$ into the distribution expectation of the $P_{Y|H}^{(\Psi)}(y|h)$ by scaling method. This representation is similar to cross-entropy, but the known distribution of conditional PDF is still the encoded information h, which is still difficult to solve. Since there exists a relationship as

$$P_{Y|T}(y|h) = P_{Y|P_{\hat{Y}|H}(\hat{y}|h)}\left(y|P_{\hat{Y}|H}(\hat{y}|h)\right)$$
$$= P_{Y|\hat{Y}}(y|\hat{y}). \tag{3.121}$$

Thus, we can replace the known information h with the output $\hat{y}$, which is produced by the network with a non-optimal parameter set φ. Then it can be further expressed as

$$H_2\left(P^{\varphi}_{Y|H}(y|h)\right) \Leftarrow -E_{P^{\omega}_{Y|\hat{Y}}(Y|\hat{Y})}\left[E_{P^{\omega}_{Y|\hat{Y}}(Y|\hat{Y})}\left(\log P^{\varphi}_{Y|\hat{T}}(Y|\hat{Y})\right)\right]$$
$$= - E_{P^{\omega}_{Y|\hat{Y}}(Y|\hat{Y})}\left[\log P^{\varphi}_{Y|\hat{T}}(Y|\hat{Y})\right]. \tag{3.122}$$

This representation can be approximated by cross-entropy, so we get the equation as $H_2\left(P^{\varphi}_{Y|H}(y|h)\right)=L_{CE}\left(P^{\varphi}_{Y|\hat{T}}(Y|\hat{Y})\right)$, where $L_{CE}(\cdot)$ denotes the cross-entropy. When $P^{\varphi}_{Y|H}(y|h)$ is equal to the optimal mapping $P^{\varphi}_{Y|H}(y|h)$, the inequality can be converted to the above equation. When KL divergence has the expression as $D_{KL}\left(P^{\omega}_{Y|H}(y|h)||P^{\varphi}_{Y|H}(y|h)\right) \to 0$, there exists a formulation as $H_2\left(P^{\omega}_{Y|H}(y|h)\right) \to L_{CE}\left(P^{\varphi}_{Y|\hat{H}}(Y|\hat{Y})\right)$. Since the distance between φ and ω can be trained to be reduced, the entropy can approach its upper bound. It should be noted that the calculation of the marginal distribution H is difficult. Thus, this study employs the variational distribution $R(h)$ to approximate $P_{H|X}(h|x)$ according to the variational IB theory. Due to the non-negative KL divergence, the equation is obtained as

$$D_{KL}\left(P_{H|X}(H|X)||R(H)\right) = \sum_h P_{H|X}(h|x)\log P_{H|X}(h|x)$$
$$- \sum_h P_{T|X}(h|x)\log R(H)$$
$$\geq 0. \tag{3.123}$$

Then we have the following formulation

$$I(X; H) \leq \sum_{x,h} P_{H|X}(h|x)\log P_{H|X}(h|x)$$
$$- \sum_h P_{H|X}(h|x)\log R(H)$$
$$= \sum_{x,h} P(x)P_{H|X}(h|x)\log P_{H|X}(h|x)$$
$$- \sum_h P(x)P_{H|X}(h|x)\log R(H)$$
$$= \frac{1}{N}\sum_{n=1}^{N} P_{H|X}(h|x_n)\log \frac{P_{H|X}(h|x_n)}{R(H)}$$
$$= \frac{1}{N}\sum_{n=1}^{N} D_{KL}\left(P_{H|X}(H|X)||R(H)\right), \tag{3.124}$$

where N represents the data sample number. Since $H_2(y|h) \geq 0$ and $I(X; H) \geq 0$, the upper bound of SOIB can be expressed as

$$L_{SO}(H) \leq \left[L_{CE}\left(P_{Y|\hat{Y}}^{\varphi}(Y|\hat{Y}) \right) \right]^2$$

$$+ \beta \left[\frac{1}{N} \sum_{n=1}^{N} D_{KL}\left(P_{H|X_n}(H|X_n)||R(H) \right) \right]^2. \tag{3.125}$$

The upper bound $\overline{L}$ is further described as

$$L_{SO}(H) \leq \overline{L} = \mu \|W\|^2 + \left[L_{CE}\left(P_{Y|\hat{Y}}^{\varphi}(Y|\hat{Y}) \right) \right]^2$$

$$+ \beta \left[\frac{1}{N} \sum_{n=1}^{N} D_{KL}\left(P_{H|X_n}(H|X_n)||R(H) \right) \right]^2. \tag{3.126}$$

Therefore, the challenging problem to minimize the loss function $L_{SO}(H)$ can be considered as minimizing the upper bound L by reducing to obtain the minimum $L_{SO}(H)$. For $L_{CE}\left(P_{Y|\hat{Y}}^{\varphi}(Y|\hat{Y}) \right)$, it is a concave function of decoding mapping, which can be approximated by cross entropy. For $\beta \left[\frac{1}{N} \sum_{n=1}^{N} D_{KL}\left(P_{H|X_n}(H|X_n)||R(H) \right) \right]^2$, we utilize KDE to solve KL divergence. Gaussian KDE is employed, which can be expressed as $K(x, x_c) = \exp\left[-\|x - x_c\|^2/(2 * \sigma)^2 \right]$. In this equation x_c and σ represent kernel function center and function width respectively. $\mu \|W\|^2$ is an L2 regularization term of the network weight W with a small positive constant μ. Algorithm 5 demonstrates the detailed training pseudo code of the entire network based on SOIB scheme.

Another version of HOSIB is based on the presentation of TOIB principle. Based on the basic idea of the IB theory, we further propose TOIB principle, which can be expressed as

$$L_{TO}(H) = I(X; H)^3 - \beta I(H; Y)^3. \tag{3.127}$$

We derive an upper bound of this loss function by solving the upper and lower bound of $I(X; H)$ and $I(H; Y)$. Similarly, method of reducing upper bound is used to reduce the value of the loss function, so as to realize the training of the network. The upper bound of $I(X; H)$ is formulated as follows

$$I(X; H) = D_{KL}(p(X, H)||p(X)p(H))$$

$$= D_{KL}(p(H|X)||p(H))$$

$$= D_{KL}(p(H|X)||r(H)) - D_{KL}(p(H)||r(H))$$

$$\leq D_{KL}(p(H|X)||r(H)) \tag{3.128}$$

$$= \frac{1}{N} \sum_{n=1}^{N} D_{KL}(p(H|x_n)||r(H)).$$

The lower bound for $I(H; Y)$ is expressed as follows

$$I(H; Y) = \sum_{y,h} p(y, h) \log \frac{p(y, h)}{p(y)p(h)}$$

$$= \sum_{y,h} p(y, h) \log p(h|y) - \sum_{h} p(h) \log p(h) \qquad (3.129)$$

$$\geq \sum_{y,h} p(y, h) \log p(h|y).$$

We make two assumptions about the network weights, assuming that the optimal network weights and their corresponding outputs are ω and $Y^{\omega} = \{y_1, y_2, \ldots, y_n, \ldots, y_m\}$, while a random set of network weights and their corresponding network outputs are φ and $Y^{\varphi'} = \{y'_1, y'_2, \ldots, y'_n, \ldots, y'_m\}$. Therefore, we can draw the following conclusions as

$$p_{H|Y^{\omega}}(h|y_n) = p_{q(Y'\varphi|T)Y^{\omega}}(q(y'_n|h)|y_n)$$
$$= p_{Y'\varphi|Y^{\varphi}}(y'_n|y_n) \qquad (3.130)$$

where y'_n represents the output result of input sample, y_n denotes the correct result of the corresponding input sample. Then, the mutual information $I(H; Y)$ between Y and H can be formulated as

$$I(H; Y) \geq \sum_{y,h} p(y, h) \log p(h|y)$$

$$= \sum_{y,h} p(h|y)p(y) \log p(h|y)$$

$$= \frac{1}{N} \sum_{n=1}^{N} p(h|y_n) \log p(h|y_n) \qquad (3.131)$$

$$= \frac{1}{N} \sum_{n=1}^{N} p_{Y'\varphi|Y^{\omega}}(y'_n||y_n) \log p(y'_n|y_n).$$

Combining the above derivation, we replace $I(X; H)$ with the upper bound, while replace $I(H; Y)$ with the lower one, so that upper bound of the third-order information bottleneck loss function L_{TO} can be obtained. By gradually reducing the upper bound, the value of the loss function is gradually reduced, so that we can realize the training of the network with the loss function as

$$L_{TO} = \beta I(X; H)^3 - I(H; Y)^3$$

$$\leq \beta \left(\frac{1}{N} \sum_{n=1}^{N} D_{KL}(p(H|x_n)||r(H)) \right)^3 \qquad (3.132)$$

$$- \left(\frac{1}{N} \sum_{n=1}^{N} p_{Y'_\varphi|Y^\omega}(y'_n||y_n) \log p(y'_n||y_n) \right)^3 .$$

Similar with SOIB loss, the decoding mapping function $\frac{1}{N}\sum_{n=1}^{N} P_{y_{\varphi'}|y^\omega}(y'_n|y_n) \log p(y'_n|y_n)$ in TOIB is approximated by cross entropy. For $\beta\left(\frac{1}{N}\sum_{n=1}^{N} D_{KL}(p(H|x_n)||r(H))\right)^3$, KDE is utilized to solve KL divergence. Gaussian KDE is employed, which is consistent with SOIB loss. We have $K(x, x_c) = \exp\left[-\|x - x_c\|^2/(2 * \sigma)^2\right]$, where x_c and σ represent kernel function center and function width respectively. The training process is also consistent with SOIB as demonstrated in Algorithm 5.

We apply the proposed HOSIB framework in deep VGG9 architecture with spiking neurons. The detailed mathematical formulation of the SNN with VGG9 architecture is described in 3.4.1. The proposed training scheme is based on rate coding. Spike generator produces a Poisson spike train for each pixel in the image with frequency proportional to the pixel intensity. The weighted sum of the input signal is transmitted through the hidden layer and accumulated in the membrane potential. The spiking neurons spikes when the membrane potential is over the firing threshold. As shown in Algorithm5, we calculate SOIB loss based on the accumulated effects of the membrane potential, i.e., $L \leftarrow SOIB(V_L^T, Y)$. It means the input voltage over all time steps is accumulated for the last layer, which a softmax layer is used to output a probability distribution, which has also depicted in Fig. 2. The SOIB loss is calculated and gradients for weight of each layer are also calculated with the approximated gradient function. The membrane potential is leveraged for spike information representation in training on VGG architecture. We define the accumulated gradients at the lth hidden layer based on chain rule as

$$\Delta W = \sum_t \frac{\partial L}{\partial O_l^t} \frac{\partial O_l^t}{\partial V_l^t} \frac{\partial V_l^t}{\partial W_l^t}. \qquad (3.133)$$

For the output layer, it can be formulated as

$$\Delta W = \sum_t \frac{\partial L}{\partial V_l^T} \frac{\partial V_l^T}{\partial W_l^k}, \qquad (3.134)$$

where O_l and V_l represent output spikes and membrane potential at lth layer respectively. In addition, in order to effectively and efficiently train SNN through gradients of the mapping on ResNet architecture, we use spike representation to process the spiking information and avoid the non-differentiable problem of the partial derivative of the output with respect to the membrane potential in the process of gradient solution. For ResNet architecture, we also encode the spike trains into spike representation and train deep SNNs using rate coding scheme. Specifically, denoting by s^i the output spike information of ith layer, the representation of s^i can be expressed as

$$O^i = R\left(s^i\right) \approx \text{clamp}\left(g_{W^i} R\left(s^{i-1}\right), 0, b_i\right), \tag{3.135}$$

where $clamp(a, b, c) \triangleq \max(b, \min(a, c))$. $R\left(s^i\right)$ represents the spike representation of the spike information s^i. b^i denotes the value of $\frac{\theta^i}{\Delta t}$. g_{W^i} is the mapping result of network parameter W^i. Based on the spike representation, the weight gradient can be expressed as

$$\Delta W = \frac{\partial L}{\partial W^i} = \frac{\partial L}{\partial O^i} \frac{\partial O^i}{\partial W^i}. \tag{3.136}$$

For the calculation of classification, we use softmax function, which can be expressed as

$$y_m = \frac{e^{V_i^T}}{\sum\limits_{j=1}^{M} e^{V_j^T}}, \tag{3.137}$$

where V_i^T represents the membrane potential of the ith neuron at the Tth time step in the output layer, and y_m represents the probability of classification result for the mth class. We select the maximum probability for one class based on softmax function as the classification result. The accuracy is calculated by the ratio of the number of samples correctly classified by the classifier to the total number of samples for a given test data set. For the comprehensive exploration of the HOSIB learning performance, three static data sets are used, including Fashion-MNIST, CIFAR10 and CIFAR100. The surrogate function for the gradient is formulated as follows

$$\frac{\partial O_i^t}{\partial V_i^t} = \alpha \max\left\{0, 1 - \left|\frac{V_i^t - V_{th}}{V_{th}}\right|\right\}, \tag{3.138}$$

where α represents a damping factor for back-propagated gradients, and θ denotes the membrane threshold. We train the VGG architecture with standard SGD with momentum 0.9, weight decay 0.0005, while applying random crop and horizontal flip to input images. The base learning rate is set to 0.1, and step-wise learning rate scheduling with a decay factor 10 at 50%, 70%, and 90% of the total number of epochs is used. The total number of epochs is set to 120 in this study. We use VGG9 for Fashion-MNIST and CIFAR10, and VGG11 for CIFAR100 and DVS-CIFAR10 respectively. For ResNet architecture, we use Eq. (3.135) for spike representation, which can effectively solve the non-differentiability problem in SNN training. The experiment uses SGD optimizer with

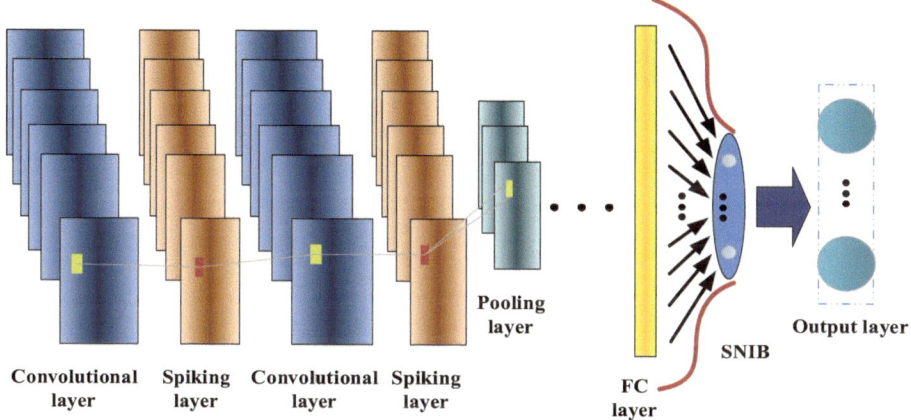

Fig. 3.31 Application of SNIB in deep SCNN architecture. FC layer represents fully connected layer [59]

total number of epochs set to 120. The learning rate is set to 0.1, and ResNet18 is used for all the data sets with ResNet architecture. The network architecture of SNIB-based convolutional SNN is shown in Fig. 3.31, which contains five-layered feature extractor and a classifier. The detailed training process of SNIB with deep VGG architecture is demonstrated in Table 3.9.

3.5.3 Performance Evaluation

(1) Learning performance of SNIB model

Input pattern noise is a critical type of noise for SNNs, which is also called attribute noise. The learning robustness of SNIB is first explored with different types of input pattern noise. Gaussian noise is considered as one of the most common types of noise, which is defined with mean value $\mu = 0$ and standard variance σ as

$$f(x) = \frac{1}{\sqrt{2\pi}\sigma} e^{-\frac{(x-\mu)^2}{2\sigma^2}}. \tag{3.139}$$

where the standard variance σ is considered as the noise intensity. The SP noise uses white and black pixels based on sharp-and-sudden disturbance. The Rayleigh noise can be formulated as

$$z = \alpha_{ray} + \beta_{ray}\sqrt{-\ln[1 - u(0, 1)]} \tag{3.140}$$

Table 3.9 Training process with **SNIB** (From [59])

Input: Lagrange parameter (β); mini-batch (B); label set (Y); timestep (T); training epoch (p); learning rate (r); max iteration(MT)

Output: Encoding scheme $P(h \,|\, x)$ Encoding scheme from input information X to encoded information H; updated network weights (W);

1: **for** $i \leftarrow 1$ to MT **do**

2:      fetch a mini batch B

3:      **for** $t \leftarrow 1$ to T **do**

4:          $O \leftarrow$ PoissonGenerator(X)

5:          **for** $l \leftarrow 1$ to L-1 **do**

6:              $V_l^t \leftarrow (V_l^{t-1},\ SNN_{v^t}(W_t, o_{l-1}^{t-1}))$

7:          **end for**

8:      **end for**

9:      %Calculate the nonlinear IB loss function and back-propagate it

10:     $L \leftarrow \sum_{L-2}^{L} SNIB(V_l^T, Y)$

11:     Update the network weight parameters

12:     %Calculate the classification accuracy for X

13:     Acc(X)$\leftarrow$SNIB(X)

14: **end for**

---

where α_{ray} and β_{ray} represent the critical parameters of the Rayleigh noise. The parameter β_{ray} is defined as the noise density.

The robustness of SNIB is tested on the Fashion-MNIST and CIFAR10 datasets. Image pixels were converted to spike timing using an encoding rule that maximizes spike time separation. The Fashion-MNIST dataset consists of ten types of images with dimensions of 28×28 pixels, while CIFAR10 contains 60,000 RGB color images based on 32×32 pixels. In Table 3.10, the learning accuracy of SIB, CIB, and QIB is tested on both Fashion-MNIST and CIFAR10 datasets, under various noise conditions, including SP, Gaussian, and Rayleigh noise. The performance is measured based on learning accuracy and the relative improvement compared to the VGG9 architecture without the SNIB framework (w.o. SNIB) and with the conventional IB framework. The results show that the learning robustness of SNIB outperforms the networks without SNIB and with IB strategies in all four noisy situations. Interestingly, no single strategy clearly outperforms others of SNIB under SP noise. However, for Gaussian noise, QIB demonstrates the best performance on both Fashion-MNIST and CIFAR10 datasets in all situations. This suggests that SNIB with high non-linearity contributes significantly to improving the robustness of the conventional IB framework.

The robustness of the SNIB framework is further tested using the more complex CIFAR-100 dataset. In Fig. 3.32, the performance is demonstrated on CIFAR-100 with

Table 3.10 Classification accuracy (%) with VGG architectures on Fashion-MNIST and CIFAR-10 data sets with different types of noise. The dark areas mean the best results in the corresponding situation (From [59])

Noise type	Noise intensity	Data set	Accuracy (%)				
			w.o.SNIB	IB	SIB	CIB	QIB
Salt-and-pepper	1	Fashion-MNIST	89.12	89.64	**89.79**	89.72	89.87
	4	Fashion-MNIST	45.84	48.87	**49.54**	48.89	48.40
	1	CIFAR10	88.79	89.49	**89.45**	89.58	88.95
	4	CIFAR10	44.77	50.92	**52.26**	49.07	49.81
Gaussian	0.1	Fashion-MNIST	89.19	89.81	**89.98**	89.49	90.17
	0.5	Fashion-MNIST	82.35	83.85	**83.34**	83.42	83.67
	0.1	CIFAR10	88.83	89.60	**89.04**	89.62	89.89
	0.5	CIFAR10	82.19	83.57	**83.62**	83.67	83.97
Rayleigh	0.5	Fashion-MNIST	87.29	87.56	**87.72**	87.85	87.59
	4	Fashion-MNIST	57.74	61.17	**63.23**	61.55	61.95
	0.5	CIFAR10	86.48	86.62	**87.36**	86.99	87.04
	4	CIFAR10	58.26	60.32	**62.28**	61.54	64.31

SIB, CIB, QIB, IB, and the method in [41], under Gaussian noise with normalized noise intensity (NNI) set to 0.2 and 1.0, respectively. The results reveal that SNIB can significantly enhance learning robustness compared to IB and the method in [41] when contaminated by Gaussian noise. Moving on, Fig. 3.33 and Fig. 3.34 showcase the SNIB performance under SP, and Rayleigh noise on CIFAR-100, respectively. The figures illustrate that SNIB consistently outperforms the method in [41] and IB in all these three conditions. This observation suggests that SNIB is well-suited for improving the robustness of models when dealing with more complex datasets.

Neuromorphic systems always suffer from hardware noise in operating environments. The learning robustness of SNIB is explored for neuromorphic hardware non-ideality. For simulating neuromorphic synaptic noise, Gaussian noise is added on the updating process of synaptic weight during training, which is formulated as follows:

$$w_{upd} = \Delta w + N\left(0, \sigma_{SN}^2\right) \tag{3.141}$$

where w_{upd} represents the updated noisy synaptic weight, and Δw represents the updating value of the synaptic weight. The Gaussian noise $N\left(0, \sigma_{SN}^2\right)$ is defined by mean value 0 and standard variance σ_{SN}. SIB, CIB, and QIB are compared with IB and method in [41] along with neuromorphic synaptic noise with NNI = 1, 3, 5, and 15 respectively. As demonstrated in Fig. 3.35, the robustness of SNIB (including SIB, CIB, and QIB) is superior to IB and method in [41], and the robustness improvement is larger as NNI

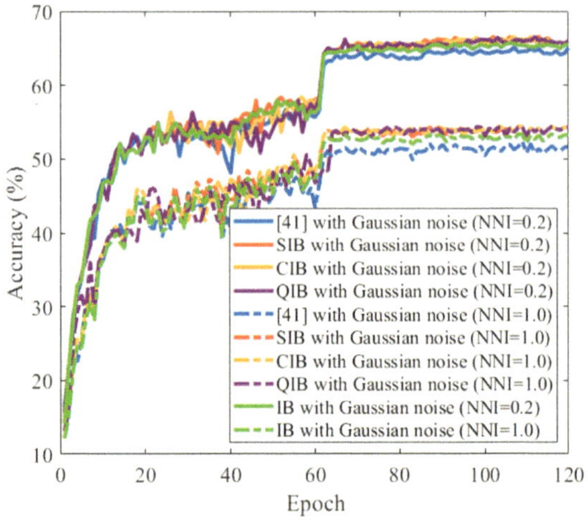

Fig. 3.32 Performance comparison on CIFAR100 data set among SIB, CIB, QIB, and the method in [41] with Gaussian noise using NNI = 0.2 and 1.0 respectively [59]

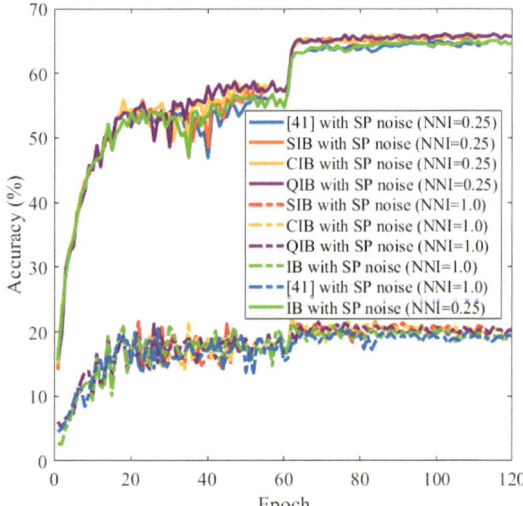

Fig. 3.33 Performance comparison on CIFAR100 data set among SIB, CIB, QIB, and the method in [41] with SP noise using NNI = 0.2 and 1.0 respectively [59]

increasing. It suggests SNIB is an effective approach to improve the learning robustness when facing neuromorphic synaptic noise.

The robustness of SNIB with neuromorphic membrane potential noise is tested, which is induced by background noise. Gaussian noise $N\left(0, \sigma_{MPN}^2\right)$ is employed to simulate the membrane potential noise, which is expressed as follows

$$V_{upd} = \Delta V + N\left(0, \sigma_{MPN}^2\right) \tag{3.142}$$

where V_{upd} represents the updated noisy membrane potential, and ΔV represents the updating value of the membrane potential. Figure 3.36 shows the learning robustness

Fig. 3.34 Performance comparison on CIFAR100 data set among SIB, CIB, QIB, and the method in [41] with Rayleigh noise using NNI = 0.25 and 2.00 respectively [59]

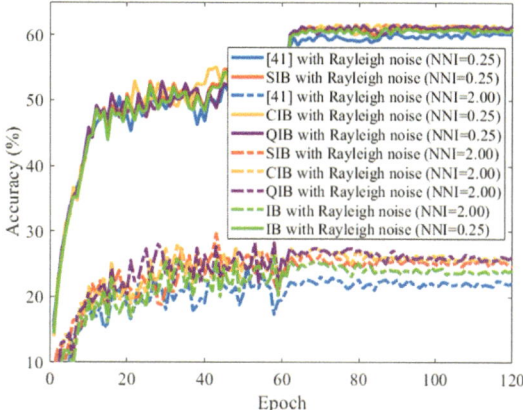

Fig. 3.35 Learning robustness with synaptic noise existing on neuromorphic hardware. The comparison is among SIB, CIB, QIB, and the method in [41]. This figure is from [59]

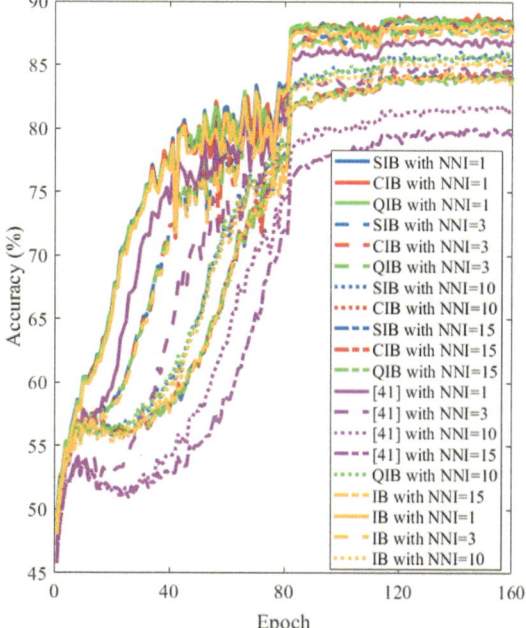

of SIB, CIB, and QIB in comparison with method in [41] and IB. It shows the SNIB can demonstrate is superiority on learning robustness when contaminated with high-level neuromorphic membrane potential noise, such as NNI = 3 and 6. Therefore, SNIB is a better solution to improve the learning robustness with neuromorphic background noise on both neural synapse and membrane potential.

Further exploration of the parameter sensitivity is conducted on the learning robustness of SNIB. For this investigation, the QIB strategy is selected as a representative of

Fig. 3.36 Learning robustness with membrane potential noise existing in neuromorphic hardware. The comparison is among SIB, CIB, QIB, and the method in [41]. This figure is from [59]

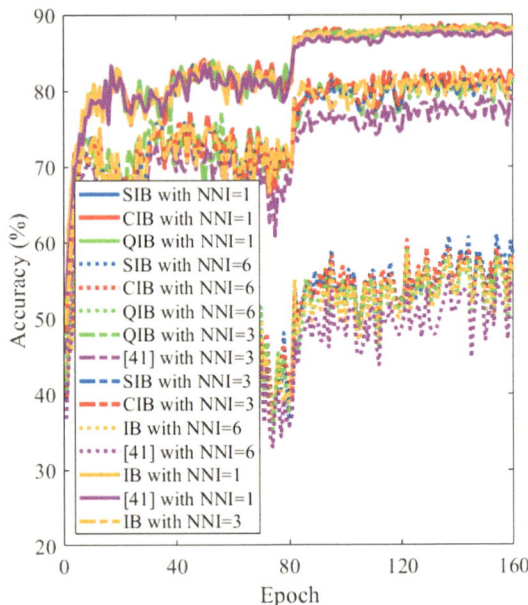

the SNIB framework. Figure 3.37 presents the performance of QIB with varying values of hyperparameters, including the Lagrange parameter and NNI. Four types of noise are considered, namely Gaussian, Rayleigh, neuromorphic synaptic noise, and neuromorphic membrane potential noise. The results reveal that QIB is more sensitive to Gaussian and neuromorphic synaptic noise, while the impact of the Lagrange parameter is less pronounced in Rayleigh noise. Interestingly, for QIB, both low and high levels of the Lagrange parameter induce higher learning robustness across different noise intensities. On the other hand, a medium value of β tends to reduce the learning robustness. Overall, this analysis sheds light on the parameter sensitivity of SNIB, highlighting the importance of selecting appropriate values for the Lagrange parameter and NNI to achieve optimal learning robustness in different noise scenarios.

SNIB is applied to a recurrent shallow SNN architecture based on SAM neuron [59]. The SAM model consists of three parts, a spiking body, an excitatory spiking dendrite, and an inhibitory spiking dendrite. The variable $\Psi_j(t)$ is also introduced as a learnable parameter. Its discrete membrane potential equation formula is as follows:

$$\begin{cases} V_j(t + \Delta t) = \mu V_j(t) + (1 - \xi)\alpha_m I_j(t) + (1 - \xi)V_j^{den,i}(t) \\ \quad +(1 - \xi)V_j^{den,e}(t) - \Psi_j(k)z_j(t)\Delta t \\ V_j^{den,e}(t + \Delta t) = \xi V_j^{den,e}(t) + (1 - \xi)\alpha_m^e I_j^e(t) \\ V_j^{den,i}(t + \Delta t) = \xi V_j^{den,i}(t) + (1 - \xi)\alpha_m^i I_j^i(t) \end{cases} \quad (3.143)$$

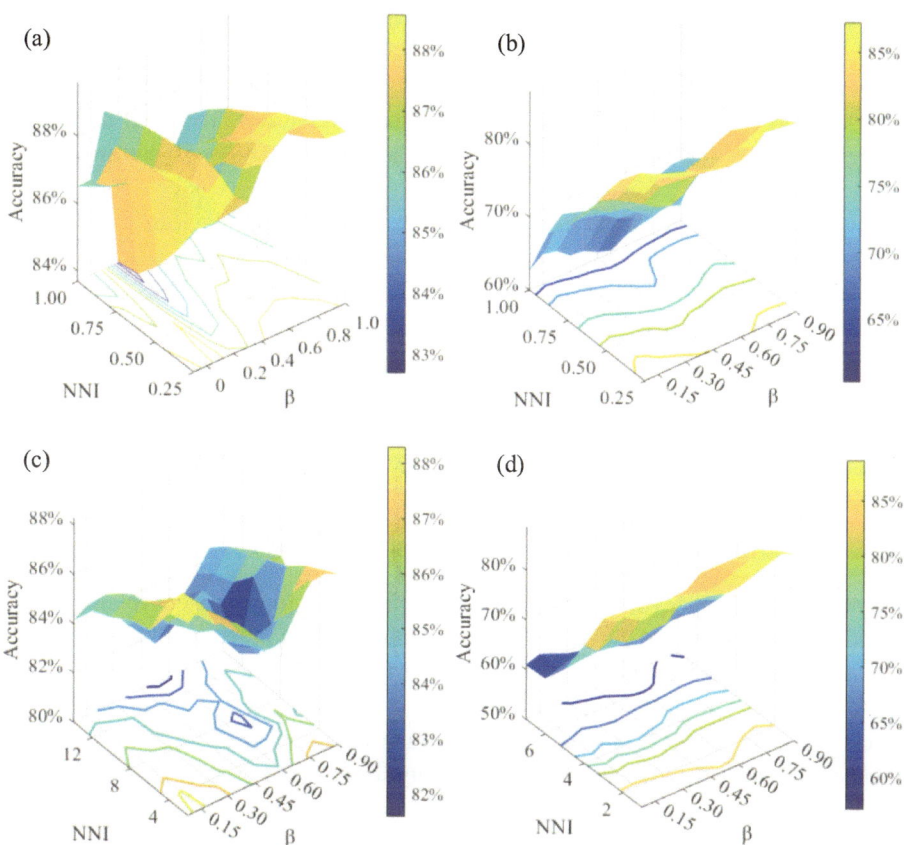

Fig. 3.37 Parameter sensitivity analysis of Lagrange parameter β of the QIB strategy along with the NNI increasing [59]. **a** Effects of β on QIB performance with Gaussian noise. **b** Effects of β on QIB performance with Rayleigh noise. **c** Effects of β on QIB performance with neuromorphic synaptic noise. **d** Effects of β on QIB performance with neuromorphic membrane potential noise

where $\mu = \exp(-\Delta t / \tau)$. Variable $z_j(t)$ denotes the spike train of the jth neuron in $\{0, 1/\Delta t\}$. The dynamics of $\Psi_j(t)$ represent the firing rate of neuron j, and is updated with spikes. It can be written as:

$$\Psi_j(t) = \psi_j^0 + \varepsilon \cdot \psi_j(t) \tag{3.144}$$

$$\psi_j(t + \Delta t) = \mu_j \psi_j(t) + (1 - \mu_j) z_j(t) \tag{3.145}$$

where η represents a constant that scales the deviation $\psi_j(t)$ from the baseline ψ_j^0. $\lambda_j = \exp(-\Delta t / \tau_{a,j})$ and $\tau_{a,j}$ represents the adaptation time constant. If $V_j(t) > \Psi_j(t)$, the SAM neuron omits a spike.

In the hidden layer, SAM neurons are interconnected with lateral inhibitory connections, as depicted in Fig. 3.38. To achieve a balance between information usage and compression, SNIB is employed as the loss function. The effects of SNIB on the learning robustness of the SAM network architecture are investigated. To assess the learning robustness, a classification task on the sequential MNIST dataset is performed. This involves applying five categories of input noise, where noisy pixels of each handwritten image are sequentially input into SAM. Figure 3.39 illustrates the learning robustness of the SAM-based RSNN, showcasing that all SIB, CIB, and QIB models can enhance the robustness of the deep SCNN with VGG architecture. Notably, QIB demonstrates superiority with Gaussian noise, while CIB outperforms with uniform noise. As the NNI increases, the learning accuracy experiences a more significant reduction. Consequently, the improvement effects of the SNIB strategy, including SIB, CIB, and QIB, become increasingly prominent. Furthermore, Fig. 3.39 indicates that SNIB can enhance learning robustness, even for shallow SNN architectures, providing valuable insights into the benefits of employing SNIB in the SAM network.

(2) Power consumption of SNIB

Another advantage of SNIB is its low power consumption. Suppose a convolutional layer l with weight tensor $\mathbf{W}^l \in \mathbb{R}^{k^l \times k^l \times C_i^l \times C_o^l}$ using input activation tensor $\mathbf{A}^l \in \mathbb{R}^{H_i^l \times W_i^l \times C_i^l}$, with H_i^l, W_i^l, k^l, C_i^l and C_o^l as the input height, width, filter height and width, channel size, and filter number. Table 3.11 shows the FLOPs consumption for an ANN and SNN for the corresponding layer to generate an output activation tensor $\mathbf{O}^l \in \mathbb{R}^{H_o^l \times W_o^l \times C_o^l}$. The variable ζ^l denotes the associated spiking activity for layer l. For an L-layer SNN model with rate-coded and direct input, the inference computation energy can be formulated as

$$E_{SNN}^{rate} = \left(\sum_{l=1}^{L} FL_{SNN}^l \right) \cdot E_{AC} \tag{3.146}$$

$$E_{SNN}^{direct} = FL_{SNN}^l \cdot E_{MAC} + \left(\sum_{l=2}^{L} FL_{SNN}^l \right) \cdot E_{AC} \tag{3.147}$$

where E_{AC} and E_{MAC} denote the power consumption of AC and MAC operation respectively. The evaluation is performed using a 45 nm CMOS neuromorphic chip, TrueNorth, as a reference. Considering that SNIB employs spike-based coding and learning schemes, it significantly reduces energy consumption compared to conventional ANNs. The calculation of energy consumption is based on the data provided in Tables 3.11 and 3.12, allowing us to demonstrate the substantial energy savings achieved by SNIB over traditional ANN-based models.

The neural spiking rate at the lth layer can be defined as

$$x_{SR}(l) = \frac{x_{SPN}}{x_{NEU}} \tag{3.148}$$

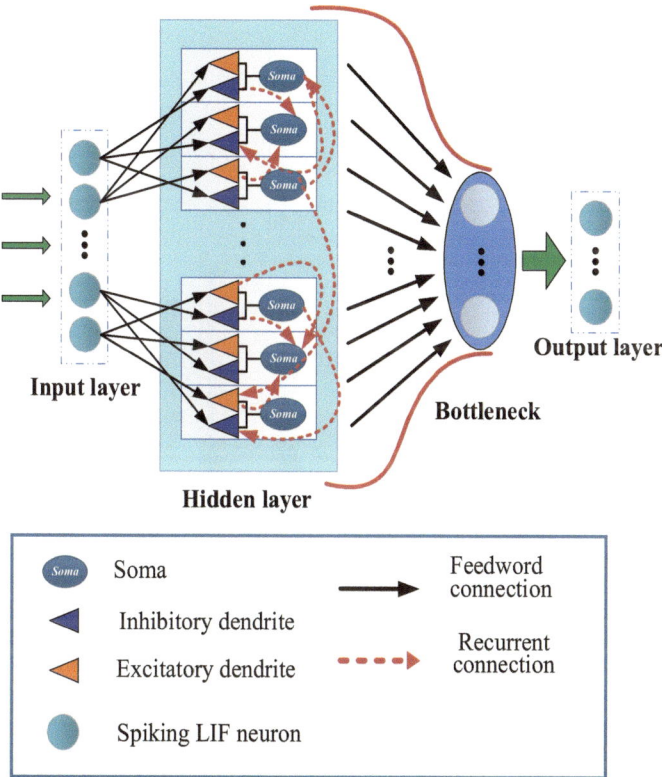

Fig. 3.38 Application of SNIB in SAM-based recurrent neural network architecture for spike-driven online learning [59]

where x_{SPN} and x_{NEU} represent the neural spikes of layer r over all time steps and neuron number at layer l respectively. Table 3.13 illustrates the critical settings of the tested SNN model with the SIB strategy. The calculation of energy consumption is demonstrated in Table 3.14. Figure 3.40 visualizes the normalized energy consumption in each layer for ASC, conventional SNN without the IB strategy, and SIB methods. It becomes evident that the energy consumption using SIB is significantly lower than the ASC and SNN methods. The overall results clearly indicate that the SNIB framework can substantially cut down the energy consumption on neuromorphic hardware. In conclusion, the SNIB approach offers promising energy efficiency benefits, making it an attractive choice for implementation on neuromorphic hardware systems.

2. Learning performance of HOSIB

We first investigate learning accuracy of simple static data set Fashion-MNIST, and the comparison with other works is presented in Table 3.15. The proposed HOSIB framework

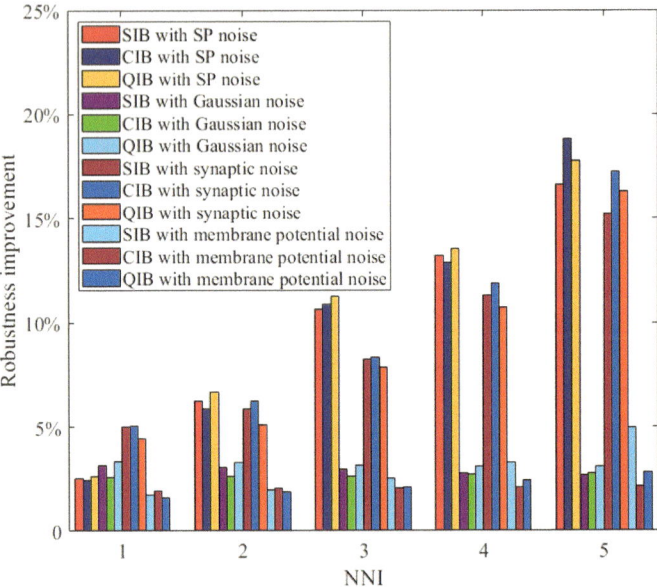

Fig. 3.39 Robustness improvement of SAM-based RSNN on MNIST data set with different types of noise [59]

Table 3.11 FLOPs number of convolutional layer for ANN and SNN models (From [59])

Model	FLOPs of a convolutional layer	
	Variable	Value
ANN	FL^l_{ANN}	$\left(k^l\right)^2 \times H^l_O \times W^l_O \times C^l_O \times C^l_i$
SNN	FL^l_{SNN}	$\left(k^l\right)^2 \times H^l_O \times W^l_O \times C^l_O \times C^l_i \times \zeta^l$

Table 3.12 Energy consumption for different operations in a 45 nm CMOS chip (for example, TrueNorth) at 0.9 V (From [59])

Operation	Energy consumption (pJ)	
	32-bit INT	32-bit FP
32-bit multiplication	3.1	3.7
32-bit addition	0.1	0.9
32-bit MAC	3.2	4.6
32-bit AC	0.1	0.9

Table 3.13 Critical settings of the tested SNN model with the SIB strategy (From [59])

Layers	Kernel size	Input/output channel number	Output mapping size
Conv#1	3×3	3,64	32
Conv#2	3×3	64,128	16
Conv#3	3×3	128,256	8
Conv#4	3×3	256,256	8
Conv#5	3×3	256,512	4
Conv#6	3×3	512,512	4
Conv#7	3×3	512,512	2
FC	3×3	512,512	2

Table 3.14 [59] Spiking number, neuron number, and spiking rate of SNIB in each layer.

Layers	Spikes number	Neuron number	x_{SR} of SNIB
Conv#1	26,913.1	126,208	0.213
Conv#2	6828	32,768	0.2084
Conv#3	14,410.2	16,384	0.8795
Conv#4	3325.3	16,384	0.2030
Conv#5	6416.9	8192	0.7833
Conv#6	5888.1	8192	0.7188
Conv#7	1433.1	2048	0.6998
FC	454.6	2048	0.2220

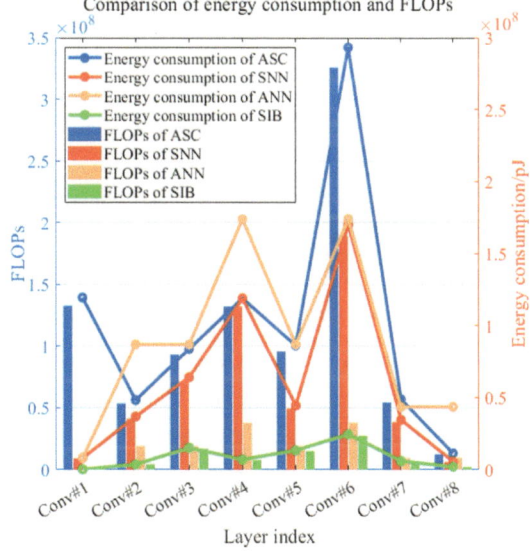

Fig. 3.40 Comparison of normalized energy consumption between SIB, ANN and ASC, SCNN. FC denotes fully connected layer accordingly. The gray line represents the actual energy consumption (EC) of ASC method in each layer for a reference [59]

can achieve accuracy of 91.6% and 92.6% with SOIB and TOIB strategies with VGG9 architecture respectively. The classification accuracy of SOIB and TOIB strategies with ResNet18 architecture can reach 94.2% and 93.6% respectively, which is only slightly lower than Fang et al. [77] and superior to other works. More complex data sets are further used for performance evaluation, including CIFAR10 and CIFAR100 data sets. Table 3.16 lists the testing performance of HOSIB on CIFAR10 and CIFAR100. To convert image pixels to spike timing, encoding rule $t_i = \gamma_{en} p_i$ is employed to maximize spike time separation. There are 60,000 images using RGB color with 32×32 pixels in ten classes contained in CIFAR-10, which includes 50,000 and 10,000 for training and testing separately. Besides, CIFAR100 consists of 100 categories of images with RGB color. For VGG architecture, the classification accuracy of TOIB is just lower than Rathi et al. [78], Han et al. [79], and Sengupta et al. [80] on CIFAR10 and CIFAR100 data sets. The latter two methods are both based on ANN-SNN conversion, which have the limitations to deal with event-based neuromorphic data set. For ResNet architecture, both SOIB and TOIB achieve the state-of-the-art accuracy. Especially for CIFAR100 data set, TOIB achieves 2.7%, 3.2%, 6.5%, and 6.5% higher than Rathi et al. [78], Li et al. [87], Han et al. [79], and Sengupta et al. [80] respectively. It further reveals that the proposed HOSIB strategy is more suitable for dealing with more complex data sets with more deep and complicated network architectures.

In addition, we investigate the learning performance of HOSIB on the neuromorphic DVS-CIFAR10 data set using the VGG11 architecture. The DVS-CIFAR10 data set is

Table 3.15 Classification accuracy (%) on Fashion-MNIST data set (From [59])

Studies	Network architecture	Acc. (%)
S4NN [71]	784–1000-10	88.0
BS4NN [72]	784–1000-10	87.3
Hao et al. [73]	784–6000-10	85.3
Zhang et al. [74]	784–400-400–10	89.5
Ranjan et al. [75]	28×28-32C3-32C3-P2-128–10	89.0
STDBP [76]	784–1000-10	88.1
STDBP [7]	28×28-16C5-P2-32C5-P2-800–128-10	90.1
Guo and Lin [54]	32C5-P2-64C5-P2-1024	87.9
Guo and Lin [54]	VGG8A	88.3
Fang et al. [77]	3Conv, 2Down, 2FC	94.4
HOSIB (ours)	**VGG9 with SOIB**	**91.6**
HOSIB (ours)	**VGG9 with TOIB**	**92.6**
HOSIB (ours)	**ResNet18 with SOIB**	**94.2**
HOSIB (ours)	**ResNet18 with TOIB**	**93.6**

Table 3.16 Comparison of classification accuracy (%) with the state-of-the-art SNNs with VGG and ResNet architectures on CIFAR-10 and CIFAR-100 data sets (From [59])

Studies	Data set	Training method	Architecture	Acc. (%)
Cao et al. [81]	CIFAR 10	ANN-SNN Conversion	3Conv, 2Linear	77.4%
Sengupta et al. [80]	CIFAR 10	ANN-SNN Conversion	VGG16	91.5%
Lee et al. [82]	CIFAR 10	Surrogate Gradient	VGG9	90.4%
Rathi et al. [83]	CIFAR 10	Hybrid	VGG16	92.0%
Han et al. [79]	CIFAR 10	ANN-SNN Conversion	VGG16	93.6%
Garg et al. [84]	CIFAR 10	STDB	VGG9	89.9%
Hunsberger and Eliasmith [85]	CIFAR 10	ANN-SNN Conversion	VGG9	83.5%
Kim and Panda [86]	CIFAR 10	Surrogate Gradient	VGG9	90.5%
Kim and Panda [86]	CIFAR 10	Surrogate Gradient	VGG9	90.3%
Zheng et al. [36]	CIFAR 10	STBP-tdBN	ResNet19	93.2%
Li et al. [87]	CIFAR 10	Dspike	ResNet18	94.3%
Rathi et al. [78]	CIFAR 10	DIET-SNN	VGG6	90.1%
Rathi et al. [78]	CIFAR 10	DIET-SNN	VGG16	93.4%
Rathi et al. [78]	CIFAR 10	DIET-SNN	ResNet20	92.5%
Deng et al. [88]	CIFAR 10	TET	ResNet19	94.5%
Fang et al. [77]	CIFAR 10	Spike-based BP	3Conv, 2Down, 2FC	93.5%
Bu et al. [89]	CIFAR 10	ANN-SNN Conversion	ResNet20	92.4%
SOIB (ours)	**CIFAR 10**	Surrogate Gradient	**VGG9**	**90.4%**
TOIB (ours)	**CIFAR 10**	Surrogate Gradient	**VGG9**	**91.3%**
SOIB (ours)	**CIFAR 10**	Spike Representation	**ResNet18**	**94.4%**
TOIB (ours)	**CIFAR 10**	Spike Representation	**ResNet18**	**94.6%**
Sengupta et al. [80]	CIFAR 100	ANN-SNN Conversion	VGG16	70.9%
Rathi et al. [83]	CIFAR 100	Hybrid	VGG16	67.8%
Han et al. [79]	CIFAR 100	ANN-SNN Conversion	VGG16	70.9%
Kim and Panda [86]	CIFAR 100	Surrogate Gradient	VGG11	66.6%
Kim and Panda [86]	CIFAR 100	Surrogate Gradient	VGG11	65.8%

(continued)

Table 3.16 (continued)

Studies	Data set	Training method	Architecture	Acc. (%)
Li et al. [87]	CIFAR 100	Dspike	ResNet18	74.2%
Rathi et al. [78]	CIFAR 100	DIET-SNN	VGG16	69.7%
Rathi et al. [78]	CIFAR 100	DIET-SNN	ResNet20	64.1%
Deng et al. [78]	CIFAR 10	TET	ResNet19	74.7%
SOIB (ours)	**CIFAR 100**	Surrogate Gradient	**VGG11**	**68.3%**
TOIB (ours)	**CIFAR 100**	Surrogate Gradient	**VGG11**	**66.8%**
SOIB (ours)	**CIFAR 100**	Spike Representation	**ResNet18**	**77.0%**
TOIB (ours)	**CIFAR 100**	Spike Representation	**ResNet18**	**77.4%**

derived from CIFAR-10 and is converted into a neuromorphic data set using an event-based sensor. The data set comprises 10 categories with 1,000 images each, totaling to 10,000 event-based images with a resolution of 128×128 pixels. The data set is divided into 20 slices, and each slice's events are integrated into a frame. This integration method is based on the transformation method outlined in [77]. We use the SpikingJelly framework [90] to implement the integration from event to frame and the data separation for training and testing. We use data augmentation techniques based on random cropping, and the spatial resolution is reduced from 128×128 to 48×48 pixels. We partition the data set into 9,000 training images and 1,000 testing images. Our experimental results presented in Table 3.17 demonstrate that the HOSIB framework outperforms other representative studies that use direct training methods.

We conduct extensive ablation study to evaluate the proposed HOSIB strategy with SOIB and TOIB loss functions. In this experiment, we train the SNNs based on the same

Table 3.17 Comparison of classification accuracy (%) with the SOTA direct training methods of deep SNNs on DVS-CIFAR10 data set (From [59])

Studies	Data set	Architecture	Accuracy
Wu et al. [91]	DVS-CIFAR10	VGG7	62.5%
Wu et al. [92]	DVS-CIFAR10	7-layer	65.6%
Wu et al. [93]	DVS-CIFAR10	7-layer	60.5%
Zheng et al. [36]	DVS-CIFAR10	ResNet19	67.8%
Fang et al. [77]	DVS-CIFAR10	7-layer	74.8%
Li et al. [87]	DVS-CIFAR10	ResNet-18	75.4%
Fang et al. [54]	DVS-CIFAR10	SEW ResNet	74.4%
HOSIB (ours)	**DVS-CIFAR10**	**VGG11 with SOIB**	**78.2%**
HOSIB (ours)	**DVS-CIFAR10**	**VGG11 with TOIB**	**75.7%**

Table 3.18 Ablation study on different data sets for the evaluation of SOIB and TOIB loss (From [59])

Loss	Fashion-MNIST /Architecture	CIFAR10 /Architecture	CIFAR100 /Architecture
Cross-entropy	93.0%/VGG9	86.6%/VGG9	58.4%/VGG11
IB	93.0%/VGG9	86.2%/VGG9	62.2%/VGG11
SOIB	93.1%/VGG9	90.4%/VGG9	68.3%/VGG11
TOIB	**94.0%/VGG9**	**91.3%/VGG9**	**66.8%/VGG11**
Cross-entropy	93.7%/ResNet18	94.3%/ResNet18	75.8%/ResNet18
IB	94.1%/ResNet18	94.4%/ResNet18	76.4%/ResNet18
SOIB	**94.2%/ResNet18**	94.4%/ResNet18	77.0%/ResNet18
TOIB	94.1%/ResNet18	**94.6%/ResNet18**	**77.4%/ResNet18**

network architecture and training details but with cross-entropy loss, common IB loss, SOIB, and TOIB respectively, and compare the test accuracy. As shown in Table 3.18, the proposed SOIB and TOIB strategies outperform the conventional cross-entropy and IB loss for all types of data sets with both VGG and ResNet architectures. The highest improvement is in the case of TOIB strategy with VGG11 architecture for CIFAR100 data set, which achieves 8.4% improvement of learning accuracy compared to cross-entropy loss and 4.6% improvement compared to conventional IB loss. In addition, the learning accuracy of TOIB is 4.7% and 5.1% higher than cross-entropy and IB loss for CIFAR 10 with VGG9 architecture. We also test the performance improvement for neuromorphic DVS-CIFAR10 data set. The learning accuracies with cross-entropy and IB loss are 74.5% and 75.4%. Therefore, performance improvements for DVS-CIFAR10 with SOIB loss are 3.7% and 2.8% higher compared to cross-entropy and IB loss respectively.

We first investigate the situations of five different noise categories that added on the input images. Gaussian noise is the normal noise with the statistical characteristics dependent on the normal distribution. The Gaussian noise with mean value $\mu = 0$ and standard variance σ can be expressed as

$$f(x) = \frac{1}{\sqrt{2\pi}\sigma} e^{-\frac{(x-\mu)^2}{2\sigma^2}}. \tag{3.149}$$

The noise density means the standard deviation. The SP noise is generated with sparsely white and black pixels, which is produced by sharp and sudden disturbances in the image signal. The uniform noise can be formulated as

$$z(x) = \alpha_{uni} + (\beta_{uni} - \alpha_{uni})u(0, 1), \tag{3.150}$$

where $u(0, 1)$ is uniformly distributed between 0 and 1. The parameter β_{uni} is defined as the noise density, and the parameter $\alpha_{uni} = 0$. The Gamma noise is represented by n

exponentially distributed noise, which is expressed by

$$z = z_1 + z_2 + \cdots + z_n, \tag{3.151}$$

where $z_i = -\frac{1}{\alpha} \ln[1 - u(0, 1)]$. The gamma noise can be regarded as the exponential noise when $n = 1$. For this noise, $1/\alpha$ is defined as the noise density.

Figure 3.41 shows the learning robustness of SOIB and TOIB compared to the SNN model without (w.o.) IB (i.e., cross-entropy), common IB, and non-BP IB (NIB) that is introduced by previous study [54]. Four types of noise are added to SOIB and TOIB models. We normalize the noise intensity of SP, Gaussian, uniform, and Gamma noise by the parameters $\alpha_1 = 0.0017$, $\alpha_2 = 1.6667$, $\alpha_3 = 16.6667$, and $\alpha_4 = 1.6667$. The fair comparison between cross-entropy (w.o. IB), common IB and the proposed HOSIB loss that follow the exact same setting (network structure, training details, etc.) is included. As shown in Fig. 3.41, it is surprising to find that the learning accuracy of both SOIB and TOIB can maintain almost the same accuracy along with the noise intensity of SP noise increasing from 100 to 600, while the accuracy is reduced by 21.5% and 20.7% based on w.o.IB and NIB method respectively. In terms of Gaussian noise, it shows that TOIB achieves the highest accuracy along with the noise intensity of Gaussian noise from 0.1 to 0.6. When the noise intensity is 0.6, the learning accuracy of TOIB is 3.67%, 2.38%, and 5.21% much higher than SOIB, NIB, and w.o.IB respectively. It should be noted that the proposed SOIB performs better than w.o.IB model, but is less robust than the NIB model. The learning accuracy of TOIB is 2.18%, 2.63%, and 4.93% higher than SOIB, NIB, common IB, and w.o.IB respectively when the noise intensity of uniform noise is at 0.01, and 9.81%, 32.46%, and 35.06% higher with the noise intensity of uniform noise at 0.06. Learning robustness with Gamma noise is demonstrated in Fig. 3.43, which shows that the learning accuracy of TOIB is superior to the other three methods along with the noise intensity increasing. TOIB is 2.18% and 9.81% higher than SOIB respectively when the noise intensity of Gamma noise is 0.1 and 0.6. Thus, it reveals that the learning robustness of TOIB is superior to other models, and it can maintain a high level of accuracy regardless the noise intensity of SP noise and uniform noise increasing.

Furthermore, we explore the robustness improvement on CIFAR-100 data set. The relative error of improvement is defined as the criteria of robustness improvement, i.e., the y-axis of Fig. 3.42, by comparing with w.o.IB (i.e. cross-entropy loss) model as the basis. As shown in Fig. 3.42, SOIB has superior robustness than TOIB on CIFAR-10 with SP and Gamma noise. In terms of CIFAR-10 with uniform noise, the learning robustness of TOIB is higher than SOIB conversely. SOIB has higher robustness than TOIB with a low noise-to-signal level of Gaussian noise, but is less robust than TOIB when facing larger level of Gaussian noise, such as the normalized noise intensity equaling to 1. Both the SOIB and TOIB method are more robust than NIB and w.o.IB. Therefore, it suggests SOIB is the best choice with SP, Gamma, and low Gaussian noise, but TOIB is more preferable with high level of Gaussian noise as well as uniform noise.

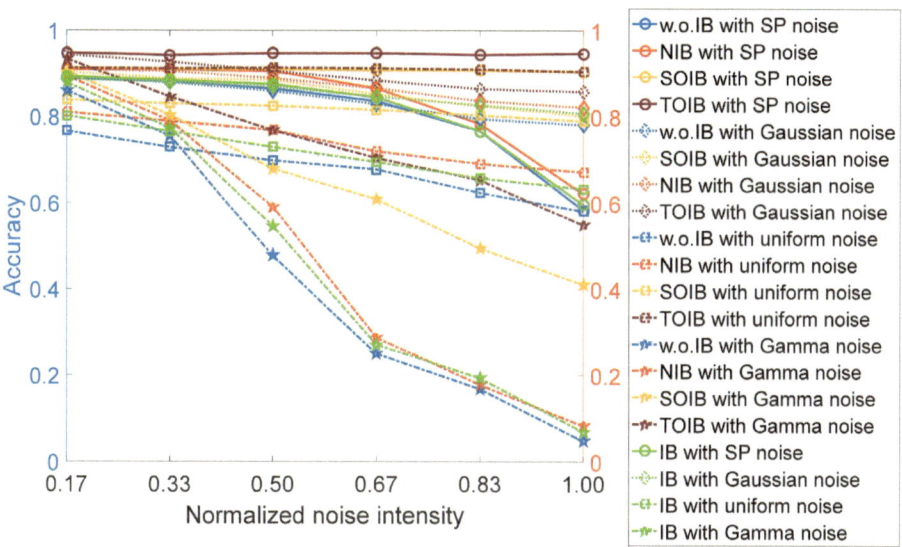

Fig. 3.41 Learning robustness test on the average accuracy of common IB, NIB, SOIB and TOIB along with the normalized noise intensity increasing [70]

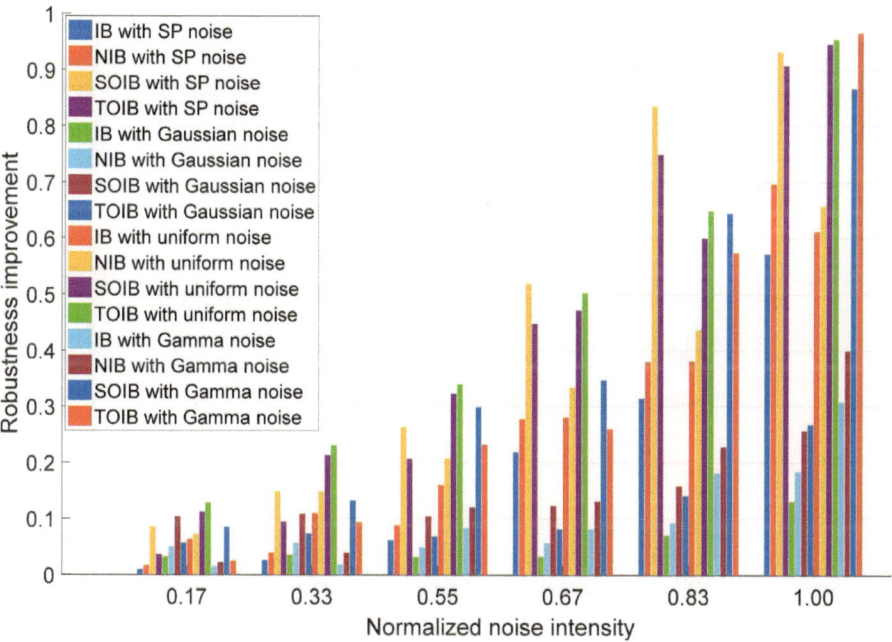

Fig. 3.42 Robustness improvement of average accuracy on CIFAR-100 data set using common IB, NIB, SOIB, and TOIB with different types of noise along with the noise intensity increasing [70]

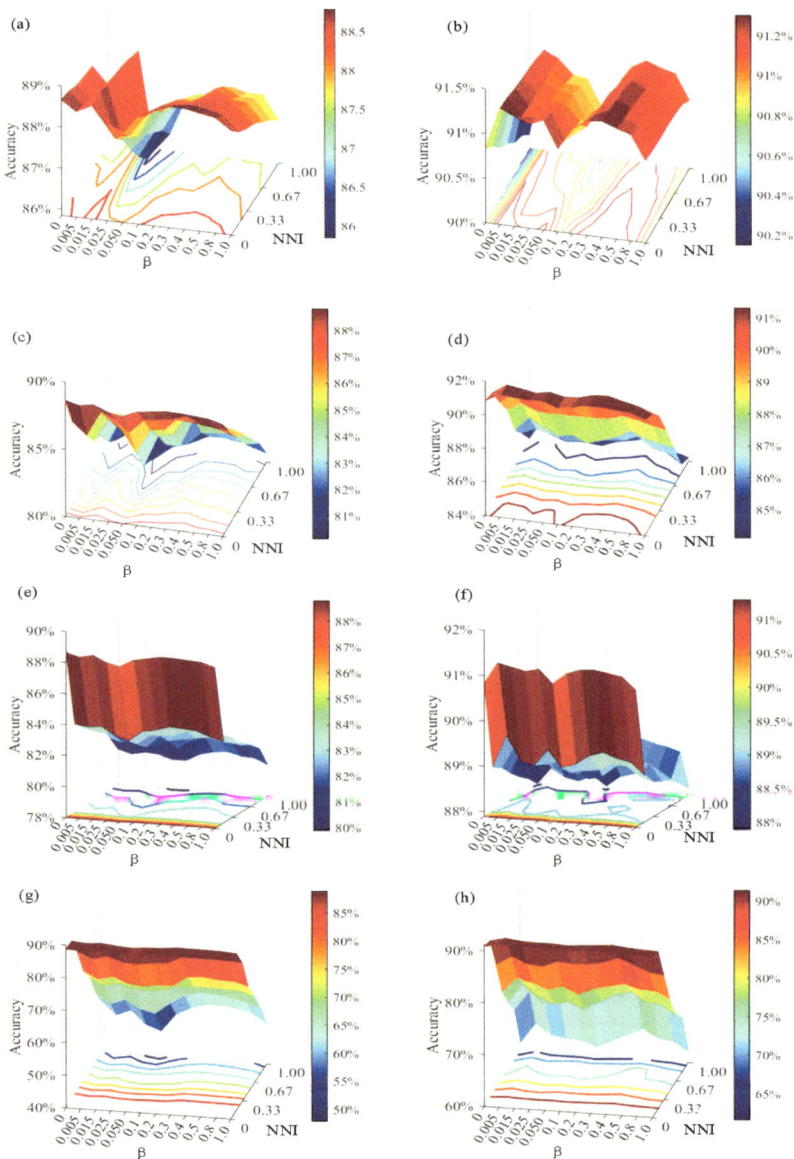

Fig. 3.43 Parameter sensitivity analysis of Lagrange parameter β in the proposed HOSIB framework along with the NNI changing. **a** Effects of β on SOIB performance with SP noise. **b** Effects of β on TOIB performance with SP noise. **c** Effects of β on SOIB performance with Gaussian noise. **d** Effects of β on TOIB performance with Gaussian noise. **e** Effects of β on SOIB performance with uniform noise. **f** Effects of β on TOIB performance with uniform noise. **g** Effects of β on SOIB performance with Gamma noise. **h** Effects of β on TOIB performance with Gamma noise [70]

For the selection of coefficient value, the optimal β value is selected based on multiple sets of experiments. For example, we set β to be 0.015 in general. Although we do not conduct fine tuning for the coefficient parameter in HOSIB framework, the experiments results have obviously demonstrated the superiority of the application of HOSIB learning strategy. We further investigate the parameter sensitivity analysis in the range of [0, 1] in Fig. 3.43, which can be employed to be a concrete guidance for parameter value selection. We test network robustness performance by tuning the Lagrange parameter β from {0, 0.005, 0.015, 0.025, 0.050, 0.1, 0.2, 0.3, 0.4, 0.5, 0.8, 1.0} and changing the NNI from {0, 0.17, …, 1}. Figure 3.43 shows the performance of SOIB and TOIB schemes with different values of hyper parameters, i.e., NNI and β, on CIFAR10 data set. A promising accuracy is achieved when the value of β falls into a certain range. After it becomes larger, a new balance of compression and exploitation is achieved, which enables a higher performance. Finally, when the value close to 1, the accuracy decreases due to the information compression. Furthermore, too larger NNI produces lower performance, and the Gamma noise has the heaviest influence on the information representation of HOSIB with both SOIB and TOIB schemes.

(4) Power Consumption of HOSIB

The energy consumption of HOSIB on digital neuromorphic hardware is further evaluated. Total energy consumption is calculated based on the total number of FLOPs or matrix–vector multiplication (MWM) operations. The FLOPs for the lth convolutional layer in the HOSIB framework can be expressed as

$$FLOPs(l) = x_{KS}^2 \times x_{OFM}^2 \times x_{IC} \times x_{OC}, \tag{3.152}$$

where x_{KS} denotes the kernel size, and x_{OFM} denotes the output feature map size. The variables x_{IC} and x_{OC} represent the input and output channel numbers respectively. The FLOPs for the lth linear layer in HOSIB can be defined as

$$FLOPs(l) = x_{IC} \times x_{OC}. \tag{3.153}$$

The neural spiking rate at the lth layer can be defined as

$$x_{SR}(l) = \frac{x_{SPN}}{x_{NEU}}, \tag{3.154}$$

where x_{SPN} and x_{NEU} represent the neural spikes of layer r over all time steps and neuron number at layer l respectively.

In addition, the FLOPs of SNN model can be calculated based on the spiking rate $x_{SR}(l)$ and FLOPs, which is expressed as

$$FLOP_{SNN}(l) = FLOPs(l) \times x_{SR}(l). \tag{3.155}$$

The MAC operation can be represented by AC operation because SNN is based on 0/1 binary spike-driven computation. The total energy consumption of ANNs and SNNs can be formulated as

$$E_{ANN} = \sum_l FLOPs(l) \times E_{MAC}, \qquad (3.156)$$

$$E_{ANN} = \sum_l FLOP_{SNN}(l) \times E_{AC}. \qquad (3.157)$$

We further demonstrate the results of FLOPs and energy consumption by comparing SOIB and TOIB with SNN and ASC models. The calculation of energy consumption is based on Table 3.19. Figure 3.44 depicts the FLOPs and energy consumption in each layer based on ASC, conventional SNN without IB rule, SOIB, and TOIB methods. ASC method induces higher spiking rate for each layer since it forwards spike trains with a large number of time steps in comparison with other methods. Even though direct training method of SNN significantly cut down the firing rate and number of time steps, it still requires hundreds of spikes for each layer. Since the proposed SOIB and TOIB strategies can effectively cut down the firing rate in each layer of deep SNNs, the FLOPs and energy consumption using SOIB and TOIB are significantly lower than ASC and SNN methods. In comparison with ASC method, the energy reduction percentages of SOIB and TOIB are 87.70% and 82.46% respectively. The FLOPs and energy consumption of TOIB are 3.6025 slightly higher than SOIB, because the spiking rate in each layer of TOIB is higher than SOIB accordingly. The proposed SOIB and TOIB schemes can significantly cut down the firing rate of spiking neurons in each layer of deep SNNs, since the spike information can be represented in a more efficient and compressed manner based on the HOSIB learning strategy. Therefore, it reveals that the proposed HOSIB framework with both SOIB and TOIB schemes can significantly cut down the energy consumption on neuromorphic hardware.

Table 3.19 [59] Spiking rate of SOIB and TOIB in each layer.

Layers	Neuron number	x_{SR} of SOIB	x_{SR} of TOIB
Conv#1	126,208	0.3053	0.3870
Conv#2	32,768	0.2357	0.3250
Conv#3	16,384	1.0356	1.4093
Conv#4	16,384	0.2356	0.3198
Conv#5	8192	0.9287	1.3696
Conv#6	8192	0.8673	1.2891
Conv#7	2048	0.9512	1.3477
FC	2048	0.3560	0.4463

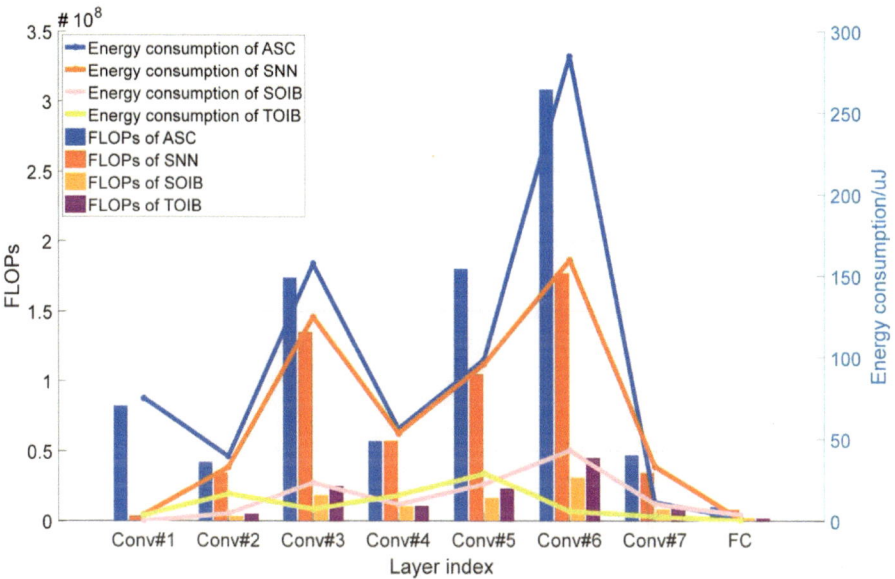

Fig. 3.44 Comparison of energy consumption and FLOPs between SOIB, TOIB, and conventional ANN [70]

3.6   Summary

This chapter introduces spike-based learning with ITL theories. Firstly, it presents ITL criteria. It elaborates on the mathematical derivations and the significance of these ITL criteria. Next, it introduces the robust spike-driven few-shot learning algorithm, HES-FOL, which employs an ensemble loss function. It provides insights into HESFOL's network model architecture, neuron models, and performance evaluation. Furthermore, we introduce the robust spike-driven meta-learning algorithm, MeMEE, along with its performance test results. Subsequently, it presents the spike-based learning algorithm, SIBoLS, based on the IB theory, significantly enhancing the robustness of spike-driven learning. Finally, it introduces the SNIB algorithm based on nonlinear IB, and the HOSIB algorithm based on high-order IB, further improving the robustness and reducing power consumption in deep SNN learning. The content of this chapter demonstrates that ITL provides a powerful approach for the direct training of SNNs, laying important theoretical foundations for designing future SNN learning algorithms with enhanced generalization and robustness.

References

1. Achille A, Soatto S. Emergence of invariance and disentanglement in deep representations. J Mach Learn Res. 2018;19(1):1947–80.
2. Shamir O. Learning and generalization in neural networks: a statistical physics view [J]. Adv Neural Inf Process Syst. 2010;23:1–9.
3. Wainwright MJ. High-dimensional statistics: A non-asymptotic viewpoint [M]. Cambridge University Press, 2019.
4. Tishby N, Pereira FC, Bialek W. The information bottleneck method [J]. arXiv preprint physics/0004057, 2000
5. Cover TM. Elements of information theory [M]. Wiley, 1999
6. Principe JC. Information theoretic learning: Renyi's entropy and kernel perspectives [M]. Springer Science & Business Media, 2010
7. Liu W, Pokharel PP, Principe JC. Correntropy: Properties and applications in non-Gaussian signal processing [J]. IEEE Trans Signal Process. 2007;55(11):5286–98.
8. Zhang L, Liu Y, Zhang D. Robust principal component analysis via mixture maximum correntropy criterion [J]. IEEE Trans Cybern. 2016;46(12):2906–18.
9. Zheng Y, Chen B, Wang S, et al. Mixture correntropy-based kernel extreme learning machines [J]. IEEE Trans Neural Netw Learn Syst. 2020;33(2):811–25.
10. Natarajan BK. On minimizing the maximum error entropy criterion [J]. IEEE Trans Signal Process. 1995;43(2):392–5.
11. Zhang L, Liu Y, Liu X. Minimum error entropy criterion for robust beamforming [J]. Signal Process. 2011;91(12):2888–96.
12. Wen J, Wu Y, Zhang B. A minimum error entropy approach for fault detection in networked control systems [J]. Automatica. 2016;68:81–9.
13. Li Y, Chen B, Yoshimura N, et al. Restricted minimum error entropy criterion for robust classification [J]. IEEE Trans Neural Netw Learn Syst. 2021;33(11):6599–612.
14. Liu Y, Zhang L, Liu X. Restricted mixture maximum correntropy criterion for robust regression [J]. IEEE Trans Neural Netw Learn Syst. 2014;25(11):2083–93.
15. Zhang L, Liu Y, Liu X. Restricted mixture maximum correntropy criterion for robust principal component analysis [J]. IEEE Signal Process Lett. 2014;21(8):956–9.
16. Chen B, Xing L, Xu B, et al. Insights into the robustness of minimum error entropy estimation [J]. IEEE Trans Neural Netw Learn Syst. 2016;29(3):731–7.
17. Chen B, Zhu P, Principe JC. Survival information potential: a new criterion for adaptive system training [J]. IEEE Trans Signal Process. 2011;60(3):1184–94.
18. Chechik G, Sharma V, Bengio S. The information bottleneck revisited [J]. In Advances in neural information processing systems, 2005: 233–240.
19. Alemi AA, Fischer I, Dillon JV, et al. Deep variational information bottleneck [J]. arXiv preprint arXiv:1612.00410, 2016
20. Yang S, Gao T, Wang J, et al. Efficient spike-driven learning with dendritic event-based processing [J]. Front Neurosci. 2021;15: 601109.
21. Jiang R, Zhang J, Yan R, et al. Few-shot learning in spiking neural networks by multi-timescale optimization [J]. Neural Comput. 2021;33(9):2439–72.
22. Santoro A, Bartunov S, Botvinick M, et al. Meta-learning with memory-augmented neural networks[C]. International conference on machine learning. PMLR, 2016: 1842–1850
23. Koch G, Zemel R, Salakhutdinov R. Siamese neural networks for one-shot image recognition[C]. ICML deep learning workshop. 2015, 2(1)

24. Merolla PA, Arthur JV, Alvarez-Icaza R, et al. A million spiking-neuron integrated circuit with a scalable communication network and interface [J]. Science. 2014;345(6197):668–73.
25. Qiao N, Mostafa H, Corradi F, et al. A reconfigurable on-line learning spiking neuromorphic processor comprising 256 neurons and 128K synapses [J]. Front Neurosci. 2015;9:141.
26. Esser S, Merolla P, Arthur J, et al. Convolutional networks for fast, energy-efficient neuromorphic computing. arXiv 2016 [J]. arXiv preprint arXiv:1603.08270
27. Rodrigues CF, Riley G, Luján M. SyNERGY: An energy measurement and prediction framework for Convolutional Neural Networks on Jetson TX1[C]. Proceedings of the International Conference on Parallel and Distributed Processing Techniques and Applications (PDPTA). The Steering Committee of The World Congress in Computer Science, Computer Engineering and Applied Computing (WorldComp), 2018: 375–382
28. Roy K, Jaiswal A, Panda P. Towards spike-based machine intelligence with neuromorphic computing [J]. Nature. 2019;575(7784):607–17.
29. Panda P, Roy K. Learning to generate sequences with combination of hebbian and non-hebbian plasticity in recurrent spiking neural networks [J]. Front Neurosci. 2017;11:693.
30. Soures N, Kudithipudi D. Deep liquid state machines with neural plasticity for video activity recognition [J]. Front Neurosci. 2019;13:686.
31. Wijesinghe P, Srinivasan G, Panda P, et al. Analysis of liquid ensembles for enhancing the performance and accuracy of liquid state machines [J]. Front Neurosci. 2019;13:504.
32. Wang J, Hafidh B, Dong H, et al. Sitting posture recognition using a spiking neural network [J]. IEEE Sens J. 2020;21(2):1779–86.
33. Luo S, Guan H, Li X, et al. Improving liquid state machine in temporal pattern classification[C]. 2018 15th International Conference on Control, Automation, Robotics and Vision (ICARCV). IEEE, 2018: 88–91
34. Al Zoubi O, Awad M, Kasabov NK. Anytime multipurpose emotion recognition from EEG data using a Liquid State Machine based framework [J]. Artif Intell Med. 2018;86:1–8.
35. Ding J, Yu Z, Tian Y, et al. Optimal ann-snn conversion for fast and accurate inference in deep spiking neural networks [J]. arXiv preprint arXiv:2105.11654, 2021
36. Zheng H, Wu Y, Deng L, et al. Going deeper with directly-trained larger spiking neural networks[C]. Proceedings of the AAAI Conference on Artificial Intelligence. 2021, 35(12): 11062–11070
37. Kim Y, Panda P. Revisiting batch normalization for training low-latency deep spiking neural networks from scratch [J]. Front Neurosci. 2021: 1638
38. Kim Y, Chough J, Panda P. Beyond classification: Directly training spiking neural networks for semantic segmentation [J]. Neuromorphic Comput Eng. 2022;2(4): 044015.
39. Kim Y, Panda P. Visual explanations from spiking neural networks using inter-spike intervals [J]. Sci Rep. 2021;11(1):19037.
40. Venkatesha Y, Kim Y, Tassiulas L, et al. Federated learning with spiking neural networks [J]. IEEE Trans Signal Process. 2021;69:6183–94.
41. Guo S, Lin T. An Efficient non-Backpropagation Method for Training Spiking Neural Networks, in 2021 IEEE 33rd International Conference on Tools with Artificial Intelligence (ICTAI), pp. 192–199, 2021
42. Bellec G, Kappel D, Maass W, et al. Deep rewiring: Training very sparse deep networks[J]. arXiv preprint arXiv:1711.05136, 2017.
43. Kingma DP, Ba J. Adam: a method for stochastic optimization [J]. arXiv preprint arXiv:1412.6980, 2014
44. Schulman J, Wolski F, Dhariwal P, et al. Proximal policy optimization algorithms [J]. arXiv preprint arXiv:1707.06347, 2017

45. Wolff MJ, Jochim J, Akyürek EG, et al. Dynamic hidden states underlying working-memory-guided behaviour [J]. Nat Neurosci. 2017;20(6):864–71.
46. Tschannen M, Djolonga J, Rubenstein P K, et al. On mutual information maximization for representation learning [J]. arXiv preprint arXiv:1907.13625, 2019
47. Yang S, Gao T, Wang J, et al. SAM: a unified self-adaptive multicompartmental spiking neuron model for learning with working memory [J]. Front Neurosci. 2022, 16
48. Simonyan K, Zisserman A. Two-stream convolutional networks for action recognition in videos [J]. Advances in neural information processing systems, 2014, 27.
49. Parisi GI, Kemker R, Part JL, et al. Continual lifelong learning with neural networks: a review [J]. Neural Netw. 2019;113:54–71.
50. Serra J, Suris D, Miron M, et al. Overcoming catastrophic forgetting with hard attention to the task[C]. International Conference on Machine Learning. 2018: 4548–4557PMLR, 2018: 4548–4557
51. Yang S, Tan J, Chen B. Robust spike-based continual meta-learning improved by restricted minimum error entropy criterion [J]. Entropy. 2022;24(4):455.
52. Yang S, Linares-Barranco B, Chen B. Heterogeneous ensemble-based spike-driven few-shot online learning [J]. Front Neurosci. 2022, 16.
53. Wei D-S, Mei Y-A, Bagal A, et al. Compartmentalized and binary behavior of terminal dendrites in hippocampal pyramidal neurons [J]. Science. 2001;293(5538):2272–5.
54. Guo S, Lin T. An efficient non-backpropagation method for training spiking neural networks[C]. 2021 IEEE 33rd International Conference on Tools with Artificial Intelligence (ICTAI). IEEE, 2021: 192–199
55. Horowitz M. Computing's energy problem (and what we can do about it) [C]. 2014 IEEE International Solid-State Circuits Conference Digest of Technical Papers (ISSCC). IEEE, 2014: 10–14.
56. Alemi AA, Fischer I, Dillon JV, et al. Deep variational information bottleneck [J]. ar**v preprint ar**v:1612.00410, 2016
57. Sun Q, Li J, Peng H, et al. Graph structure learning with variational information bottleneck[C]. Proceedings of the AAAI Conference on Artificial Intelligence. 2022, 36(4): 4165–4174
58. Kim Y, Venkatesha Y, Panda P. Privatesnn: Fully privacypreserving spiking neural networks [J]. arXiv preprint arXiv:2104.03414, 2021
59. Yang S, Chen B. SNIB: improving spike-based machine learning using nonlinear information bottleneck [J]. IEEE Trans Syst Man Cybern Syst
60. Yang S, Wang H, Chen B. SIBoLS: Robust and energy-efficient learning for spike-based machine intelligence in information bottleneck framework. IEEE Trans Cognit Dev Syst, 2023.
61. He R, Hu BG, Zheng WS, et al. Robust principal component analysis based on maximum correntropy criterion [J]. IEEE Trans Image Process. 2011;20(6):1485–94.
62. He R, Zheng WS, Hu BG. Maximum correntropy criterion for robust face recognition [J]. IEEE Trans Pattern Anal Mach Intell. 2010;33(8):1561–76.
63. Li R, Liu W, Principe JC. A unifying criterion for instantaneous blind source separation based on correntropy [J]. Signal Process. 2007;87(8):1872–81.
64. Singh A, Principe JC. A loss function for classification based on a robust similarity metric[C]. The 2010 International Joint Conference on Neural Networks (IJCNN). IEEE, 2010: 1–6
65. Seth S, Príncipe JC. Compressed signal reconstruction using the correntropy induced metric[C]. 2008 IEEE International Conference on Acoustics, Speech and Signal Processing. IEEE, 2008: 3845–3848
66. Singh A, Principe JC. Using correntropy as a cost function in linear adaptive filters[C]. 2009 International Joint Conference on Neural Networks. IEEE, 2009: 2950–2955

67. Zhao S, Chen B, Principe JC. Kernel adaptive filtering with maximum correntropy criterion[C]. The 2011 International Joint Conference on Neural Networks. IEEE, 2011: 2012–2017

68. Wu Z, Shi J, Zhang X, et al. Kernel recursive maximum correntropy [J]. Signal Process. 2015;117:11–6.

69. Shi L, Lin Y. Convex combination of adaptive filters under the maximum correntropy criterion in impulsive interferenc e [J]. IEEE Signal Process Lett. 2014;21(11):1385–8.

70. Yang S, Chen B. Effective surrogate gradient learning with high-order information bottleneck for spike-based machine intelligence [J]. IEEE Trans Neural Netw Learn Syst, 2023

71. Kheradpisheh SR, Masquelier T. Temporal backpropagation for spiking neural networks with one spike per neuron [J]. Int J Neural Syst. 2020;30(06):2050027.

72. Kheradpisheh SR, Mirsadeghi M, Masquelier T. BS4NN: binarized spiking neural networks with temporal coding and learning [J]. Neural Process Lett. 2022;54(2):1255–73.

73. Hao Y, Huang X, Dong M, et al. A biologically plausible supervised learning method for spiking neural networks using the symmetric STDP rule [J]. Neural Netw. 2020;121:387–95.

74. Zhang W, Li P. Temporal spike sequence learning via backpropagation for deep spiking neural networks [J]. Adv Neural Inf Process Syst. 2020;33:12022–33.

75. Ranjan JAK, Sigamani T, Barnabas J. A novel and efficient classifier using spiking neural network [J]. J Supercomput. 2020;76(9):6545–60.

76. Zhang M, Wang J, Wu J, et al. Rectified linear postsynaptic potential function for backpropagation in deep spiking neural networks [J]. IEEE Trans Neural Netw Learn Syst. 2021;33(5):1947–58.

77. Fang W, Yu Z, Chen Y, et al. Incorporating learnable membrane time constant to enhance learning of spiking neural networks[C]. Proceedings of the IEEE/CVF international conference on computer vision. 2021: 2661–2671

78. Rathi N, Roy K. Diet-snn: A low-latency spiking neural network with direct input encoding and leakage and threshold optimization [J]. IEEE Trans Neural Netw Learn Syst. 2021

79. Han B, Srinivasan G, Roy K. Rmp-snn: Residual membrane potential neuron for enabling deeper high-accuracy and low-latency spiking neural network[C]. Proceedings of the IEEE/CVF conference on computer vision and pattern recognition. 2020: 13558–13567

80. Sengupta A, Ye Y, Wang R, et al. Going deeper in spiking neural networks: VGG and residual architectures [J]. Front Neurosci. 2019;13:95.

81. Cao Y, Chen Y, Khosla D. Spiking deep convolutional neural networks for energy-efficient object recognition [J]. Int J Comput Vision. 2015;113:54–66.

82. Lee JH, Delbruck T, Pfeiffer M. Training deep spiking neural networks using backpropagation [J]. Front Neurosci. 2016;10:508.

83. Rathi N, Srinivasan G, Panda P, et al. Enabling deep spiking neural networks with hybrid conversion and spike timing dependent backpropagation [J]. arXiv preprint arXiv:2005.01807, 2020

84. Garg I, Chowdhury SS, Roy K. Dct-snn: Using dct to distribute spatial information over time for learning low-latency spiking neural networks [J]. arXiv preprint arXiv:2010.01795, 2020

85. Hunsberger E, Eliasmith C. Training spiking deep networks for neuromorphic hardware[J]. arXiv preprint arXiv:1611.05141, 2016

86. Kim Y, Panda P. Revisiting batch normalization for training low-latency deep spiking neural networks from scratch [J]. Front Neurosci. 2021, 15: 773954

87. Li Y, Guo Y, Zhang S, et al. Differentiable spike: Rethinking gradient-descent for training spiking neural networks [J]. Adv Neural Inf Process Syst. 2021;34:23426–39.

88. Deng S, Li Y, Zhang S, et al. Temporal efficient training of spiking neural network via gradient re-weighting [J]. arXiv preprint arXiv:2202.11946, 2022

89. Bu T, Fang W, Ding J, et al. Optimal ANN-SNN conversion for high-accuracy and ultra-low-latency spiking neural networks [J]. arXiv preprint arXiv:2303.04347, 2023

90. Fang W, Chen Y, Ding J, et al. SpikingJelly: An open-source machine learning infrastructure platform for spike-based intelligence [J]. Sci Adv. 2023, 9(40): eadi1480

91. Wu H, Zhang Y, Weng W, et al. Training spiking neural networks with accumulated spiking flow[C]. Proceedings of the AAAI conference on artificial intelligence. 2021, 35(12): 10320–10328

92. Wu J, Chua Y, Zhang M, et al. A tandem learning rule for effective training and rapid inference of deep spiking neural networks [J]. IEEE Trans Neural Netw Learn Syst, 2021.

93. Wu Y, Deng L, Li G, et al. Spatio-temporal backpropagation for training high-performance spiking neural networks [J]. Front Neurosci. 2018;12:331.

Scalable Architectures of Neuromorphic Computing

4

4.1 Scalability of Neuromorphic Architectures

To implement neuromorphic intelligence, the development of a novel computing architecture that overcomes the limitations of the von Neumann architecture and enhances the scalability of neuromorphic systems is imperative [1]. The von Neumann architecture, characterized by the separation of computation and storage, constitutes the primary bottleneck in traditional computing paradigms [2]. Firstly, this architecture typically converts high-dimensional information into one-dimensional data for processing, which hampers real-time processing capabilities for complex intelligence tasks [3]. Secondly, the division of information processing between memory and computing units that are physically segregated leads to substantial energy wastage and constrains system performance [4]. Thirdly, it fails to mimic the operational principles of the human brain, rendering it unsuitable for neuromorphic computing implementations [5]. Consequently, the design of a non-von Neumann computing architecture that draws inspiration from the human brain's mechanisms is pivotal, as underscored by the preceding analysis.

While the von Neumann architecture has undergone numerous optimizations, it has not fundamentally evolved. Fueled by Moore's Law, computers built upon the von Neumann architecture have experienced exponential performance growth, serving as foundational platforms for artificial intelligence applications [6]. However, around 2004–2005, it became widely accepted that Moore's Law would eventually reach its limits after 50 years, necessitating a reevaluation of computer architectures.

As depicted in Fig. 4.1a, the classic von Neumann architecture perpetuates the separation of computation and storage, a primary hindrance in traditional computing architectures [1–3]. Even though the Harvard architecture, shown in Fig. 4.1b, represents an improvement over the conventional model, it still cannot entirely alleviate the separation of computation and storage, despite being an enhanced version of the von Neumann

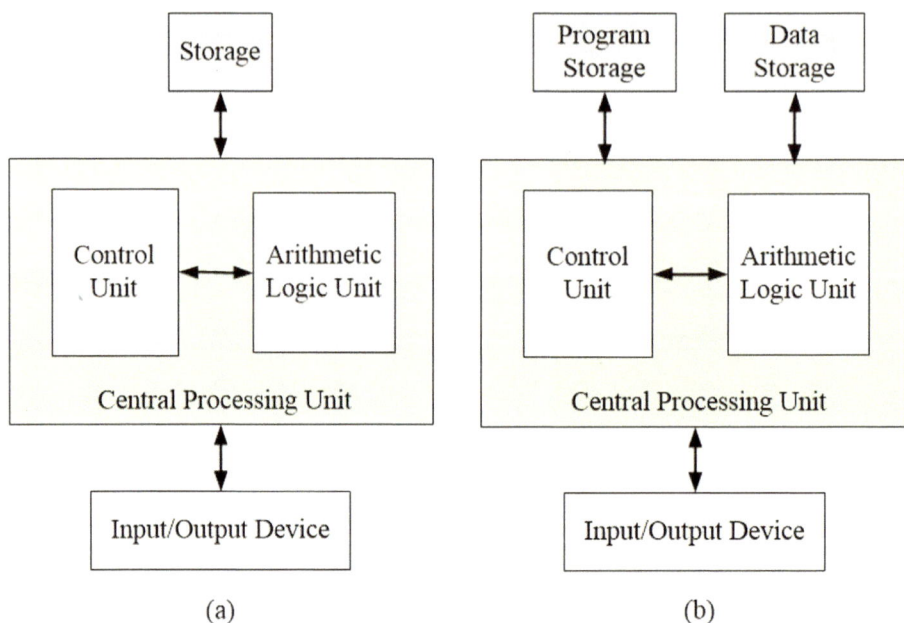

Fig. 4.1 Traditional computing architectures. **a** Von Neumann architecture. **b** Harvard architecture

architecture. To address this issue, researchers have dedicated themselves to proposing non-von Neumann computing architectures that align with the human brain's information processing mechanisms. Such architectures can surmount the bottlenecks associated with traditional designs and simulate the computational architecture of the human brain, laying the foundation for realizing neuromorphic computing.

In alignment with the information processing architecture of the human brain, neuromorphic engineering employs a low-power asynchronous event-driven approach to process information [7–18]. The human brain's neural network simultaneously manages information storage and complex nonlinear calculations. Thus, neuromorphic engineering leverages localized storage methods and integrates computing and storage units. The brain's cortex utilizes multiple regions with distinct perceptual and cognitive functions to maximize functionality while minimizing reliance on routing and storage units. Consequently, neuromorphic computing necessitates more complex systems, featuring distributed, heterogeneous multi-core, hierarchical architectures for computation.

NoC represents a novel approach to on-chip information communication and processing [19], finding extensive application in scalable neuromorphic architectures. The realm of NoC research encompasses multiple facets, including the investigation of topology architectures and routing algorithms. With the rapid advancement of semiconductor technology, more modules can be integrated onto a single chip, enhancing information processing capabilities. Consequently, system-on-chip (SoC) devices face a multitude of design and manufacturing challenges. Bus architectures, in particular, grapple with issues like transmission speed, power consumption, and throughput. NoC emerges as a novel on-chip data communication solution, aiming to integrate diverse electronic system components, such as central processing unit (CPU), I/O ports, and buses, into a unified entity capable of handling complete data flow communication functions [20]. Ogras et al. introduced the concept of the NoC design space and distilled key NoC research concerns into three categories: infrastructure, communication mechanisms, and mapping optimization [21]. Designers must select the fundamental NoC architecture based on specific application requirements, taking into account factors like application feature maps and constraints including time, implementation costs, and technological reserves. Subsequently, they need to address function-related challenges within the design space of these NoCs and embark on a journey of optimization to explore optimal implementation strategies [22].

NoC has established itself as the future communication platform mainstream, boasting superior transmission speed, data throughput, and computational velocity. NoC topologies should enable data exchange between any two resource nodes. However, in cases where a perfect topology map is unavailable, an outstanding routing algorithm assumes the role of determining the communication path between these resource nodes. Thus, NoC-based routing algorithms emerge as pivotal technologies within NoC research, as they not only select the path for packet transmission but also significantly influence factors like latency, network link utilization, and communication performance. Research on routing algorithms within the NoC domain boasts a long history, with the classical XY algorithm being the first to surface. The XY algorithm can be categorized as a form of dimensional order routing, known for its simplicity and deadlock-free implementation. However, its performance degrades considerably as network injection rates increase [23]. Subsequent research efforts aimed to enhance the XY algorithm, leading to designs like the turn model routing model, which balances simplicity and adaptability [24]. Nevertheless, network congestion concerns, coupled with the challenges of routing through curved paths, highlight the need to introduce virtual channels. This approach, while addressing congestion issues, escalates challenges including routing power consumption and logic complexity. Evaluating and optimizing routing algorithms entails considering factors such as deadlock occurrence rates, hardware implementation costs, and logic complexity [25]. Neuromorphic computing, based on NoC, necessitates routing algorithms capable of delivering flexible routing performance for multi-core computing architectures housing numerous distributed processing units. Consequently, devising a high-performance architecture suitable for the neuromorphic computing model and enhancing the stability, scalability, and flexibility of NoC routing algorithms become pivotal in boosting neuromorphic computing performance.

4.2   Mesh and Torus Architectures

Neuromorphic systems [26, 27] strive to emulate the functionality and behavior of bio-logical neurons and synapses to perform cognitive tasks more efficiently than traditional computing methods. One approach to crafting neuromorphic systems involves adopting a mesh architecture, which entails connecting numerous simple processing nodes in a grid-like pattern. This configuration enables parallel processing and distributed compu-tation, both crucial traits of the brain's information processing system. Within a mesh architecture, each processing node assumes responsibility for performing elementary com-putational tasks, such as multiplication or addition. These nodes connect to one another via a network of interconnections, which can be either fixed or dynamically reconfig-urable. This network of interconnections facilitates simultaneous communication and computation throughout the system, promoting efficient parallel processing. An exem-plary neuromorphic system employing a mesh architecture is the SpiNNaker system. SpiNNaker is a large-scale parallel computing platform designed to real-time simulate SNNs. It comprises numerous simple processing nodes connected via a two-dimensional mesh network with dynamically reconfigurable interconnections, allowing flexible routing of signals and computations across the system.

Another instance of a neuromorphic system employing a mesh architecture is the TrueNorth chip. TrueNorth, a custom-designed neuromorphic chip, houses over a million spiking neurons and more than 250 million synapses. It relies on a mesh architecture to enable efficient parallel processing and distributed computation. Mesh architecture stands as a crucial approach for crafting efficient and scalable neuromorphic systems, facilitating parallel processing and distributed computation, in alignment with the brain's information processing principles. The SpiNNaker and TrueNorth systems serve as prime examples of neuromorphic systems leveraging mesh architectures to attain high-performance and energy-efficient computing.

Torus architecture [28, 29] serves as a network topology beneficial for interconnect-ing a multitude of processing nodes efficiently within neuromorphic systems. Under a torus architecture, each processing node connects to its neighboring nodes in a two-dimensional grid, forming a toroidal shape. This topology fosters efficient com-munication and computation throughout the system, enabling signals to travel via the shortest paths between nodes. One application of torus architecture in neuromorphic sys-tems is the creation of large-scale SNNs. These networks, modeling biological brains, utilize spiking neurons and synapses to process information, demanding substantial com-putation and communication. Therefore, efficient interconnectivity is pivotal for their performance. The BrainScaleS system serves as an instance, using torus architecture to efficiently interconnect processing nodes. It relies on a torus architecture to facilitate effi-cient communication and computation, implementing interconnections for low-latency, energy-efficient communication.

4.3 IBFT Architecture

4.3.1 Large-Scale Scalable Architecture

As depicted in Fig. 4.2, a novel three-dimensional NoC architecture, employing an improved butterfly fat tree (IBFT) architecture at each layer, is presented, showcasing excellent scalability for large-scale neuromorphic computing [30]. This hierarchical architecture comprises two components: a horizontal layer and a vertical connection layer, illustrated in Fig. 4.2a. The horizontal IBFT layers can be extended limitlessly, with each horizontal IBFT layer being FPGA-based. To facilitate vertical connections, the architecture employs the high-speed Terasic connector (HSTC) found on the DE3 development board. For data buffering between layers, a First in First Out (FIFO) module is utilized. Data communication, both within layers and across layers, relies on the same packet type. The transmitted data stream comprises 26-bit packets containing 8-bit address event representation (AER) data, 3-bit chip address, 3-bit layer address, 6-bit node address, and 6-bit timestamp. The AER communication protocol determines the route address by consulting the synaptic routing table in on-chip memory, facilitating synaptic connections in a dynamically reconfigurable manner. Ultimately, it maps the presynaptic source address to the postsynaptic destination address using synaptic parameters.

In Fig. 4.2b, the planar architecture of the IBFT topology within a three-dimensional NoC architecture is illustrated. The number of multi-compartment neuron processors (MNPs) is determined by available hardware resources. For instance, when employing 64 MNPs, a three-level router within the IBFT architecture becomes necessary.

Figure 4.3 displays a router architecture comprised of three modules: a buffer module, a judgment module, and a crossbar switch module connecting the I/O ports. Within the buffer module, AER data is stored, and it decides the output direction and sequence for each packet. In contrast to the traditional router in the BFT topology, which features six I/O ports connected via a cross-switch module and a multiplexer, the IBFT topology reduces the number of I/O ports in the top-level router, resulting in hardware resource savings.

Figure 4.3a shows the router architecture in the IBFT architecture. It contains 6 input ports and 6 output ports. Each port has six virtual channels, each serving as a FIFO buffer. The selection of virtual channels is determined by the judgment module. The IBFT architecture router algorithm is a deterministic non-shortest path router algorithm. The routing path obtained from this algorithm is not guaranteed to be the shortest path. The node code of the router is compared to the node code of the target node. The pseudo code of IBFT routing algorithm is shown in Fig. 4.3b, where *NAD* represents the node address of the destination and the node of current router is encoded as $C\ (i, j)$. To reduce the routing latency, the local load pattern is chosen, which means that when the distance between the two nodes increases, the probability of information routing between the source and destination nodes decreases. As shown in Fig. 4.4, the first-level router provides six internal

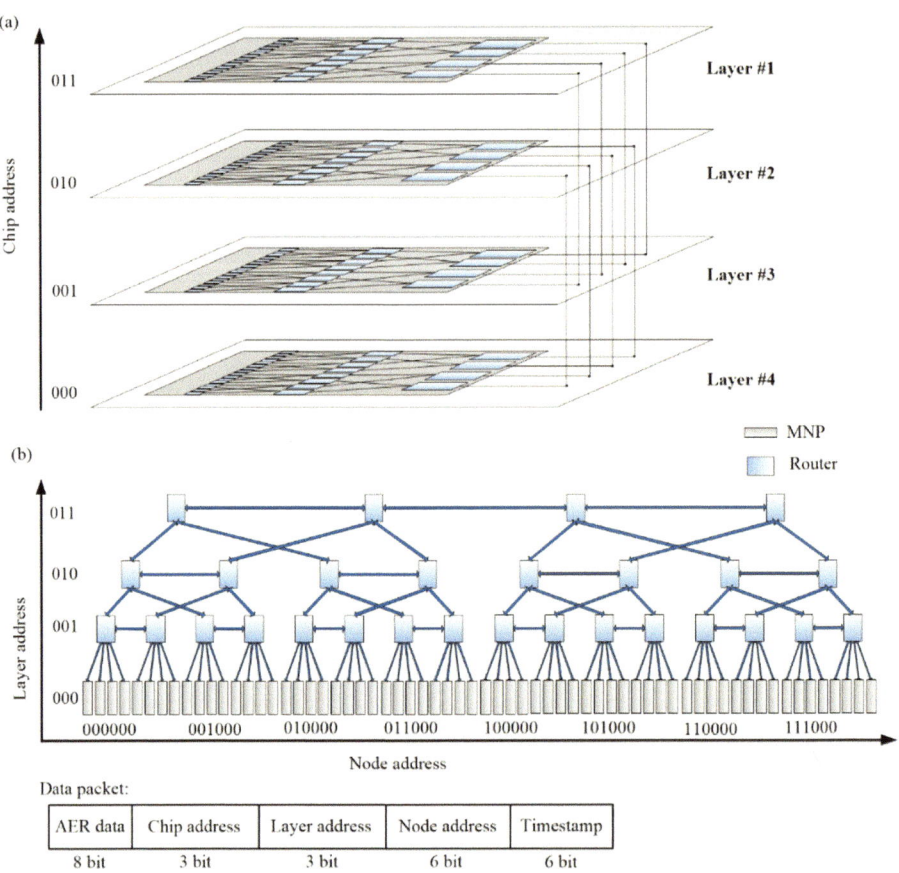

Fig. 4.2 Three-dimensional layered extensible architecture. **a** Connection architecture of extensible system. **b** NoC architecture [30]

input ports for interface connections to adjacent routers or MNPs, and each input port is equipped with a virtual channel with six FIFOs per channel. These virtual channels are controlled by virtual channel arbitrators, which use counters to sequentially select the appropriate FIFO in each virtual channel. During the process of routing, when the first-level router receives a spike event from MNP, the spike wrapper module initiates a packet consolidation process. The spike wrapper module is located behind the virtual channel and uses a configuration processor to process single or multiple spike events from MNP into valid 26-bit AER spike packets. The packet is divided into five fields: AER data, chip address, layer address, node address, and timestamp.

All information about the NoC architecture configuration is stored in the configuration processor, which can be found in each first-level router. These configuration processors can be reconfigured at any time, depending on the neural connection specifications

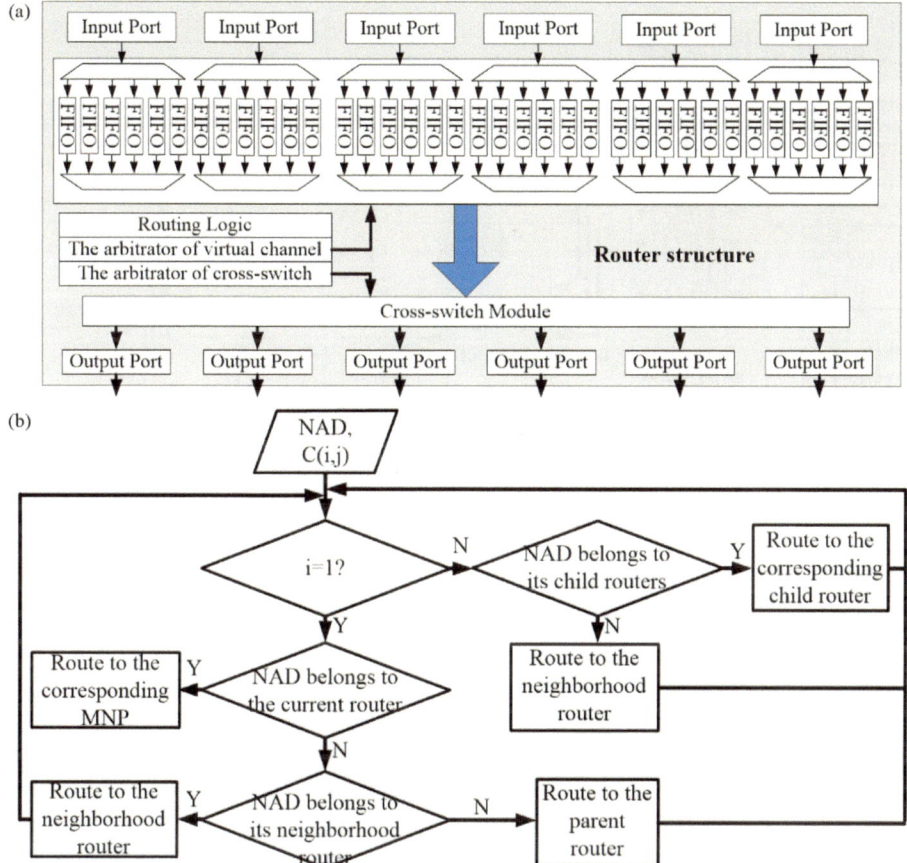

Fig. 4.3 Improved routing of BFT architecture. **a** Router architecture. **b** Pseudo code of Routing algorithm [30]

required for application execution, including the number of neurons per layer and the type of neuron connections. The configuration processor contains four types of registers, namely chip address registers, layer address registers, node address registers, and timestamp registers. The chip address register and layer address register hold 3 bits of value and are used to select the corresponding chip and layer on each chip, respectively. The 6-bit data is stored in the node address register to select the target node on each layer. Finally, the 6-bit timestamp data is stored in the timestamp register. When a valid ARE spike packet passes a spike event, the routing logic module processes the event according to the routing algorithm shown in Fig. 4.5. Another important component of a first-layer router is a cross-switch implemented using a multiplexer. It is controlled by the order of the routing logic modules.

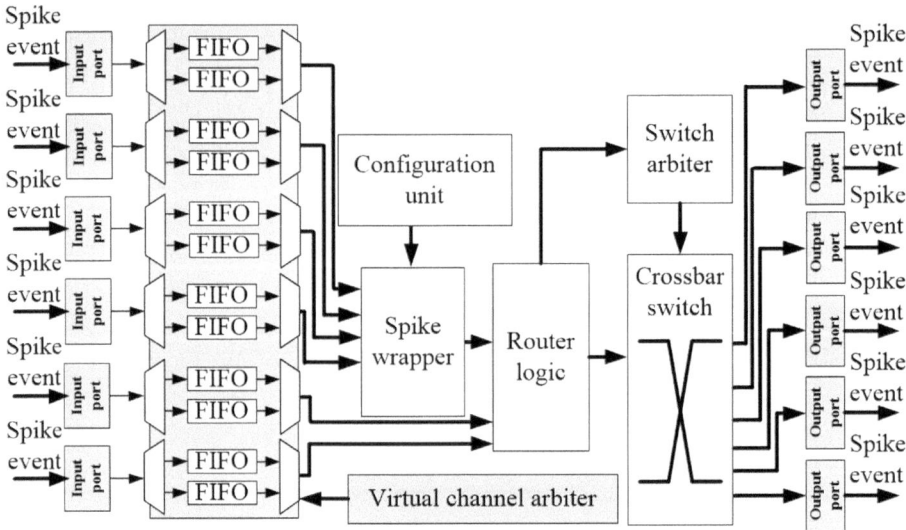

Fig. 4.4 Design and implementation of the first level [30]

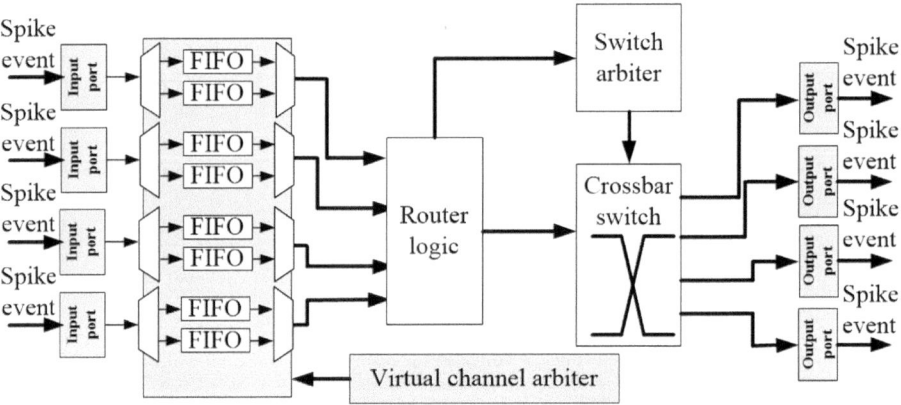

Fig. 4.5 Design and implementation of the second and third level [30]

   The detailed implementation of the second and third-level routers is shown in Fig. 4.5, and it consists of four internal input and output ports and interfaces with the other four routers. The design and implementation of these routers is similar to that of the first-level routers. In this routing architecture, the spike wrapper in the first-layer router is removed because there is no spike event input from MNP. The upstream and downstream data path includes a 4×4 crossover switch that sends NoC packets from the four input ports connected to the router to the output ports that communicate with other routers. If two spike events are judged to be sent to the same output port, one of them will be sent to the other output port to avoid data communication conflicts.

The detailed routing algorithm in the router is shown in Fig. 4.6. Using segmented encoding, it consists of a 3-bit chip address, a 3-bit layer address, and a 6-bit node address. To avoid deadlocks, when the algorithm operates the routing logic, it checks the node addresses of all connected routers and compares the current router with the node addresses of the connected routers. The destination node is denoted as *dest* and the node address of the current router is $S(l, r)$. The addresses of the source and target nodes are represented by *src* and *dest*, respectively. The number of IP cores is N. The first step is to enter the addresses of the source and destination nodes $(src, dest)$ and the addresses of current router (l, r). If $l = 1$, you need to calculate the values of the current router S, which is called *lowRange* and *highRange*. It is necessary to determine whether the *dest* value is in the range of [*lowRange, highRange*]. If the value of *dest* falls into this range, the packet will be transmitted to the destination node. If it is not in this range, the packet is transmitted to any parent node of the current node and this routing logic is repeated again. After that, it needs to calculate the range of cross-connected router nodes for the current router, which is denoted as [*lowRange, highRange*]. If the value of *dest* belongs to a cross-connected router, the packet is transmitted to it through the cross-connect link and the routing logic repeats again. If the value of *dest* does not belong to a cross-connected router, the relationship between *src* and *dest* should be judged. If $dest - src < (N - 1)/2$ and $dest - src > -(N - 1)/2$, the packet is sent to the right neighbor node, otherwise it is routed to the left neighbor node, and the routing logic is repeated again.

The address code consists of three parts, as shown in Table 4.1. Firstly, the information is processed using segmented encoding, using 3-bit *y_dest* address encoding and m-bit *x_dest* address encoding, respectively. The parameter m is determined based on the total number of MNPs. Secondly, the layer address is set to 000 and the node address to be assigned from minimum to maximum. Finally, the router encoding uses the binary form

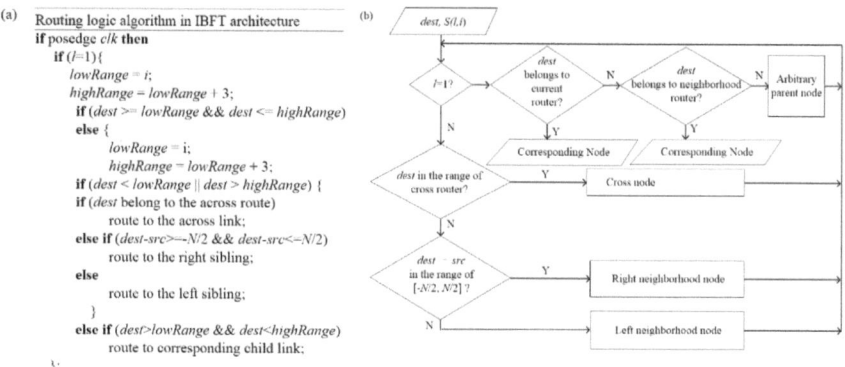

Fig. 4.6 Routing algorithm. **a** Pseudo code of routing. **b** Logic flowchart of routing [30]

Table 4.1 Address of child nodes, neighbor nodes, and parent nodes of the current router

Current router	Child nodes	Neighbor nodes	Parent nodes	MNP
001,0000	–	001,0100	010,0000	0000; 0001,0010; 0011
001,0100	–	001,0000	010,0000	0100; 0101,0110; 0111
001,1000	–	001,1100	010,1000	1000; 1001,1010; 1011
001,1100	–	001,1000	010,1000	1100; 1101,1110; 1111
010,0000	001,0000; 001,0100	010,1000	–	–
010,1000	001,1000; 001,1100	010,0000	–	–

of its layer encoding, and the node address of the router uses the minimum value of the router's child node address.

4.3.2 Performance Evaluation

To assess the performance enhancements achieved by the architecture, a comparison is made between the IBFT architecture and the flat architecture proposed in earlier studies. The flat architecture necessitates the provision of a unique global address for each source node, a requirement that significantly limits the system's capacity to accommodate a large number of neurons. In test experiments, Poisson spike generators are employed to dispatch 40,000 neuronal events at a rate of 40 M SynOps. As illustrated in Fig. 4.7a, the delay is measured based on 64 MNP data for a flat architecture, revealing a marked reduction in maximum delay when employing the IBFT architecture. The graph in Fig. 4.7c displays the measured average latency against synaptic event rates for both the previous flat and IBFT architectures. Notably, the average latency of the IBFT architecture amounts to approximately 12.5% of the average latency observed in the flat architecture. Furthermore, the IBFT architecture offers scalability, a feature absent in the flat architecture. This scalability facilitates the expansion of neuromorphic computing, enabling simulations of synaptic connections and network scales akin to the human central neural system.

From the perspective of system architecture, the coding methods of routers and MNPs, and the router mapping algorithms will affect the latency, hardware resource overhead, and power consumption of neuromorphic computing systems. Therefore, the NoC design of neuromorphic computing needs to increase resource utilization, and the cost function of buffer utilization is defined as follows

$$CF_{BUR} = \frac{\sum_{i=1}^{t_{tot}} \left[\sum_{j=1}^{N_R} N_{BBS}^{i,j} + \sum_{j=1}^{N_{IP}} N_{BSN}^{i,j} \right]}{N_{VC} \cdot N_{PPB} \cdot (N_{PL} + N_{IP}) \cdot t_{tot}} \tag{4.1}$$

where N_{VC} and N_{PPB} describe the number of virtual channels and packets per buffer, and N_{PL} represents the total number of physical links. N_{IP} is the total number of IP cores

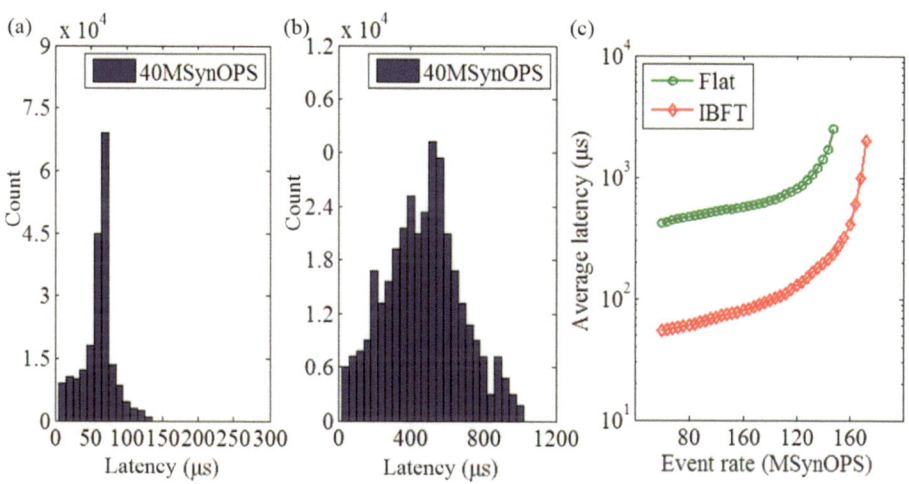

Fig. 4.7 Comparison of the IFBT architecture with the flat architecture. **a** Latency of IBFT architecture. **b** Latency of flat architecture. **c** Comparison of the IBFT and flat architecture [30]

in NoC architecture. t_{tot} describes the total time from the first packet to the last packet, and N_R represents the total number of routers in NoC architecture. $N_{BBS}^{i,j}$ and $N_{BSN}^{i,j}$ represent the buffers used in the nodes and routers, respectively. Assuming that count represents the buffer utilization, Fig. 4.8a shows the simulated hardware performance for a comparison of BFT and IBFT in aspect of buffer utilization. When the event rate is low, the utilization is same for both architectures. As the event rate increases, the utilization of the IBFT architecture becomes higher than that of the BFT architecture. It shows that IBFT architectures have higher utilization than traditional BFT architectures.

To demonstrate the computational efficiency and scalability, the previous studies are compared with software simulations. The cost function for computational efficiency can be described as

$$Q = t_{bio}/t_{exp} \qquad (4.2)$$

where t_{exp} and t_{bio} are the experimental calculation time and the biological activity time of the simulated system, respectively. Software emulation based on the von Neumann architecture runs on the Intel Core2 2.4 GHz CPU. As shown in Fig. 4.8b, comparing to other hardware versions, it has higher computing speed and scalability. When the number of neurons increased to 10^6, the computational efficiency Q reach 204.35 times higher than the previous study in reference and 5351 times higher than in software-based simulations. Despite the significant increase in network size, the neuromorphic computational efficiency will not be affected and maintain a high level of real-time computing.

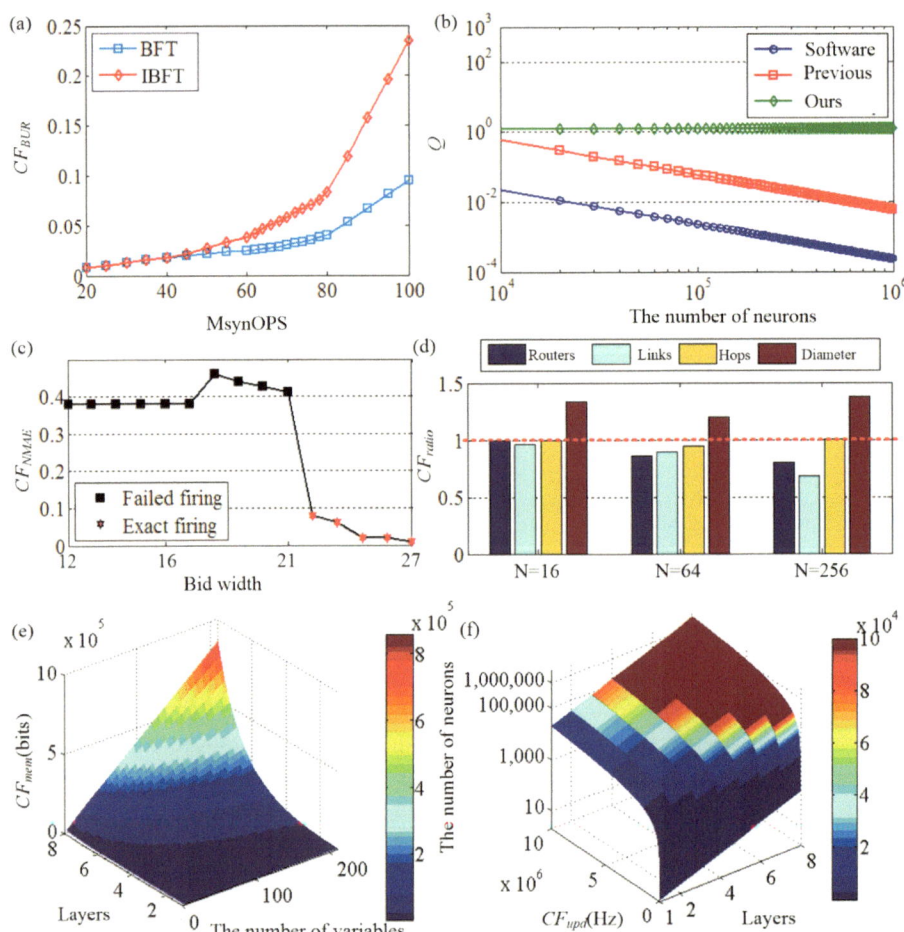

Fig. 4.8 Simulated hardware performance. **a** Comparison of BFT of the utilization with the improved architecture. **b** Comparison of the computational efficiency of other methods. **c** Accuracy analysis. **d** Evaluation of NoC performance. **e** Memory consumption analysis. **f** Analysis of the total number of neurons achieved [30]

The hardware implementation consists of integers and decimal places. In order to present the accuracy analysis, the integer part is determined according to the range of data values, and the fractional part is determined by the calculation accuracy. For a real value x, the number of digits of the integer portion a is defined as

$$a = \log_2(ceil|x| + 1) \tag{4.3}$$

where $ceil|x|$ is the upward value of a number x. The signed bit adds an extra 1 bit to avoid overflow of all variables and parameters in the model.

The fractional part of the fixed-point representation can be considered the accuracy of the system. Figure 4.8b shows the accuracy analysis between software simulation and the results calculated by hardware. Precision analysis guarantees the accurate reproduction of neural dynamics. Accurate analysis results are obtained through simulations of different fixed-point representations, and the calculation error is calculated using the index characterization CF_{NMAE}. When the bit length exceeds 21 bits, accurate discharge dynamics can be obtained. According to the precision analysis of Fig. 4.8c, in order to acquire accurate biological behavior, the fractional part of the fixed-point calculation contains 24 bits. Although higher bit widths enable more optimal computational accuracy, further increases in bit widths will result in higher hardware resource costs, which will limit network size and increase power consumption on a single chip.

As shown in Fig. 4.8d, in order to further compare the NoC performance of the IBFT architecture and the BFT architecture, CF_{ratio} is used to represent the ratio of IBFT to BFT two different NoC characteristics, including routers, links, average hops and the number of NoC network diameter. The reduction of the number of these NoC features means that the complexity of the NoC architecture is reduced. Compared to the original BFT architecture, the performance of IBFT is excellent taking into account the number of routers, the number of links, and the average hop, and the NoC network diameter remains the same. On-chip storage is a key factor in the large-scale implementation of SNN. To assess the scale of the network on the chip, the cost of memory in the IBFT architecture is further studied. Figure 4.8e shows the relationship between the number of layers of IBFT CF_{mem} and the number of compartment neuron models. The maximum number of IBFT architectural layers is regarded as 8 because hardware constraints on on-chip resources restrict the implementation of more IBFT layers. In fact, there is a conflict between the cost of memory resources, the number in the neuron model, and the number of NoC layers of IFBT. Figure 4.8f shows the relationship between achievable network scale, the speed of network update CF_{upd} and the number of computational architecture layers, and it is obviously that it is possible to implement a large-scale biologically significant network using appropriate digital neuromorphic computing. The number of layers determines the parallelism of neuromorphic computing system, the update speed CF_{upd}. When the local communication load becomes high, communication bottleneck occurs. Therefore, it is necessary to effectively deal with bottlenecks caused by higher local communication loads.

In Fig. 4.9a, a comparison of the NoC network throughput is shown between the original BFT architecture and the IBFT architecture under local load conditions. When the network's response rate remains below 100 million synaptic operations per second (MSynOps), both architectures exhibit consistent throughput rates. However, as the response rate exceeds 120 MSynOps, the BFT architecture experiences buffer congestion, leading to packet loss and limiting the communication throughput. In contrast, the IBFT architecture features shorter inter-node distances compared to BFT, reducing data transmission time,

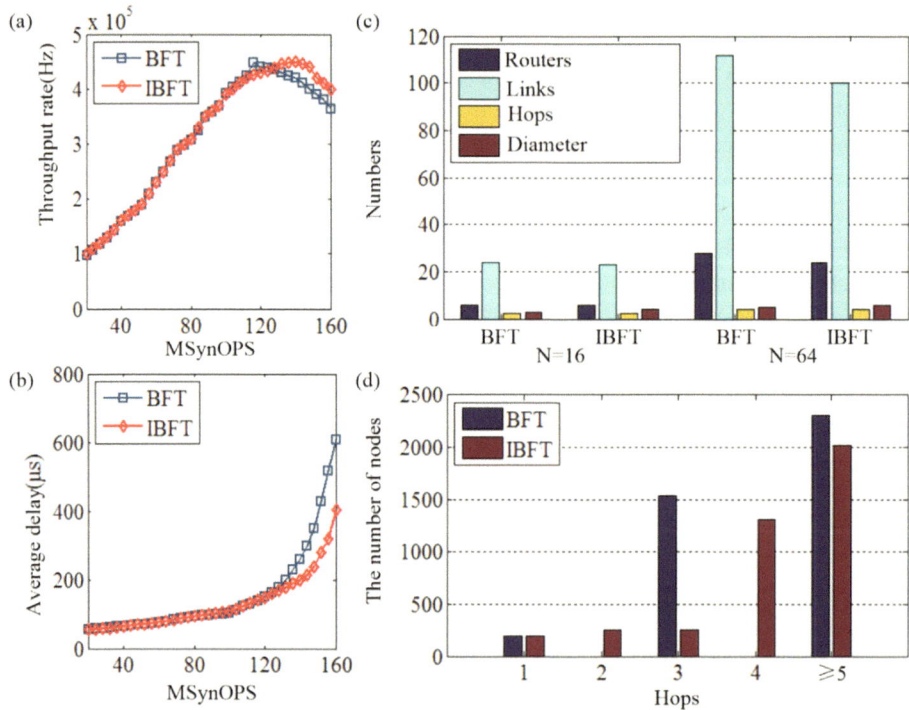

Fig. 4.9 Comparison of BFT with the improved architecture. **a** Throughput rate. **b** Average delay. **c** Resource cost. **d** Distance between nodes [30]

supporting higher data transfer speeds, and mitigating packet loss to some extent. Consequently, when dealing with greater event emissivity, the IBFT architecture outperforms the traditional BFT architecture in terms of throughput. Moving to Fig. 4.9b, a comparison of communication delays is examined. The latency of both BFT and IBFT architectures remains consistent with each other. However, as throughput rates increase and surpass 120 MSynOps, communication latency escalates due to increased data waiting in the cache. In the IBFT architecture, communication nodes are more closely situated, implying the utilization of fewer routers in data transmission. In Fig. 4.9c, for a NoC network with N = 16 nodes, the IBFT architecture exhibits a reduction of 14.3% in the number of routers and a 10.7% reduction in the number of links compared to BFT. In networks with N = 64, these reductions increase to 20% and 32.1%, respectively. Consequently, the IBFT architecture effectively lowers NoC complexity and hardware resource costs, with the average hop count remaining consistent for both NoC architectures. Furthermore, the network diameter of IBFT exceeds that of BFT, resulting in increased latency during global communication scenarios. As indicated in Fig. 4.9d, the communication distance in BFT can be categorized into three parts: 1 hop, 3 hops, and 5 hops or more, with corresponding node pair counts of 192, 1536, and 2304, respectively. In contrast, the IBFT architecture offers more hop categories: 1 hop (192 pairs of points), 2 hops (256 node pairs), 3 hops (256 node pairs), 4 hops (1312 node pairs), and more than 5 hops (2016 node pairs). This additional granularity in hop categories proves beneficial when dealing with high local loads, addressing data communication challenges in NoC designs. Consequently, IBFTs effectively enhance architectural performance in neuromorphic computing under such conditions.

4.4   3D Mesh Architecture

4.4.1   Large-Scale Scalable Architecture

The digital topology of the NoC plays a pivotal role in determining the system's simulation performance, making it one of the most critical aspects of system design and implementation. Building upon the concepts presented in [31], the three-dimensional NoC architecture is strategically divided into two integral components: the vertical beam column and the horizontal network layer. This architectural synergy harnesses the benefits of three-dimensional integrated circuits while mitigating the potential deadlock scenarios that can affect edge nodes and vertices. In this setup, each layer of horizontal neuron networks is meticulously implemented using FPGA technology, while the vertical beams and columns are skillfully realized through the utilization of high-speed Terasic connectors (HSTC). These connectors are seamlessly integrated into DE3 development boards, facilitating high-speed interconnects and configurable I/O standards. By effectively reducing the number of hops between nodes, this architectural configuration not only enhances data throughput but also significantly reduces latency. To optimize performance further, the hardware architecture is meticulously designed to adhere to a $6 \times 6 \times 6$ framework, ensuring robust and efficient operation.

As illustrated in Fig. 4.10a, the inter-chip data communication employs four HSTC connectors, with each HSTC connector capable of transmitting 120 bits of data per clock cycle. Notably, the data transfer width from the core processor to other layers is set at 24 bits, as depicted in Fig. 4.10b. Furthermore, a multicast mesh architecture is thoughtfully implemented at each layer of the network, encompassing a total of 36 core processors. This architectural design enables bidirectional communication through dedicated data communication interfaces at each network layer.

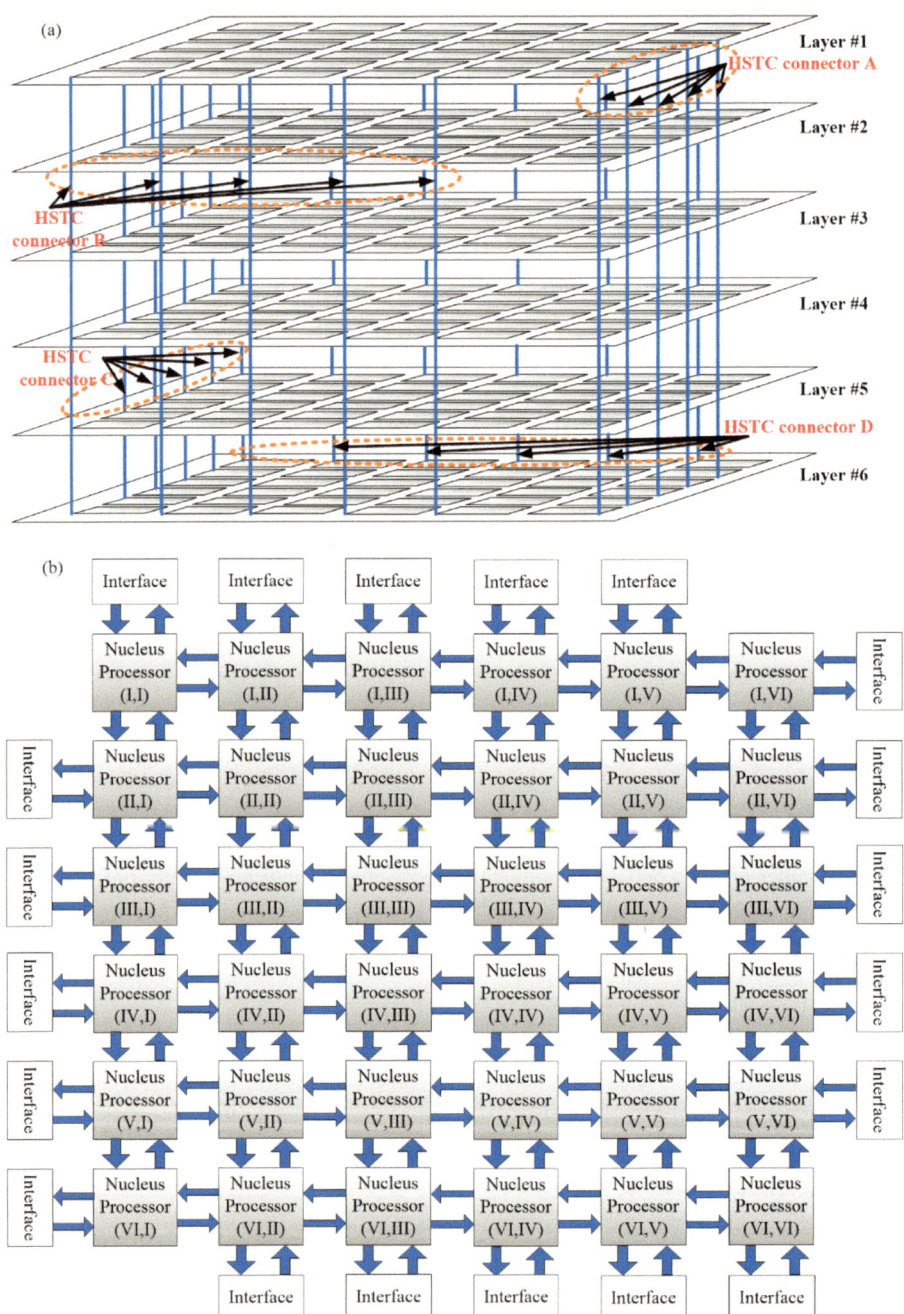

Fig. 4.10 Three-dimension hardware architecture of neuromorphic computing. **a** Top-level design for multi-core networks. **b** Architecture of multicast mesh type in AER [31]

4.4.2   Performance Evaluation

Path cost is the time it takes for a packet to travel from source node to the destination node, which can be measured by clock cycles. In LaCSNN system, the transition and restore between neuron information and packet only takes one clock cycle. The path cost of the intermediate block can be described as

$$f_{PCI} = (f_{hop} - 2) \cdot (f_R + 2 \cdot f_{FIFO}) \tag{4.4}$$

where the cost function f_{hop} represents the number of nodes that a packet needs to pass through during transmission, including the source and destination nodes. The cost functions f_R and f_{FIFO} represent the average of all the time a packet spends in the router and the time the packet spends in the FIFO in the transmission path, respectively. The path cost for the source and target nodes can be defined as

$$f_{PCSD} = 2f_R + 4f_{FIFO} + f_{src} + f_{dst} \tag{4.5}$$

The cost functions f_{src} and f_{dst} represent the time when the synaptic information is converted to a packet and the time the packet is restored to synaptic information, respectively. In fact, packet translation requires a clock cycle to convert synaptic information into packets, so the value of f_{src} and f_{dst} should be 1. The path cost on the 3D multicast network NoC can be described as

$$f_{PC} = \begin{cases} f_{hop} \cdot (f_R + 2 \cdot f_{FIFO}) + f_{src} + f_{dst}, & f_{hop} \geq 2 \\ f_R + 2 \cdot f_{FIFO} + f_{src} + f_{dst}, & f_{hop} = 1 \end{cases} \tag{4.6}$$

The relationship between the cost function f_{PC}, f_{hop} and f_{FIFO} is shown in Fig. 4.11a. For 3D NoC, the maximum distance between any two nodes depends on the network width X_{max}, network length Y_{max}, and network height Z_{max}, which can be described as

$$\forall n > 2, \exists f_{hop}, 1 \leq f_{hop} \leq ((X_{max} + Y_{max} + Z_{max} - 3) - 1) \tag{4.7}$$

In Fig. 4.11b, the average delay time is presented as a function of the network scale, with measurements taken at a synaptic discharge rate of 100 Hz. It's noteworthy that all synaptic rates fall below 100 Hz, remaining within the specified network performance parameters. As the network size expands, both the median and the latency range increase. However, it's essential to highlight that all packet transmission times remain within 100 clock cycles, which is the time required to update the network's state. Importantly, there is no packet loss observed at any input measurement frequency, underscoring the network's precise capability for calculating dynamics.

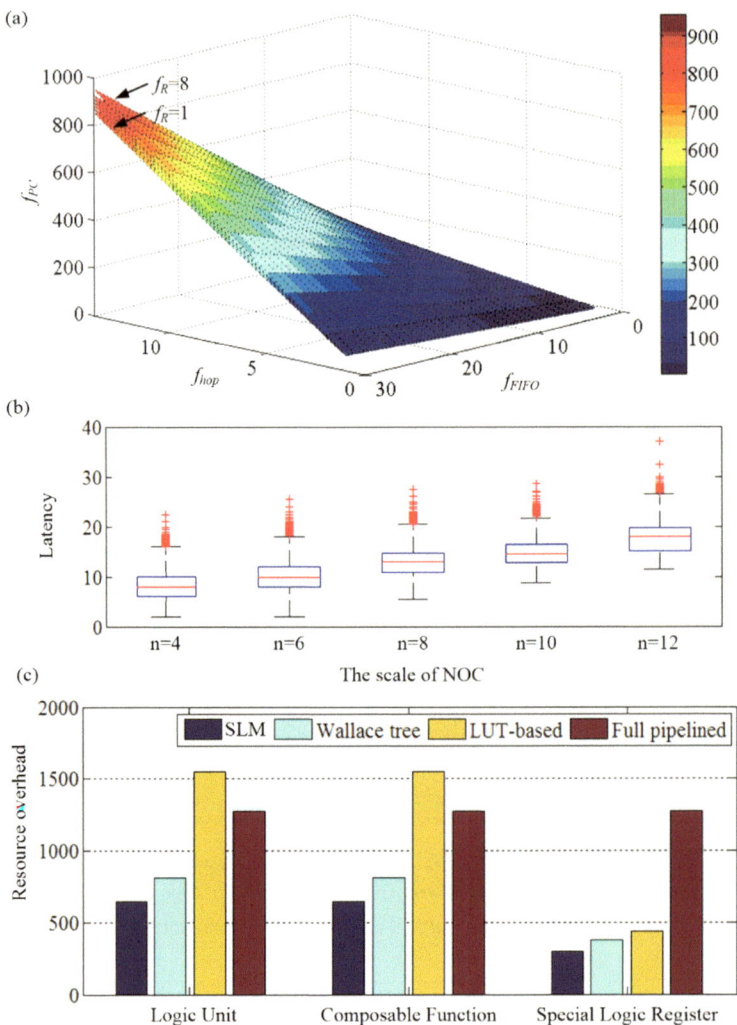

Fig. 4.11 Analysis of performance for multi-nuclei neuromorphic computing. **a** Relationship between path cost, hop count and FIFO delivery time. **b** Latency under different conditions of NoC size. **c** Comparison with SLM multiplier and other multiplier designs [31]

The use of multipliers can significantly consume computing resources and power, and it's advisable to minimize their usage during implementation. SLM multipliers, as introduced in Chapter IV, are employed. For the sake of achieving high-precision results, the SLM multipliers are designed with a 10-bit integer part and a 10-bit fractional part. As depicted in Fig. 4.11c, a comparison is made between the SLM multiplier and other conventional multipliers when applied to the Altera Cyclone-IV EP4CE115 FPGA chip. The compared multipliers include the Wallace tree multiplier, the lookup table (LUT)-based multiplier, and the fully pipelined multiplier. Notably, the comparison reveals that the SLM multiplier exhibits lower power consumption.

To enable large-scale computing, a multi-nuclei neuromorphic computing method and the LaCSNN neuromorphic computing system, based on a multi-nuclei collaboration mechanism, are introduced. To showcase the computational prowess and precision of the LaCSNN system, its performance is compared with three alternative methods: CPU, GPU, and multi-core bus. The cost function for computational efficiency is defined as

$$Q = t_{exp}/t_{bio} \tag{4.8}$$

where t_{exp} and t_{bio} represent the experimental calculation time and physiological action time of the simulation system, respectively. Intel Core2 2.4 GHz CPU and NVIDIA GTX 280 GPU are used. The multi-core bus mode is implemented by FPGA. The computational efficiency is assessed by previous research results [32–36]. LaCSNN with scalable 3D NoC architectures exhibit greater scalability compared to other hardware platforms. As shown in Fig. 4.12, as the size of the network increases, the computational efficiency of the LaCSNN system is still higher than that of the other three systems.

Fig. 4.12 Computing efficiency with different numbers of neurons for different computing platforms [31]

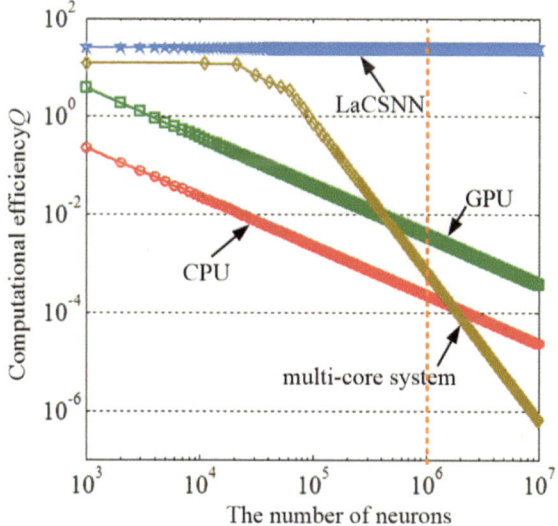

4.5   Fault-Tolerant Mesh Architecture

4.5.1   Large-Scale Scalable Architecture

The fault-tolerant strategy can be seamlessly integrated with the 3D mesh NoC topology, as illustrated in Fig. 4.13. In line with previous research [37], the mapping strategy for SNNs onto NoC-based neuromorphic systems assumes a pivotal role in neuromorphic applications. It exerts a significant influence on both the overall performance and power consumption. Within this framework, the neuron units are implemented within an 8×8 NoC architecture, which efficiently employs a mesh-based multicasting AER strategy. Notably, all three network layers are interconnected, facilitating comprehensive connectivity. The spike event packet transferred within this network encompasses 22 bits of data, encompassing a 3-bit layer ID, 1-bit AER data, 3-bit Y_dest address, 3-bit X_dest address, and 12-bit Timestamp.

The intricate digital architecture of the neuron unit is meticulously depicted in Fig. 4.14a. It comprises four core components: a neuron processor, a fault-tolerant router, a synapse unit, and a configuration unit. Notably, each router is endowed with six ports,

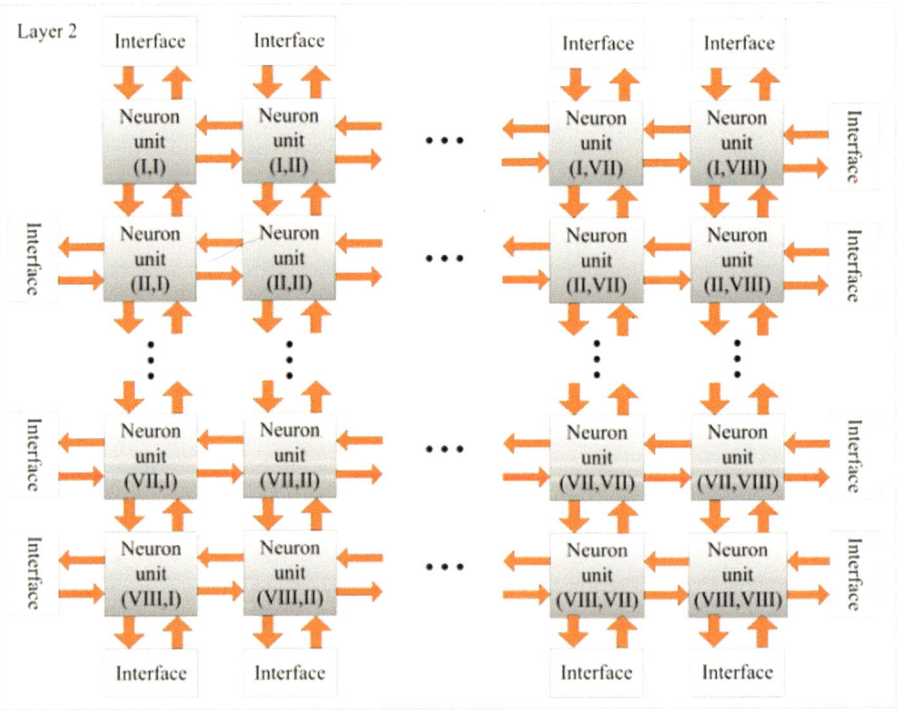

Fig. 4.13 3D NoC architecture [37]

encompassing up, down, north, west, east, and south, enabling seamless routing of AER packets to neighboring neuron units. This architecture is characterized by three key attributes: (1) Computation Utilizing Event-Based Synaptic Weighting: It excels in processing computations through event-based mechanisms that involve synaptic weighting. (2) Physical Synapse Implementation: It integrates the physical instantiation of synapses, enhancing its neural processing capabilities. (3) Fault-Tolerant Neuromorphic Routing Capability: The architecture is endowed with fault-tolerant routing capacities, rendering it particularly well-suited for neuromorphic SNN computations geared toward intricate cognitive behaviors, such as context-dependent learning.

The router holds paramount importance within the neuromorphic architecture, as it assumes pivotal roles in realizing the targeted fault tolerance. The fault-tolerant multicast 3D router architecture, implemented within the neuron units of the system, is meticulously illustrated in Fig. 4.14b. In the initial stage, spike events are received from the four neighboring nucleus processors and their packets are temporarily stored within the input buffer, awaiting processing. The spike wrapper unit, operating based on neuronal connectivity, is tasked with transforming individual spike events into valid AER spike packets. It draws upon information received from the configuration processor to accomplish this task. The configuration processor itself is equipped with four distinct types of registers, including the chip address register, layer address register, node address register, and timestamp register. Incoming spike events, along with their corresponding deliver-at times, are stored in the on-chip memory once their respective timestamps are reached. Subsequently, the source address of the packet is extracted and computed to determine the appropriate output port. The switch arbiter, utilizing the least recently served priority, is employed to expedite computations, maintain cost-effectiveness, and ensure equitable allocation of resources. Ultimately, the packet is dispatched to its intended output port via the crossbar, with the crossbar switch being under the control of the switch arbiter and implemented through multiplexers.

Euler method is used for the discretization of the neuron model since it can save hardware resources compared to Runge–Kutta method. Based on Euler method, the original equations can be transformed into the following equations:

$$
\begin{cases}
V(n+1) = \Delta t \cdot \left(\frac{1}{C}(G_l(V_i - V_{rest}) + I_k + \eta) \right) + V(n) \\
W_{ij}^{exc}(n+1) = \Delta t \cdot \left(\begin{array}{l} \left(w_{max} - W_{ij}^{exc} \right) \cdot A_+ \exp(-\Delta/\tau_+) \\ -\left(w_{min} - W_{ij}^{exc} \right) \cdot A_- \exp(+\Delta//\tau_-) \end{array} \right) + W_{ij}^{exc}(n)
\end{cases} \quad (4.9)
$$

The digital architecture of the neuron model in the information sensory, hippocampal and motor layers is shown in Fig. 4.15a. Multipliers are extravagant hardware resources in digital design, and are usually avoided as much as possible to gain energy and hardware cost benefits. Thus, in the digital architecture, shift logic multipliers (SLMs) are used to replace multipliers to realize multiplication operations. Figure 4.15b shows the detailed digital implementation of the SLM block, which is used to replace the multipliers.

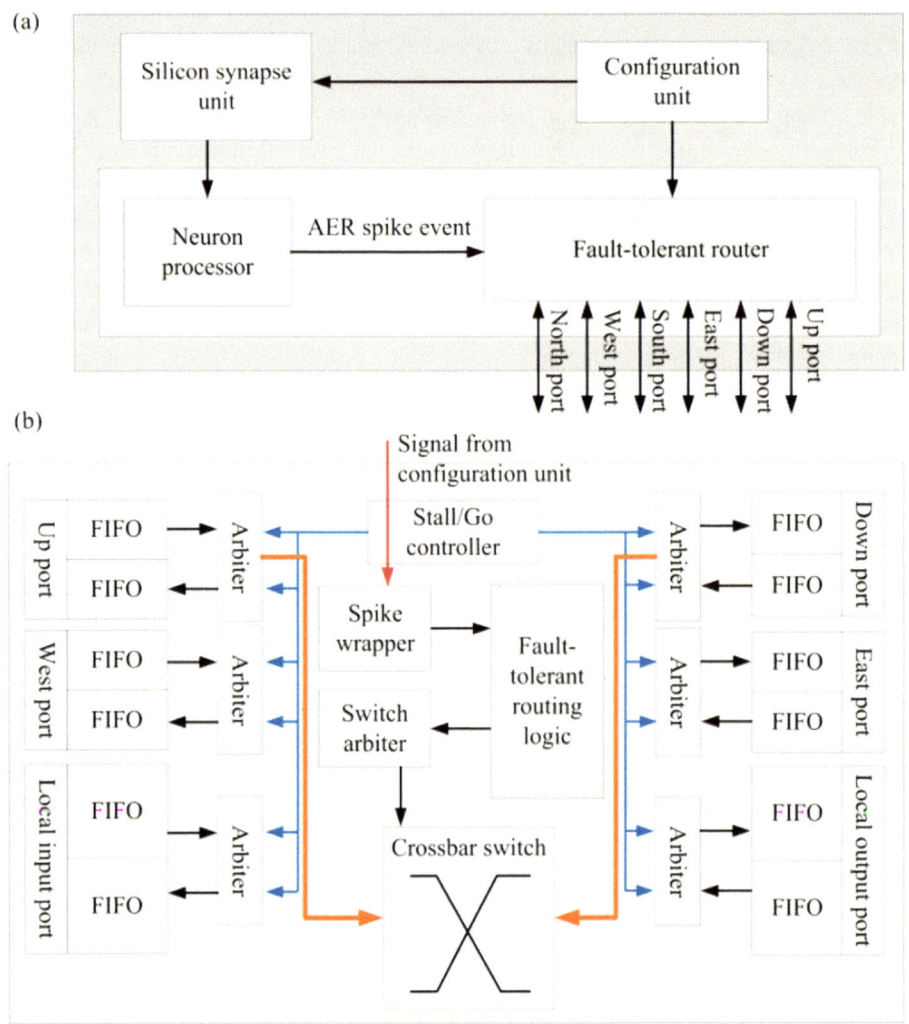

Fig. 4.14 Detailed digital neuromorphic architecture of the neuron unit and routers. **a** Digital neuromorphic architecture of the neuron unit. **b** Digital neuromorphic architecture of the fault-tolerant router [37]

The learning capability is based on STDP learning algorithm. The detailed digital implementation of this learning algorithm is shown in Fig. 4.16. As demonstrated in Fig. 4.16a, the pre-synaptic and post-synaptic spike timings "Timepre" and "Timepost" are first calculated, which are then used in the digital implementation of the STDP learning rule shown in Fig. 4.16b. Here, LUTs are used to calculate the exponential part in the STDP algorithm, and barrel shifters are used to replace the required multipliers.

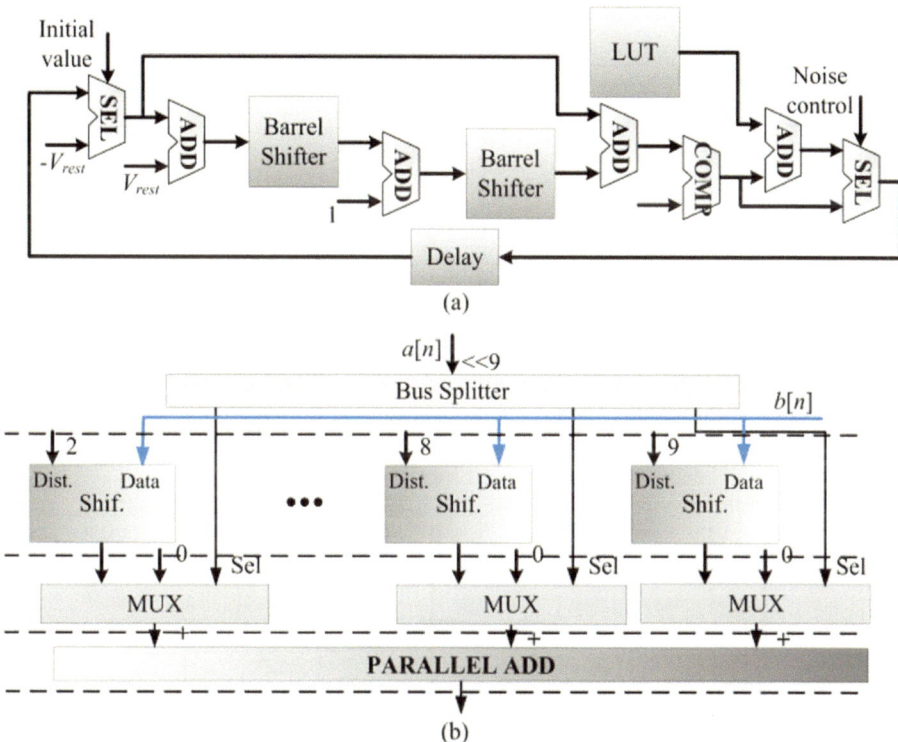

Fig. 4.15 Detailed digital architecture of the neuron processor and the SLM block. **a** Detailed digital implementation of the neuron unit. **b** Detailed digital implementation of the SLM block [37]

These components help to significantly cut down the hardware resource cost and power consumption.

In the MFTN algorithm, the XY routing strategy is employed, involving routing neural information first along the X-direction and then along the Y-direction. Consequently, the algorithm is divided into two distinct parts: routing along the X-direction and routing along the Y-direction. A prior study by Chen and Chiu introduced a fundamental fault-tolerant solution for NoC design [38]. However, Chen's algorithm comes with substantial computational requirements and occupies a large area, resulting in low node utilization. Moreover, it bears a heavy communication load and exhibits high latency. In contrast, the MFTN algorithm builds upon and expands the single-fault bypass method outlined in Zhang's algorithm [39] to address multiple-fault scenarios. In this section, it delves into the MFTN algorithm along both the X and Y directions, comparing it to Zhang's algorithm to highlight the improvements made based on Zhang's algorithm.

When considering multiple-fault-tolerant schemes, two scenarios emerge. In the first scenario, the source and destination nodes are situated on opposing sides of the fault

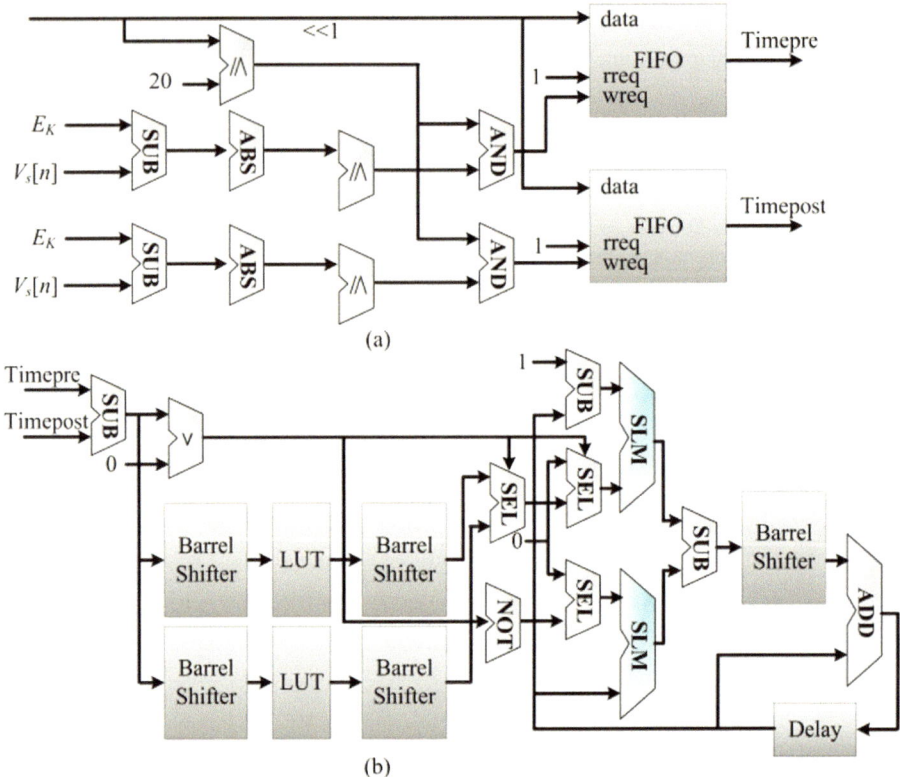

Fig. 4.16 Digital architecture of the synapse. **a** Detailed digital implementation for the computation of "Timepre" and "Timepost". **b** Digital implementation of the STDP weight updating module [37]

region, with the source node residing within one of the rows within the fault region. For example, refer to source 1 (S1) and Destination 3 (D3), or S2 and D5 in Fig. 4.17. In the second case, the source node lies within the rows of the fault region, while the destination node is located within the columns of the fault region. An instance of this case can be observed in the positions of S1 and D2 in Fig. 4.17. To determine whether the routing process traverses through the fault region, the coordinate information of the four SW, NW, SE, and NE nodes is transmitted to all the normal nodes within the corresponding column. This information is then stored in on-chip memory. In Fig. 4.17, solid arrows denote routes based on Zhang's algorithm [39], while dotted line arrows represent routes based on the MFTN routing algorithm. Compared to Zhang's algorithm, the MFTN algorithm optimizes the bypass strategy, resulting in reduced routing distances. For example, when routing from S1 to D4, Zhang's algorithm takes a detour via SE, increasing the distance. In contrast, the MFTN algorithm offers a direct route to the D4 node. The pseudo code of the MFTN algorithm is depicted in Table 4.2. In this pseudo code, C and D denote the current and destination nodes, respectively. In the XY algorithm, neural information is initially routed along the X-direction and then along the Y-direction. This approach is primarily suited for scenarios where the source and destination nodes are located on the east and west sides of the fault region, respectively. When specific conditions are met, the bypass loop is separated from the original pathway, resulting in reduced load on the bypass loop and decreased routing distances.

The scenario involving bypass along the Y-direction in the case of multiple fault nodes arises when the destination and source nodes are situated on the south and north sides of the fault region, respectively. Furthermore, the destination node is positioned within the columns of the fault region. According to Zhang's algorithm, more changes in routing

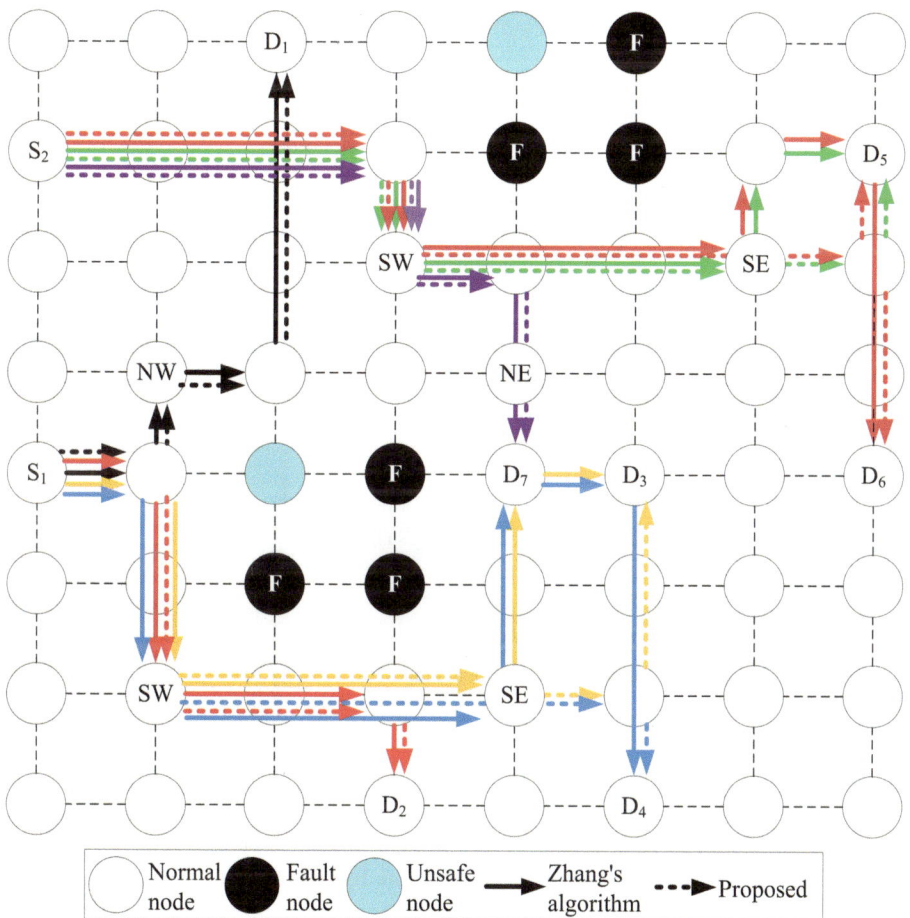

Fig. 4.17 Bypass route of the spike event along X direction [37]

Table 4.2 Pseudo code for single-fault neuromorphic tolerant algorithm along X direction (From [37])

---

if (C is destination D)

    {Processing the spike event;} // Reaching the destination node

else if (fault information memory of C is empty){

    Continue route according to XY algorithm;

    }// If the fault information is NULL, routing based on XY algorithm

else if (fault information memory of C will not affect routing from S to D){

    Continue route according to XY algorithm;

    }

    // If the fault region is not located on the pathway of XY routing, continue routing based on XY

algorithm

else if ($X\_SW <= X\_D <= X\_SE$ && $Y\_SW <= Y\_S <= Y\_NW$){

    Route according to Zhang algorithm;

    }

else if ($Y\_SW <= Y\_S <= Y\_NW$ && ($X\_D < X\_SW < X\_SE <= X\_S$ ||

$X\_S <= X\_SW < X\_SE < X\_D$)){

    // When satisfying the first case, begin optimization mode

    if (!($X\_D < X\_SW < X\_SE <= X\_C$ || $X\_C <= X\_SW < X\_SE < X\_D$)){

        Route to X_D according to Zhang algorithm;

        // Routing to the column of D based on Zhang's algorithm

        Continue route according to XY algorithm;

        // Routing spike event based on XY algorithm

    } // C and D do not locate on the both sides of the fault region

    else{

        do{

            Route to next node according to Zhang algorithm;

            // Routing to the next node based on Zhang's algorithm

        }while(!($X\_C == X\_SW < X\_S$ || $X\_C == X\_SE > X\_S$));

        Continue route according to XY algorithm;

        // Continue routing based on XY algorithm

    } // C and D locate on the both sides of the fault region

}

---

direction are necessitated along the X-direction. This leads to an increase in bypass distance, with the load primarily concentrated on the left side of the bypass loop. When the current node (C) is located on the north side of the bypass, the spike event initially follows the XY algorithm to route to the nearest corner and is then bypassed as per Zhang's algorithm. As illustrated in Fig. 4.18, if the current node is situated on the south side of the bypass loop, the spike event is first directed towards the SW node due to the restriction on turning at the NE corner. Subsequently, it follows Zhang's algorithm for the bypass. The pseudo code for the MFTN algorithm along the Y-direction is presented in Table 4.3.

The necessary and sufficient condition for any routing algorithm to be deadlock-free is the absence of loops in its corresponding component dependency graph (CDG).

In networks without any fault nodes, the CDG of the mesh-based network remains loop-free since the turn from the Y to X direction is prohibited. When the fault region is situated within the network, the MFTN algorithm introduces turns from the Y to X direction at the northwest, southwest, and southeast corners of the bypass loop, while eliminating the turn from the X to Y direction at the northeast corner. The removal of this turn ensures the absence of any cycles, rendering the algorithm deadlock-free.

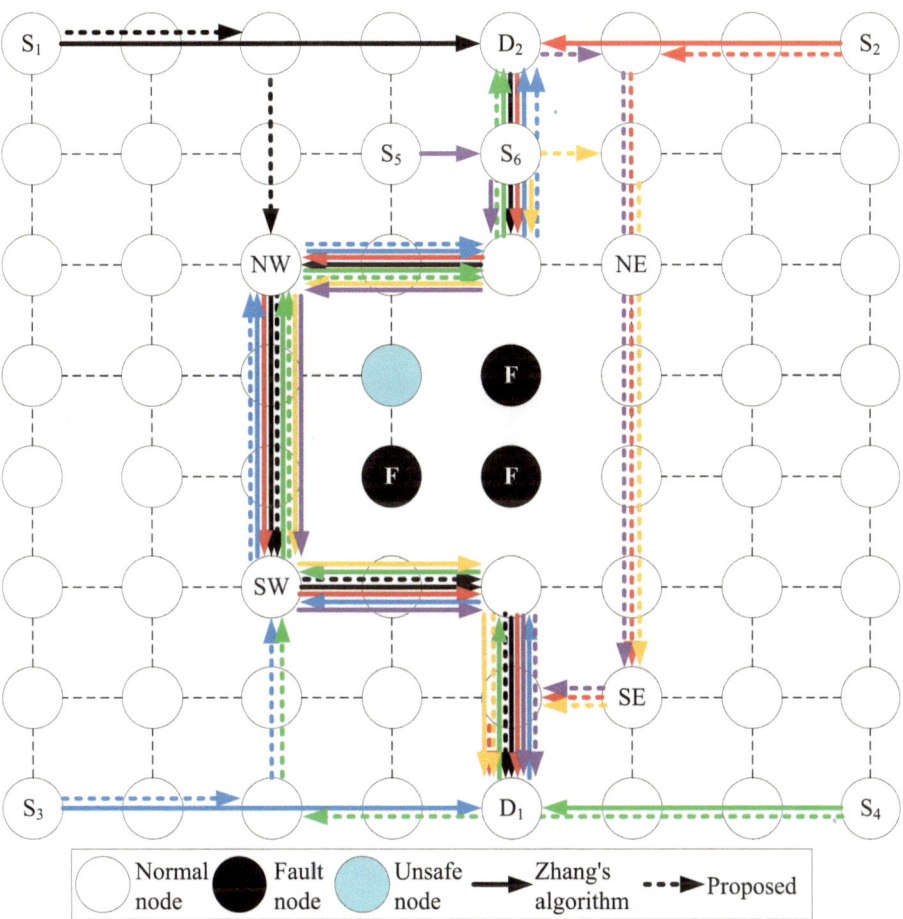

Fig. 4.18 Bypass route of the spike event along Y direction [37]

Table 4.3 The Pseudo code for single-fault neuromorphic tolerant algorithm along Y direction (From [37])

---

if (C is destination D)

    {Processing the spike event;} // Reaching the destination node

else if (fault information memory of C is empty){

    Continue route according to XY algorithm;

    }// If the fault information is NULL, routing based on XY algorithm

else if (fault information memory of C will not affect routing from S to D){

    Continue route according to XY algorithm;

    }

    // If the fault region is not located on the pathway of XY routing, routing spike event based on XY

algorithm

else if $(X\_SW \leq X\_D \leq X\_SE$ && $((Y\_S > Y\_NW$ && $Y\_D \leq Y\_SW)$ || $(Y\_S < Y\_SW$ &&
$Y\_D \geq Y\_NW)))$ {

    if (Y_C>Y_D){

        if (abs(X_C – X_NW)>abs(X_C – X_NE)){

            Route to NE according to XY algorithm;

        }

        else{

            Route to NW according to XY algorithm;

        }

    }

    else{

        Route to SW according to XY algorithm;

    }

    Route to X_D according to Zhang algorithm;

    Continue route according to XY algorithm;

}

In cases where the fault region is positioned on the edge of the mesh network, the algorithm includes turns from the Y to X direction at the vertex of the bypass loop, allowing spike events to circumvent the fault region. As the bypass loop is not a cyclic link, it does not lead to deadlock. Consequently, the MFTN algorithm is entirely deadlock-free. Furthermore, it is independent of the area and location of the fault region, as well as the scale of the NoC.

4.5.2 Performance Evaluation

The performance analysis of the fault-tolerant algorithm for multiple fault nodes is depicted in Fig. 4.19, with the northeast node on the bypass loop serving as the reference node. Figure 4.19a illustrates that the MFTN algorithm outperforms Chen's algorithm [38] across various locations of faulty regions in an (8, 8) mesh network of neurons. Both algorithms exhibit a consistent trend in Fig. 4.19, where latency is highest when the faulty region is at the network center but lowest when it's positioned at the network vertex. This is attributed to the greater number of affected nodes in the network center, which exerts the most significant impact on data transmission latency. The MFTN algorithm consistently outperforms Chen's algorithm regardless of the fault region's location. For instance, when communication latency is 80 μs, the MFTN algorithm achieves improvements of 8.3%, 1.9%, and 0.9% in the maximum event rate for fault regions at the network center (4, 4), network edge (5, 7), and network vertex (7, 2), respectively. Consequently, the MFTN algorithm performs best when the fault region is located at the network center due to the largest number of source and destination nodes satisfying optimization conditions, resulting in the most noticeable improvement. Conversely, when fault nodes are at the network vertex, the number of source and destination nodes meeting the optimization criteria is the lowest.

As the fault area expands, the algorithm provides a more significant latency advantage compared to Chen's algorithm. Figure 4.19b compares the latency of the algorithm with Chen's algorithm for fault areas of 2×2 (reference node at (4, 4)), 2×3 (reference node at (4, 4)), and 2×4 (reference node at (6, 5)). In Fig. 4.19c, the fault area size and reference node location remain the same as in part (b), but the fault regions vary along the vertical direction. The results demonstrate that, regardless of whether the fault area enlarges horizontally or vertically, the algorithm consistently exhibits lower network latency than Chen's algorithm.

When the network delay is 100 μs, compared to Chen's algorithm, the saturation injection rate increases by 5.2, 11.7, and 16.1% as the fault area expands horizontally (Fig. 4.19b). Conversely, when the fault area enlarges vertically, the saturation injection rate increases by 5.2%, 5.5%, and 6.1%, respectively (see Fig. 4.19c). This is because when data encounters the fault region along the Y direction, the algorithm cannot reduce the distance from the source node to the destination node or the transmission length on the bypass. To improve load balance on the bypass loop, it reallocates a portion of the original data intended for bypass along the left half of the fault region to the right half. Additionally, as the fault area expands horizontally, the algorithm requires fewer nodes to bypass the fault area along the X direction but more nodes to bypass it along the Y direction. Therefore, the algorithm's network performance is more pronounced when the fault area expands horizontally.

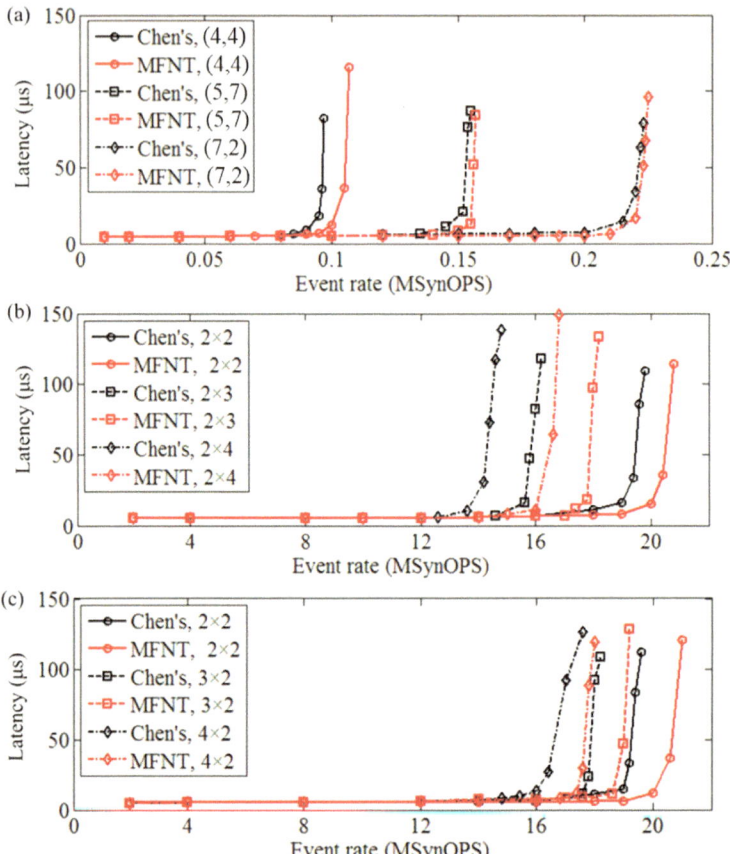

Fig. 4.19 Performance analysis of the fault tolerant algorithms for multiple fault nodes. **a** Communication latency with different locations of faulty regions in the (8, 8) Mesh network of neurons. **b** Communication latency with different areas of fault regions changing along horizontal direction. **c** Communication latency with different areas of fault regions changing along vertical direction [37]

4.6 Summary

This chapter presents scalable neuromorphic architectures designed for high-performance neuromorphic intelligence. It introduces a novel IBFT architecture that, in conjunction with an innovative routing algorithm, enhances throughput to enable large-scale neuromorphic computing. This approach ensures robust performance, biological plausibility, programmability, and computational efficiency. Additionally, the chapter introduces a novel scalable 3D mesh architecture tailored for neuromorphic computing. A fault-tolerant mesh architecture is also introduced in detail, which can solve the hardware fault problems by designing proper routing strategies.

References

1. Zhu J, Zhang T, Yang Y, et al. A comprehensive review on emerging artificial neuromorphic devices. Appl Phys Rev. 2020;7(1): 011312.
2. Wright CD, Hosseini P, iosdado JAV. Beyond von-Neumann computing with nanoscale phase-change memory devices. Adv Funct Mater. 2013;23(18):2248–2254.
3. Huang A. Architectural considerations involved in the design of an optical digital computer. Proc IEEE. 1984;72(7):780–6.
4. Ichioka Y, Tanida J. Optical parallel logic gates using a shadow-casting system for optical digital computing. Proc IEEE. 1984;72(7):787–801.
5. Sejnowski TJ. The computer and the brain revisited. Ann Hist Comput. 1989;11(3):197–201.
6. Dreslinski RG, Wieckowski M, Blaauw D, et al. Near-threshold computing: reclaiming Moore's law through energy efficient integrated circuits. Proc IEEE. 2010;98(2):253–66.
7. Ahmad AS, Sumari ADW. Cognitive artificial intelligence: brain-inspired intelligent computation in artificial intelligence. In: 2017 computing conference. IEEE;2017. p. 135–141.
8. Betzel RF, Bassett DS. Multi-scale brain networks. Neuroimage. 2017;160:73–83.
9. De Salvo B. Brain-inspired technologies: towards chips that think? In: 2018 IEEE international solid-state circuits conference-(ISSCC). IEEE;2018. p. 12–18.
10. Goertzel B, Lian R, Arel I, et al. A world survey of artificial brain projects, Part II: biologically inspired cognitive architectures. Neurocomputing. 2010;74(1–3):30–49.
11. Yu T, Park J, Joshi S, et al. 65k-Neuron integrate-and-fire array transceiver with address-event reconfigurable synaptic routing. In: Biomedical circuits and systems conference, (BioCAS), 2012. IEEE;2012. p. 21–24.
12. Schemmel J, Briiderle D, Griibl A, et al. A wafer-scale neuromorphic hardware system for large-scale neural modeling. In: Proceedings of 2010 IEEE international symposium on circuits and systems. IEEE;2010. p. 1947–1950.
13. Akopyan F, Sawada J, Cassidy A, et al. Truenorth: design and tool flow of a 65 mw 1 million neuron programmable neurosynaptic chip. IEEE Trans Comput Aided Des Integr Circuits Syst. 2015;34(10):1537–57.
14. Kranenburg T, Van Leuken R. MB-LITE: a robust, light-weight soft-core implementation of the MicroBlaze architecture. In: Proceedings of the conference on design, automation and test in Europe. European design and automation association, 2010. p. 997–1000.
15. Davies M, Srinivasa N, Lin TH, et al. Loihi: a neuromorphic manycore processor with on-chip learning. IEEE Micro. 2018;38(1):82–99.
16. Merolla PA, Arthur JV, Alvarez-Icaza R, et al. A million spiking-neuron integrated circuit with a scalable communication network and interface. Science. 2014;345(6197):668–73.
17. Park J, Kwon MW, Kim H, et al. Compact neuromorphic system with four-Terminal Si-based synaptic devices for spiking neural networks. IEEE Trans Electron Devices. 2017;64(5):2438–44.
18. Benjamin BV, Gao P, McQuinn E, et al. Neurogrid: a mixed-analog-digital multichip system for large-scale neural simulations. Proc IEEE. 2014;102(5):699–716.
19. Chaix F, Avresky D, Zergainoh NE, et al. A fault-tolerant deadlock-free adaptive routing for on chip interconnects. In: 2011 design, automation & test in Europe. IEEE;2011. p. 1–4.
20. Bjerregaard T, Mahadevan S. A survey of research and practices of network-on-chip. ACM Comput Surv (CSUR). 2006;38(1):1.
21. Sanchez D, Michelogiannakis G, Kozyrakis C. An analysis of on-chip interconnection networks for large-scale chip multiprocessors. ACM Trans Arch Code Optim (TACO). 2010;7(1):1–28.

22. Ogras UY, Bogdan P, Marculescu R. An analytical approach for network-on-chip performance analysis. IEEE Trans Comput Aided Des Integr Circuits Syst. 2010;29(12):2001–13.
23. Mehmood F, Baloch NK, Hussain F, et al. An efficient and cost effective application mapping for network-on-chip using Andean condor algorithm. J Netw Comput Appl. 2022;200: 103319.
24. Glass CJ, Ni LM. The turn model for adaptive routing. ACM SIGARCH Comput Arch News. 1992;20(2):278–87.
25. Park D, Nicopoulos C, Kim J, et al. Exploring fault-tolerant network-on-chip architectures. In: International conference on dependable systems and networks (DSN'06). IEEE;2006. p. 93–104.
26. Furber SB, Galluppi F, Temple S, Plana LA. The SpiNNaker projects. Proc IEEE. 2014;102(5):652–65.
27. Merolla PA, Arthur JV, Alvarez-Icaza R, Cassidy AS, Sawada J, Akopyan F, Modha DS. A million spiking-neuron integrated circuit with a scalable communication network and interface. Science. 2014;345(6197):668–73.
28. Schemmel J, Kriener L, Müller P, Meier K. An accelerated analog neuromorphic hardware system emulating NMDA- and calcium-based non-linear dendrites. IEEE Trans Biomed Circuits Syst. 2010;4(4):214–22.
29. Pfeil T, Grübl A, Jeltsch S, et al. Six networks on a universal neuromorphic computing substrate. Front Neurosci. 2013;7:11.
30. Yang S, Deng B, Wang J, et al. Scalable digital neuromorphic architecture for large-scale biophysically meaningful neural network with multi-compartment neurons. IEEE Trans Neural Netw Learn Syst. 2019;31(1):148–62.
31. Yang S, Wang J, Deng B, et al. Real-time neuromorphic system for large-scale conductance-based spiking neural networks. IEEE Trans Cybern. 2018;49(7):2490–503.
32. Luo J, Coapes G, Mak T, et al. Real-time simulation of passage-of-time encoding in cerebellum using a scalable FPGA-based system. IEEE Trans Biomed Circuits Syst. 2015;10(3):742–53.
33. Igarashi J, Shouno O, Fukai T, et al. Real-time simulation of a spiking neural network model of the basal ganglia circuitry using general purpose computing on graphics processing units. Neural Netw. 2011;24(9):950–60.
34. Cheung K, Schultz SR, Luk W. A large-scale spiking neural network accelerator for FPGA systems. In: International conference on artificial neural networks. Berlin, Heidelberg: Springer;2012. p. 113–120.
35. Mak TST, Sedcole P, Cheung PYK, et al. On-FPGA communication architectures and design factors. In: 2006 international conference on field programmable logic and applications. IEEE;2006. p. 1–8.
36. Petrovici MA, Vogginger B, Müller P, et al. Characterization and compensation of network-level anomalies in mixed-signal neuromorphic modeling platforms. PLoS ONE. 2014;9(10): e108590.
37. Yang S, Wang J, Deng B, et al. Neuromorphic context-dependent learning framework with fault-tolerant spike routing. IEEE Trans Neural Netw Learn Syst. 2021;33(12):7126–40.
38. Chen KH, Chiu GM. Fault-tolerant routing algorithm for meshes without using virtual channels. J Inf Sci Eng. 1998;14(4):765–83.
39. Zhang Z, Greiner A, Taktak S. A reconfigurable routing algorithm for a fault-tolerant 2D-mesh network-on-chip. In: Proceedings of the 45th annual design automation conference;2008. p. 441–446.

Large-Scale Digital Neuromorphic Systems

5

5.1 Existing Digital Neuromorphic Systems

Numerous noteworthy neuromorphic systems have showcased their potential across diverse applications, encompassing machine learning, robotics, and neuroscience research. One illustrative instance involves researchers harnessing neuromorphic systems to simulate large-scale SNNs and to realize special tasks like image recognition [1–8, 10]. This approach not only attains remarkable performance but also operates with reduced energy consumption compared to conventional computing architectures. Additionally, in the realm of robotics, neuromorphic systems have been instrumental in the creation of intelligent robots adept at executing intricate tasks, including navigation. Below, we introduce a selection of representative neuromorphic systems.

TrueNorth is a neuromorphic computing architecture developed by IBM Research that uses a massively parallel network of low-power, highly configurable digital neurons to simulate spiking neural networks [1]. The architecture is based on a custom-designed integrated circuit that can be scaled to support large-scale neural networks with billions of neurons and trillions of synapses. The digital neurons in TrueNorth are designed to mimic the behavior of biological neurons, which communicate with each other using spikes. Each TrueNorth chip contains 1 million digital neurons and 256 million synapses, and multiple chips can be interconnected to create larger neural networks. The architecture is highly configurable, allowing researchers to program the behavior of individual neurons and synapses to simulate different types of neural networks. TrueNorth has demonstrated its potential for a wide range of applications, including machine learning, robotics, and neuroscience research.

SpiNNaker is a digital hardware platform developed by researchers at the University of Manchester that uses a massively parallel architecture of low-power, highly configurable digital neurons to simulate spiking neural networks [3, 4, 9]. Each SpiNNaker chip contains 18 advanced reduced instruction set computer machine (ARM) cores and can simulate up to a million digital neurons, with each core responsible for simulating a subset of the neurons in the network. The architecture of SpiNNaker is optimized for real-time interaction with the environment, making it suitable for applications such as robotics, control systems, and sensory processing.

Loihi is a neuromorphic chip developed by Intel Labs that contains over 130,000 digital neurons and 130 million synapses [2]. Loihi's architecture supports a wide range of neural network models, including convolutional and recurrent neural networks, and also includes on-chip learning capabilities, allowing it to adapt its behavior based on changes in the environment or input data [6]. The low-power design of Loihi makes it highly energy-efficient, making it ideal for use in devices that operate on limited power, such as mobile devices and IoT devices. Tianjic is a neuromorphic computing architecture developed by Tsinghua University [5, 7]. It has been presented to develop brain-inspired intelligence by integrating the computing paradigms that are towards neuroscience and AGI.

In addition to the aforementioned work, there are many outstanding studies on digital neuromorphic systems [29], which will not be exhaustively listed here. Due to the rapid advancement in this field, more work continues to be proposed, making it challenging for this book to provide up-to-date statistics. Interested readers are encouraged to refer to relevant review papers. Below, we will focus on three digital neuromorphic systems—LaCSNN, CerebelluMorphic, and BiCoSS—all developed and implemented by the authors. Each system will be thoroughly introduced in two parts, covering both system design and experimental aspects.

5.2 LaCSNN Neuromorphic System

5.2.1 System Design

As described in [24], a large-scale conductance-based spiking neural network (LaCSNN) is implemented utilizing six Altera Stratix III 340 FPGA chips. Each of these FPGA chips is housed on a DE3 340 development board, equipped with four HSTCs for inter-board connectivity, a DDR2 memory channel for offline data storage, and a USB2.0 interface for seamless data transmission to the host computer. Extensive system testing has been conducted to validate the network's capacity to handle intricate neuron dynamics, ensure network scalability, and confirm the faithful emulation of large-scale SNN models. Notably, the system operates with a total power consumption of 10.578 W, resulting in a power density of 143.92 mW/cm^2. This remarkable efficiency surpasses that of a typical CPU, which typically falls within the range of 50–100 W/cm^2. The chosen neuron model can be effortlessly implemented in a pipelined architecture, boasting a clock frequency of 100 MHz. Leveraging the hardware resources of all six FPGA chips, the system can accommodate up to 1 million neurons in its real-time simulations, capable of handling approximately 60 million synapses at an average frequency of 50 Hz, thanks to its NoC hardware architecture.

Moreover, the system's scalability is inherently flexible. Additional FPGA development boards can be seamlessly integrated to expand the number of neurons, synapses, and overall computational power. As depicted in Fig. 5.1a and b, our architecture empowers large-scale neuromorphic computing through the collaborative efforts of high-end Altera Stratix III FPGA chips. Figure 5.1c illustrates the network's composition, comprised of conductance-based neural models meticulously designed to replicate ion channel dynamics at the subcellular level. Furthermore, Fig. 5.1d showcases the innovative digital implementation of ion channels, employing a multiplier-less and PLA approach that is pipelined to enhance operational frequency. Finally, Fig. 5.1e underscores our commitment to scalability, exemplified by the application of a 3D multicast network topology based on AER. This network topology utilizes 24-bit data packets for seamless data communication within and between chips, encompassing a 3-bit layer ID, 3-bit core ID, 6-bit AER data, 3-bit Y coordinate address, 3-bit X coordinate address, and a 6-bit timestamp.

Numerous studies have delved into the implementation of large-scale neural networks, with notable systems including Neurogrid [8], SpiNNaker [9], Truenorth [1], BrainScaleS [10], and HiAER [11]. Table 5.1 provides a comprehensive comparison of the LaCSNN system with these counterparts, factoring in crucial elements such as the biological fidelity

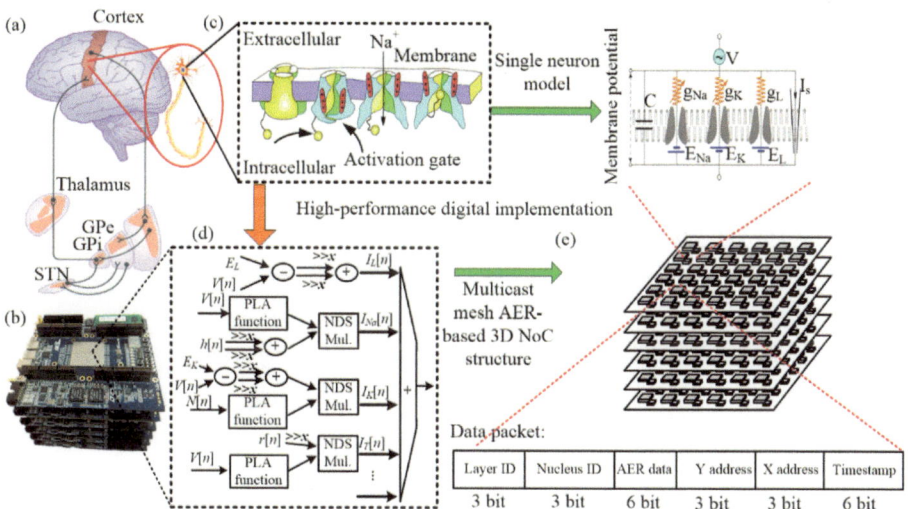

Fig. 5.1 Large-scale neuromorphic computing with multiple nuclei working together. **a** Cortico-basal ganglia-thalamus circuit in human brain. **b** Neuromorphic computing based on six DE3 340 development boards. **c** Ion channels between neuronal cell membranes. **d** Multiplier-less and memory-less digital implementations at the subcellular level. **e** 3D mesh NoC architecture based on multicast network address event routing [24]

of the neuron model, runtime plasticity, and the adaptability of the hardware infrastructure. Traditionally, the LIF neuron model is acknowledged for its relatively low biological accuracy, whereas the conductance neuron model boasts a higher level of biological fidelity. Notably, the Neurogrid project harnessed an IF neuron model for soma functionality complemented by four HH conductance models for dendrites. However, Neurogrid's limitation lies in its lack of reconfigurability, constraining its applicability to a broader spectrum of neural networks and neuron models. In contrast, the SpiNNaker project employed the Izhikevich model, offering the possibility of substituting it with the complex HH model. TrueNorth, while serving its purpose, exhibits less biological credibility and scalability compared to the model introduced in this chapter. Although SpiNNaker possesses the capability to simulate various spiking neural network models, its original design catered to the Izhikevich model, rendering it ill-suited for supporting intricate, biologically accurate neuron models. Additionally, SpiNNaker relies on ARM chips based on the von Neumann architecture, a less efficient configuration when compared to the FPGA's parallel computing capabilities featured in this chapter. TrueNorth, on the other hand, employs a LIF neuron model alongside a fixed SNN model characterized by limited programmable connections and a lack of on-chip learning capability. BrainScaleS and HiAER, while commendable, fall short in reproducing the dynamic behavior of ion currents and lack reconfigurability, given their reliance on analog circuits. In the context of large-scale applications, the NeuroDyn system is not optimally suited, and the Blue Brain model is constrained by its non-real-time nature. Considering the trifecta of biological precision, scalability, and reconfigurability, the neuromorphic computing paradigm presented in this chapter clearly excels in terms of performance. Importantly, every system mentioned in Table 5.1 possesses the capability to simulate large-scale neural networks comprising 1 million neurons in real-time.

The digital architecture of the nucleus processor is shown in Fig. 5.2, including globus pallidus externus (GPe), globus pallidus internus (GPi), subthalamic nucleus (STN) and thalamocortical (TC) processors. Each multiple synaptic information processing (MSIP) router has six ports for data communication, namely up, down, north, west, east, and

Table 5.1 Comparison with state-of-the-art neuromorphic computing systems

System	Biological plausibility	Plasticity	Reconfigurability
Neurogird	Adaptive quadratic IF with Hodgkin-Huxley channels (H)	No	No
SpiNNaker	Izhikevich (M)	Programmable	Yes
Truenorth	LIF (L)	No	Yes
BrainScaleS	AdEXP (M)	STDP	No
HiAER	IF (L)	STDP	No
LaCSNN	Conductance-based (H)	Programmable	Yes

south. Although studies have proposed a multicast AER method for CNN, it does not solve the following problems: (1) the calculation of synapse weight; (2) the realization of physical synapses; (3) the routing of mixed information. The data for the synaptic variables needs to be transmitted through the router between each core processor to calculate the synaptic current. For GPe processors and STN processors, two synaptic variables are required as outputs. Taking the GPe processor as an example, it contains two synaptic units, a GPe neuron unit, an MSIP router, and a configuration unit. The synaptic units calculate synaptic currents $I_{GPe \to GPe}$ and $I_{STN \to GPe}$, respectively. It uses synaptic variables s_{GPe} and s_{STN} from the router, and membrane potential V_{GPe} from the GPe neuronal unit. The configuration unit is responsible for setting configurable parameters for the MSIP router. Moreover, the GPe neuron unit using on-chip memory to implement 4800 virtual neurons through a physical neuron unit, and the digital implementation of other processors is similar to that of GPe processors.

Diverging from prior research that focused on CNN implementation via two-dimensional mesh routers, the current approach necessitates routers to grapple with multifaceted challenges encompassing the processing of multi-class nucleus synaptic information and inter-chip data communication. These routers are tasked with assimilating external information originating from the four neighboring core processors and

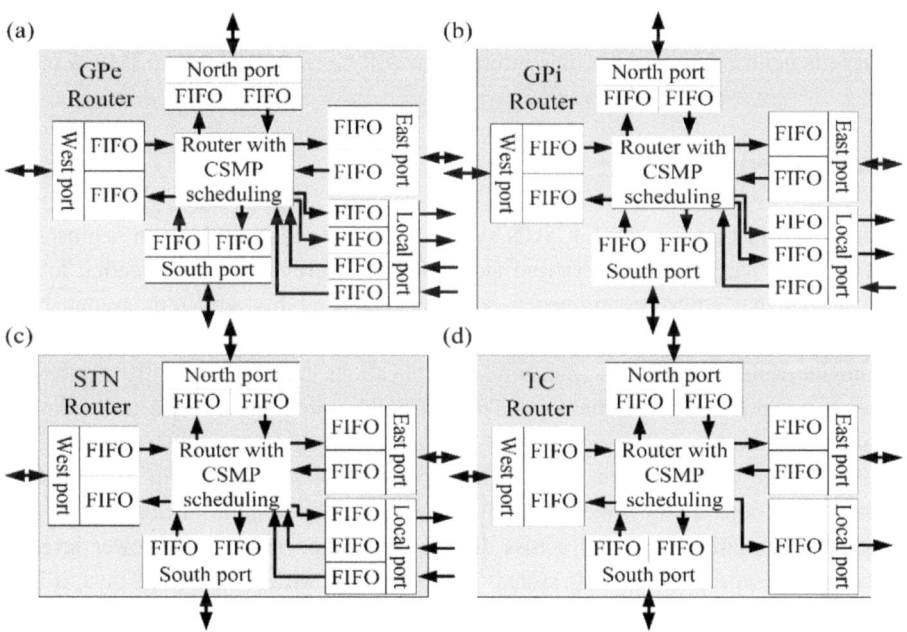

Fig. 5.2 Compute architecture of the nucleus processors. **a** GPe processor. **b** GPi processor. **c** STN processor. **d** TC processor [24]

subsequently discerning the data routing direction based on a meticulously programmed routing table within the configuration. As depicted in Fig. 5.2a, the intricacies of routing algorithms for multi-nucleus SNNs, which entail the processing of diverse synaptic information types, have prompted the development of an enhanced solution known as the improved multi-information processor (IMP) within the MSIP router. Notably, this IMP outshines conventional crossover methods by offering the distinct advantages of reconfigurability and enhanced flexibility, characteristic of digital router architectures.

According to the calculation requirements of synaptic unit, MISP routers are divided into 4 categories for different nucleus groups, namely GPe routers, GPi routers, STN routers and TC routers. It should be noted that different from the traditional application of AER mode in neuromorphic computing, the AER data in router packet transmission is synaptic information rather than spike information. The detailed architecture of GPe is given in Fig. 5.3a, and it is the most complex compared to the other 3 routers. The GPe router uses two IMPs to route the input information of two different variables. Packets are distinguished by a nucleus ID, with "000", "001", "010", "011", and "100" representing the variables s_{GPe}, s_{STN}, $s_{GPe \rightarrow GPi}$, $s_{STN \rightarrow GPe}$, and s_{GPi}, respectively. Selectors are used to select the destination node of the traffic that requires routing planning. The selection algorithm is shown in Fig. 5.3b. When the information reaches the router, the selector analyzes the nucleus ID and decides whether the information is output to IMP A or IMP B. If the nucleus ID = 000, the information is sent to IMP A, and if the nucleus group ID = 001, this information is sent to IMP B. If the information for the nucleus ID of the message is neither 000 nor 001, this information will be randomly sent to IMP A or B.

5.2.2   Experiments

As illustrated in Fig. 5.4, the LaCSNN system derives its inspiration from neuroscience principles and harnesses the advanced techniques of neuromorphic engineering to realize a design that embodies efficiency, scalability, adaptability, and reprogrammability. Despite the captivating prospects that neuromorphic engineering offers to the scientific and engineering communities, the current applications have yet to fully explore the immense potential of this technology. The LaCSNN system serves as a transformative platform, delivering both programmability and super-computing capabilities to researchers in neuroscience and cognitive science. It empowers the study of brain dynamics and cognitive functions at the subcellular level, facilitated by a multi-nucleus cooperation model. The system finds utility across diverse application tiers: at the lower levels, it can serve as a sensory processing system or a biosensor; within the middle tiers, it lends itself to environment-based brain-computer interfaces for real-time information processing. Meanwhile, at the highest echelons of application, it holds promise for deployment in decision-making processes and vision processing for humanoid robots.

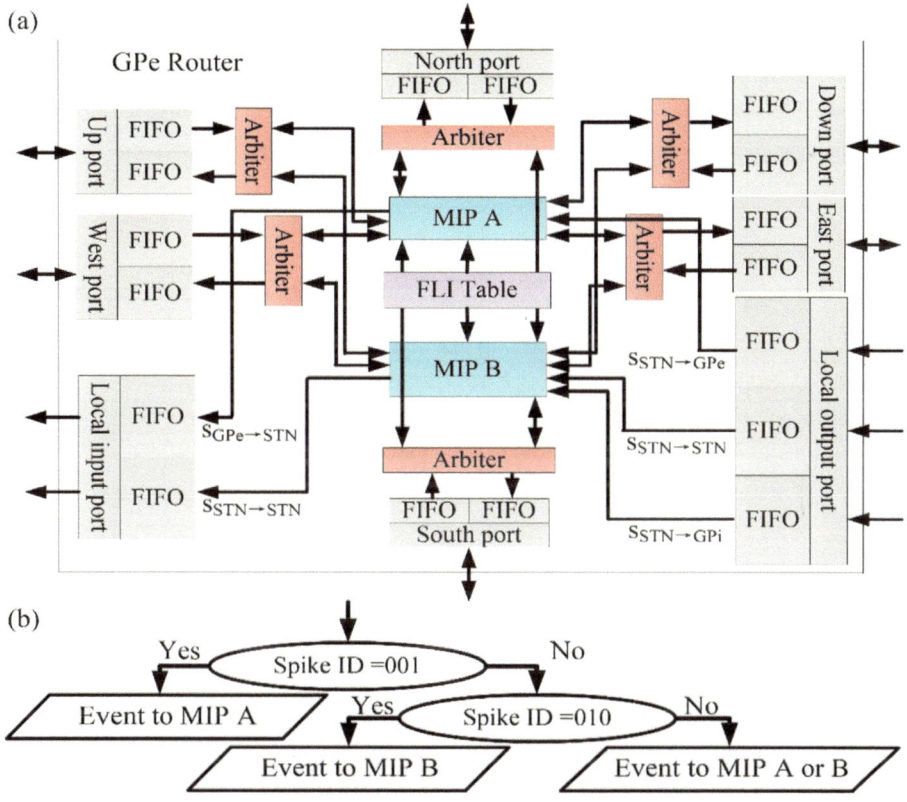

Fig. 5.3 Architecture of the nucleus processor. **a** Digital architecture of GPe. **b** Arbitration algorithm of GPe [24]

Furthermore, this system stands poised to unlock the mysteries of brain mechanisms, potentially leading to groundbreaking insights into pathology and medical discoveries. For instance, the correlation between increased irregular firing activity in the basal ganglia and dyskinesia symptoms could be meticulously investigated leveraging the LaCSNN system. In particular, the research could focus on elucidating the effects of synaptic currents originating from the GPe to other nuclei, including GPe to GPi and GPe to STN. The system could be instrumental in simulating deep brain stimulation (DBS), with the relevant equations as follows:

$$I_{DBS} = I_d H_{ea}(\sin(2\pi t/\rho_d)) \cdot [1 - H_{ea}(\sin(2\pi (t + \delta_d))/\rho_d)] \tag{5.1}$$

where $H_{ea}(\cdot)$ denotes the Heaviside function. $i_d = 300\,uA/cm^2$, $\rho_d = 1000/130 \approx 7.69\,ms$ and $\delta_d = 0.3\,ms$. Synaptic blockage was simulated by setting the corresponding synaptic current to zero.

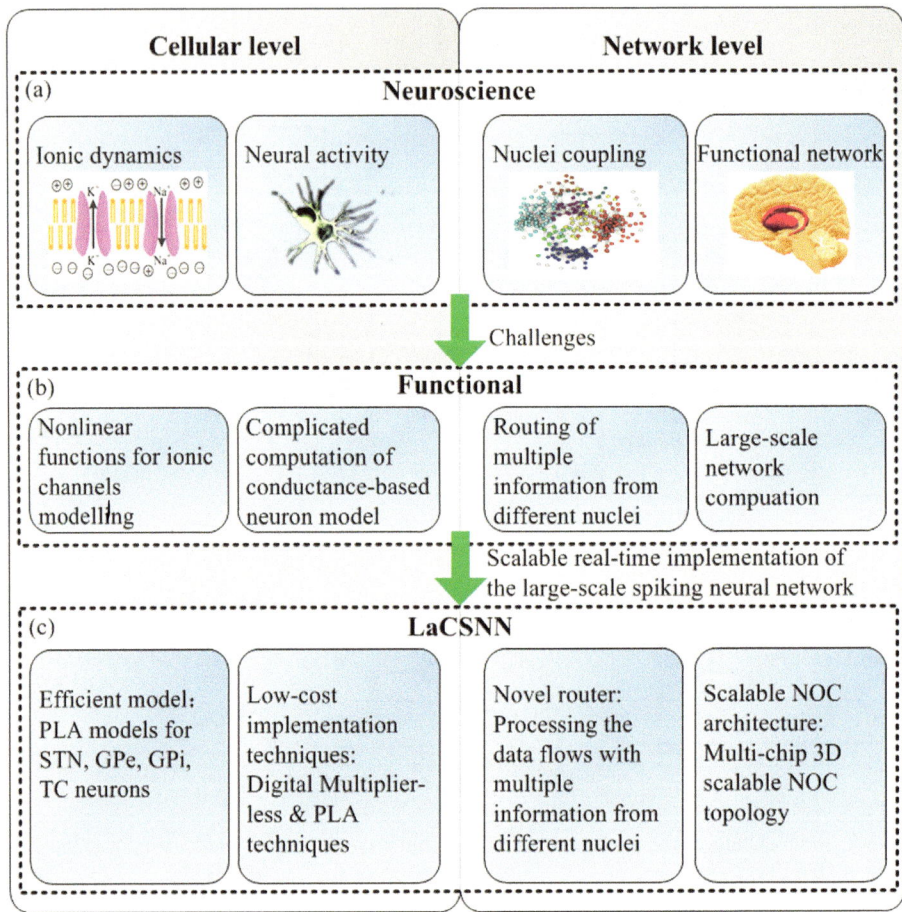

Fig. 5.4 LaCSNN system's neural heuristic concept. **a** Neuroscientific perspective. **b** Functional perspective. **c** Systems perspective [24]

In Fig. 5.5, denoted by "GPe → GPi" and "GPe → STN," these labels correspond to the respective nuclei where synaptic currents are impeded. The graphical representation in Fig. 5.5a vividly illustrates the elimination of beta-band activity within the GPi when the synaptic current originating from GPe to GPi is blocked. Complementarily, as depicted in Fig. 5.5b, the firing activity of the STN is effectively suppressed upon the blockade of synaptic input from GPe. These observations strongly suggest that synaptic currents originating from GPe to other nuclei play a pivotal role in governing the firing dynamics of the entire neural network. This insight underscores the potential significance of GPe as a viable target site for DBS.

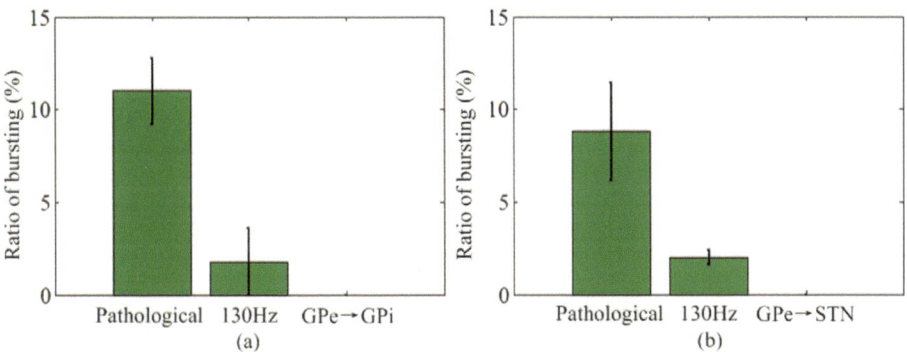

Fig. 5.5 Effect of synaptic blockade on oscillatory activity. **a** Burst firing rate of GPi nuclei under synaptic blockade. **b** Burst rate of STN nuclei under synaptic blockade [24]

The possibilities for further improvements to the LaCSNN system fall into three areas. First, there is no user interface to directly observe network dynamics and configure network parameters in real time. You can develop a graphical user interface that enables you to change model parameters. Second, the inter-chip connections of the LaCSNN system can be configured using an FPGA. Additional, FPGA chips can be used to control the vertical connections to increase system flexibility. However, this increases the hardware cost of the overall system, which is a tradeoff against system flexibility. Third, although not studied here, the LaCSNN system can be extended to any other brain region of interest. LaCSNN's high reconfigurability enables it to support a variety of complex brain-inspired computing models.

5.3 CerebelluMorphic Neuromorphic System

5.3.1 System Design

Similar to the approach outlined in [32], a large-scale SNN is established utilizing a butterfly fat tree (BFT) topology, as visually depicted in Fig. 5.6a. This extensible architecture is composed of two distinct components: the horizontal and vertical BFT layers. The horizontal BFT layer is implemented using FPGA, while the vertical layer relies on High-Speed Terasic Connectors (HSTC). Network data is transmitted in the form of 24-bit packets, encompassing 1-bit AER data, 3-bit chip address, 2-bit region address, 2-bit network address, 2-bit pad address, 2-bit region address, and a 12-bit timestamp. The encoding for a maximum of 4096 timesteps utilizes the 12-bit timestamp. Consequently, each timestamp corresponds to approximately 224 ns. LaCSNN operates at a clock frequency of 50 MHz. With each digital neuron completing its computational step within ten clock cycles (equivalent to 200 ns), the utilization of a 12-bit timestamp proves adequate for simulating time-multiplexed neurons without imposing any limitations on the maximum simulation duration. Synaptic connectivity is represented dynamically through the AER communication protocol. This representation is achieved by employing routing address events sourced from the synaptic routing table, which maps pre-synaptic source addresses to post-synaptic destination addresses. The configuration of the BFT topology within the 3D NoC architecture is vividly illustrated in Fig. 5.6b. The number of neuron units is contingent upon the available hardware resources, necessitating a three-layer BFT architecture for a network with 64 neuron units.

In the first-level router, there are six internal input ports that facilitate connections with second-level neighbors or cerebellar neuron units (CNUs). Routers are pivotal components responsible for routing and data flow control within the digital neuromorphic system architecture. Figure 5.7a illustrates the new router designed for the digital neuromorphic cerebellum. This router is equipped with six bidirectional ports connecting two parent router nodes and four child router nodes. When the first-level router receives a spike event from the CNU, it initiates the packaging process through the spike wrapper unit. The spike wrapper unit's purpose is to process a single spike event into a valid AER spike packet, utilizing information from the configuration processor for data processing. The configuration process can be reconfigured as needed, based on neural connectivity. The configuration processor encompasses four registers: a chip address register, a layer address register, a node address register, and a timestamp register. Upon reaching the *deliverat* timestamp, the incoming spike event and its corresponding *deliverat* time are stored in the on-chip memory. Figure 5.7b illustrates how the routing logic unit processes the AER packet in accordance with the routing algorithm. The crossbar switch within the first-level router is implemented using a multiplexer and is controlled by a signal from the crossbar arbiter. Subsequently, the AER spike packet is routed to the output port, with four AER spike events directed to the CNU and two events dispatched to the higher-level router. Figure 5.7b further elaborates on the detailed routing algorithm. The lay, area, network, and population serve as the source address for each router at the corresponding lay, area, network, and population levels. On the other hand, *layerd*, *regiond*, *networkd*, and *populationd* represent the router with the respective destination address. For routers operating at the population, network, and area levels, a comparison is made between the destination addresses of the population router and its underlying routers. If they match, the AER data is routed to the corresponding node in the downstream layer, which is the neuron processor. Conversely, if there is no match, the AER data is transmitted to the appropriate network router. This process is mirrored in both network and area routers. For layer routers, a comparison is made between the current router address and the destination layer address. If they align, the AER data is sent to the corresponding area router downstream. Otherwise, the data is directed to the alternate layer, contingent on the destination layer address.

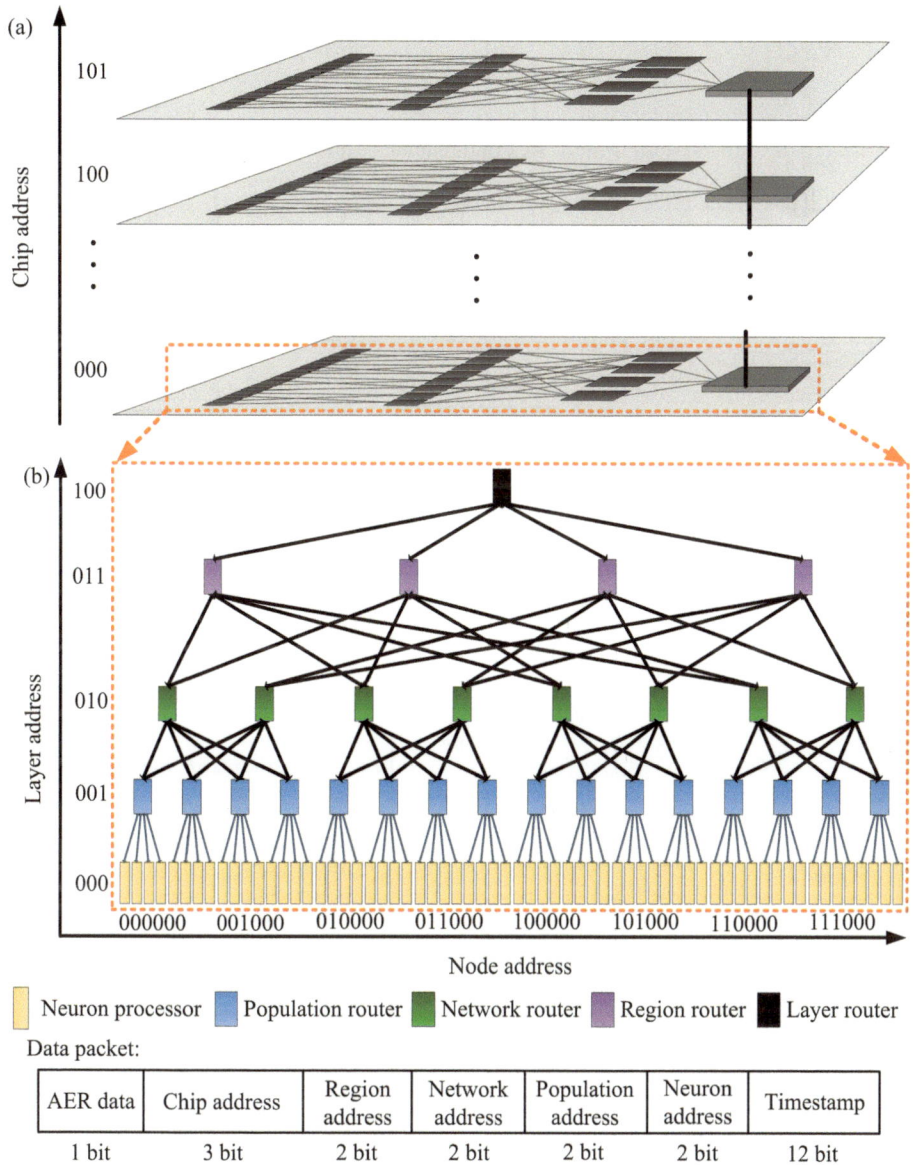

Fig. 5.6 Topology of 3-D BFT for large-scale neuromorphic implementation. **a** Scalable connection architecture for 3D NoC system. **b** Digital architecture on each chip [32]

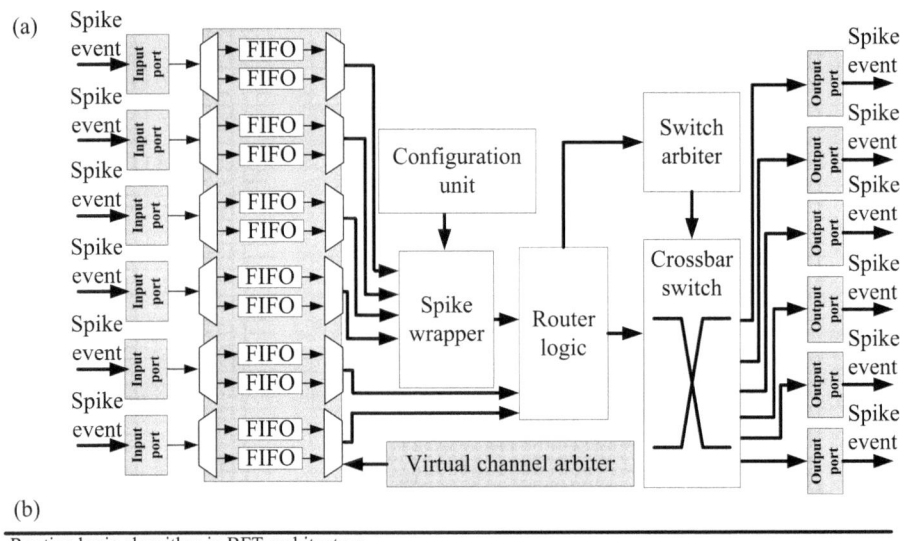

(b)

Routing logic algorithm in BFT architecture	

begin
   Neural network, population, region, layer for source and destination;
   if *Layers = Layerd* **then:**
     if *Regions= Regiond* **then:**
       if *Networks = Networkd* **then:**
         if *Populations = Populationd* **then:**
           Place spike event into respective physical neuron;
         else:
           Reach *Nd* connected to *Ps*;
           Reach *Pd*, connected to *Nd*;
           Place the spike event into respective physical neuron;
         end
       else:
         Reach *Nd*, connected to *Ps*;
         Reach *Pd*, connected to *Nd*;
         Place the spike event into respective physical neuron;
       end
     else:
       Reach *Ns*, connected to *Ps*;
       Reach respective *Rs*, connected to *Ns*;
   end
	else:
	   Reach either of two *Ns*, connected to *Ps*;
	   Reach *Rs*, connected to *Ns*;
	   Reach *Ls,* connected to *Rs*;
	   if the *address* of *Ld* > the *address* of *Ls* **then:**
	   while s<d:
	     Reach *Ls+1,*, connected to *Ls*;
	     s=s+1;
	   end
	   if the *address* of *Ld* < the *address* of *Ls* **then:**
	   while s>d:
	     Reach *Ls-1,*, connected to *Ls*;
	     s=s-1;
	   end
	   Reach *Rd*, connected to *Ld*;
	   Reach any of the two *Nd*, connected to *Rd*;
	   Reach *Pd*, connected to *Nd*;
	   Place spike event into respective physical neuron;
	 end

end	**end**

Fig. 5.7 Digital neuromorphic architecture of a routing unit. **a** Detailed digital architecture of the router. It contains FIFO blocks, virtual channel arbiters, spike wrappers, router logic, configuration unit switch arbiters and crossbar switches. It is used to efficiently route spike events from various neuronal units. **b** Pseudocode for routing logic for the BFT architecture. In the routing algorithm, the addresses of each layer, area, network and population router are considered [32]

The digital architecture of the core processor is shown in Fig. 5.8, which includes GrC, GoC, PKJ, BC, IO, and VN processors. Each router has six ports for communicating AER events to the target node of the BFT using two upstream ports and four downstream ports. AER spikes are transmitted by routers in each nucleus processor and used to calculate synaptic currents. In PKJ processor, three spike events are required in the silicon synaptic unit: β_{grc}, β_{grc}, β_{io}. The synaptic currents $I_{grc \to pkj}$, $I_{bc \to pkj}$ and $I_{io \to pkj}$ are calculated for the silicon synaptic unit. The PKJ neuron unit calculates a spike event β_{pkj} and outputs it to the router. The configuration unit is responsible for the configuration of the router and the silicon synapse unit. The PKJ neuron unit uses time-division multiplexing technology and on-chip memory to realize 9000 virtual neurons in one physical unit. Other digital implementations of the nucleus processor use the same approach as the PKJ processor, with only one silicon synapse for network computing. In order to realize the different types of neurons in the cerebellar model articulation controller (CMAC), the Euler method using numerical integration reduces the computational resources required in the digital implementation compared with the Runge–Kutta method. The digital implementation of the GrC neuron model shown in Fig. 5.9a includes a pipeline and a random-access memory (RAM) block for time multiplexing. Several RAM blocks are used to store variable values on the FPGA. The RAM module requires $SI_V * V_b$ bits of on-chip memory, where SI_V is the data size of the variable V and V_b is the bit width of each data. The delay number of the pipeline is V_{delay} which has the relationship $V_{delay} = V_{stage}$ for pipeline synchronization. The detailed digital architecture of the V-pipeline is shown in Fig. 5.9b. The addition and sub-blocks implement addition and subtraction operations, respectively. The detailed digital architecture of the "G_{ahp}" module is shown in Fig. 5.9c. The shift logic multiplier block (SLM multiplier) is a dedicated digital circuit proposed to perform multiplication calculations without embedded multiplier resources on the FPGA, as shown in Fig. 5.9d. The SLM block is used for multiplication between two variables in the neuron unit and the silicon synapse unit.

The detailed digital architecture of the synaptic unit is shown in Fig. 5.10a. There are 100 parallel sets of synaptic current processors (SCPs) in a silicon synaptic unit. The AER spike packet is input to a multiplexer with a 100 output, which is selected by a conventional counter for sequentially selected ports. The AER spike packet is processed by the decoder to obtain the event data and its corresponding timestamp. The timestamp is used as the write address of the buffer, and the read address is controlled by a counter. In each SCP, the connectivity c_{ij} is determined by the configuration unit, and all multiplication operations use SLM blocks. The cerebellar plasticity of the large-scale neuromorphic SNN is shown in Fig. 5.10b–e. The ACC block represents an accumulator having two data ports and one synchronous clear port. The MUX block represents a multiplexer that selects a data path according to a control signal. In Fig. 5.10d, the variable counter number (CN) represents the current number sequentially counted by the digital counter corresponding to the last trigger activity in the fixed time window. The LUT block represents a lookup table for looking up pre-stored values when needed. The incoming information is the

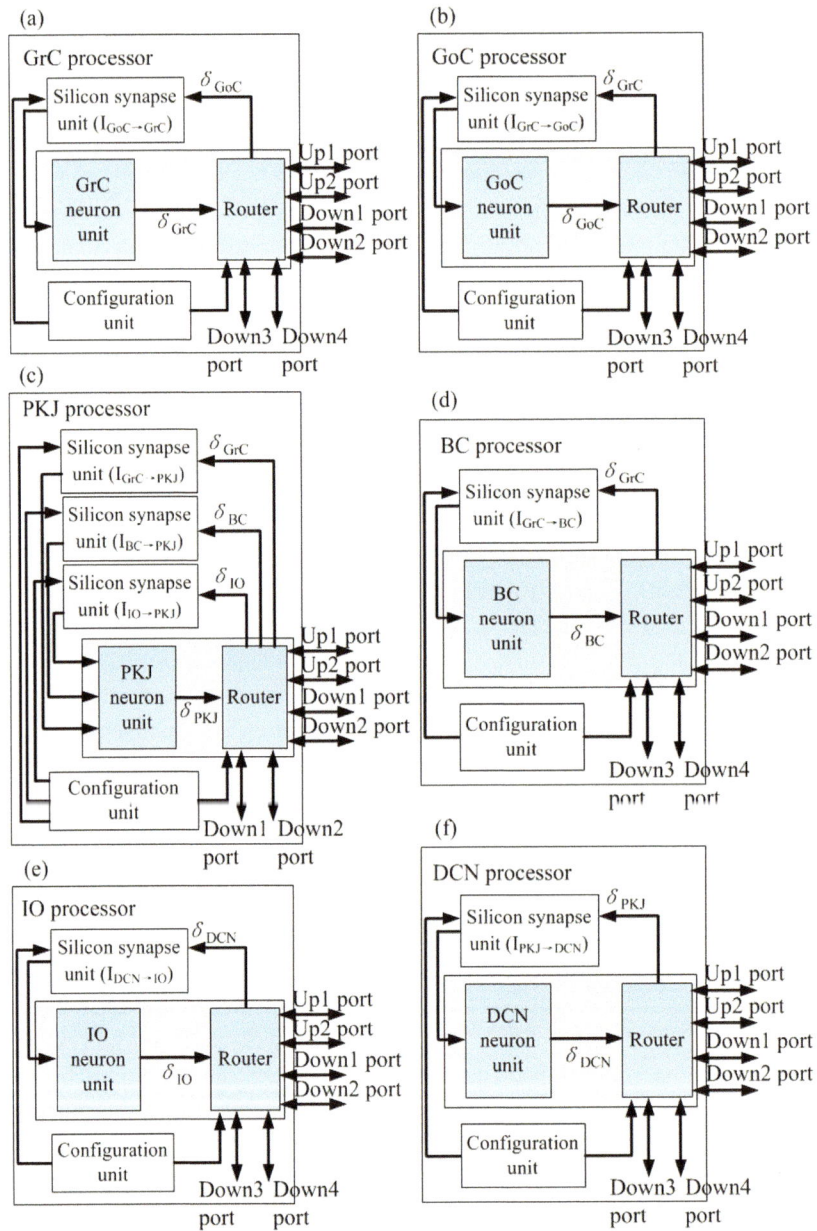

Fig. 5.8 Digital neuromorphic structure of a nucleus processor. Each nucleus processor contains one or more silicon synapse units, a neuron unit, a router, and a configuration unit. The router has ports in six directions, including two upward ports and four downward ports. The nucleus processor includes **a** a GrC processor, **b** a GoC processor, **c** a PKJ processor, **d** a BC processor, **e** an IO processor, and **f** a VN processor [32]

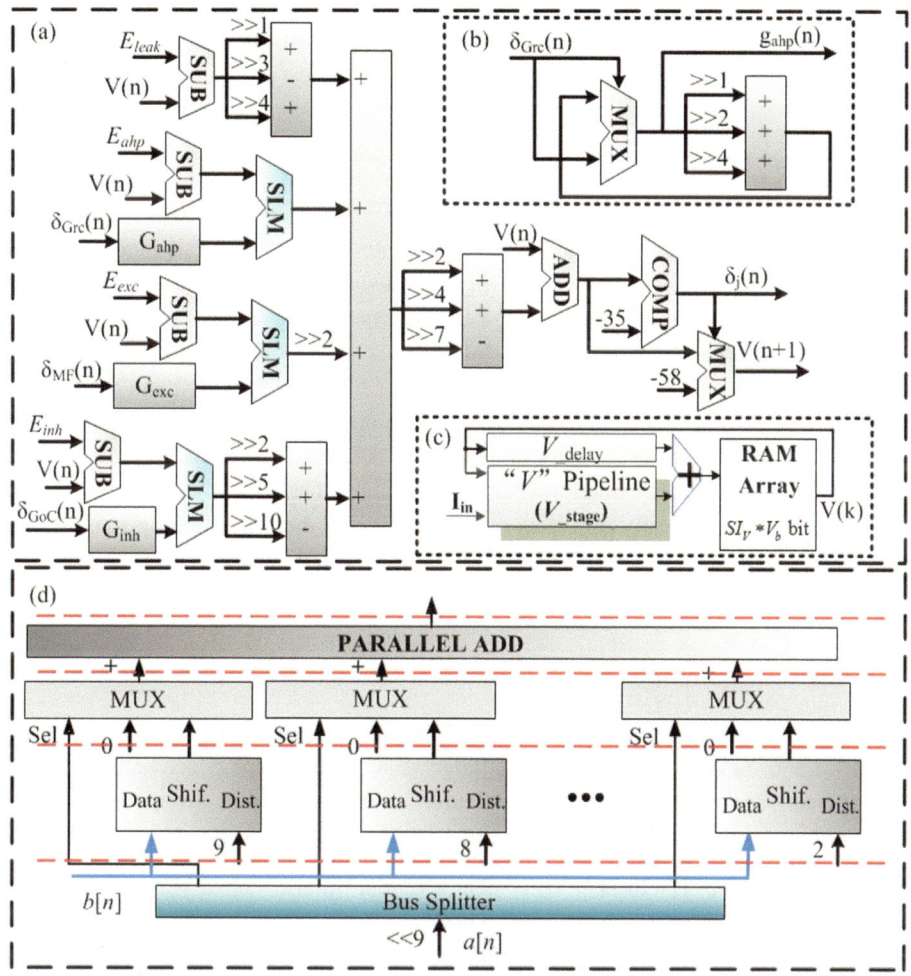

Fig. 5.9 Digital neuromorphic structure of neuron units of cerebellar GrC neurons, realized based on a multiplier-less scheme. **a** A detailed "V" pipe structure containing a RAM array for storing transient variable values. **b** General overview of the neuron unit of GrC neuron, which uses SLM modules and shifter modules instead of multipliers. **c** Detailed digital architecture of the "Gahp" module, which uses a shifter module instead of a multiplier. **d** An SLM module without a digital multiplier is implemented. It uses bus splitters, multipliers, barrel shifters and parallel adders to implement the multiplication of neural variables [32]

spike activity of the j_{th} neuron $V_{j[n]}$. The variable NCI represents network connection information. The absolute value of ABS block output input. In Fig. 5.10e, if the peak of the spike is detected, the corresponding CN is sent to the output register. The incoming information is the spike activity of the j_{th} neuron. The *sclr* signal represents the sync clear signal that resets the counter every cycle. The value of CN is obtained from the output register at the end of each time window and is calculated according to the STDP learning rule.

5.3.2 Experiments

Researchers have the capability to delve deeply into the intricate computational model of large-scale cerebellar networks with high biological plausibility. They can conduct dynamic analysis experiments on spiking networks, providing valuable insights. The core component of CerebelluMorphic is the Intel Stratix III 340 FPGA, boasting hardware resources comprising 338,000 logic elements, 16,272 kbits of memory, and a 57,618-bit 18-bit multiplier block. Cerebellar mechanisms are often studied in isolation, employing methods like OKR eye movements or Pavlovian delayed blinks [13, 14]. In the case of OKR adaptation, mossy fibers (MFs) and climbing fibers (CFs) convey retinal glide information that periodically oscillates over time. During the oscillation period, different populations of granule cells (GrCs) become active in succession. This process allows the cerebellum to learn the complete waveform indicated by the CFs.

(1) Performance evaluation of CerebelluMorphic

As depicted in Fig. 5.11a, the CerebelluMorphic system utilizes six Intel Stratix III EP3SL340 FPGAs to implement a large-scale neuromorphic cerebellar network, encompassing approximately 3.5 million neurons and 218.3 million synapses. It comprises a total of 3,411k GrC neurons, 1,024 golgi cell (GoC) neurons, 32 Purkinje (PKJ) neurons, 128 basket (BS) neurons, four inferior olive (IO) neurons, and eight vestibular nucleus (VN) neurons. These cerebellar divergence/convergence ratios closely mimic those observed in the biological cerebellum [12, 30]. To showcase the real-time computational capabilities of CerebelluMorphic, the spike activity output was sampled using an oscilloscope, as illustrated in Fig. 5.12. The discrete input spikes from neuromorphic MFs are displayed in Fig. 5.12a and are modeled as Poisson spikes. Figure 5.12b showcases the discrete spikes produced by randomly selected GrC neurons within CerebelluMorphic. Figure 5.12c presents a raster plot depicting the discrete spikes of neuron output. The evaluation criteria encompass the Root Mean Square Error (*RMSE*), Mean Absolute Error (*MAE*), Correlation Coefficient (*CORR*), and Spike Timing Error (*ERRTT*), computed as follows:

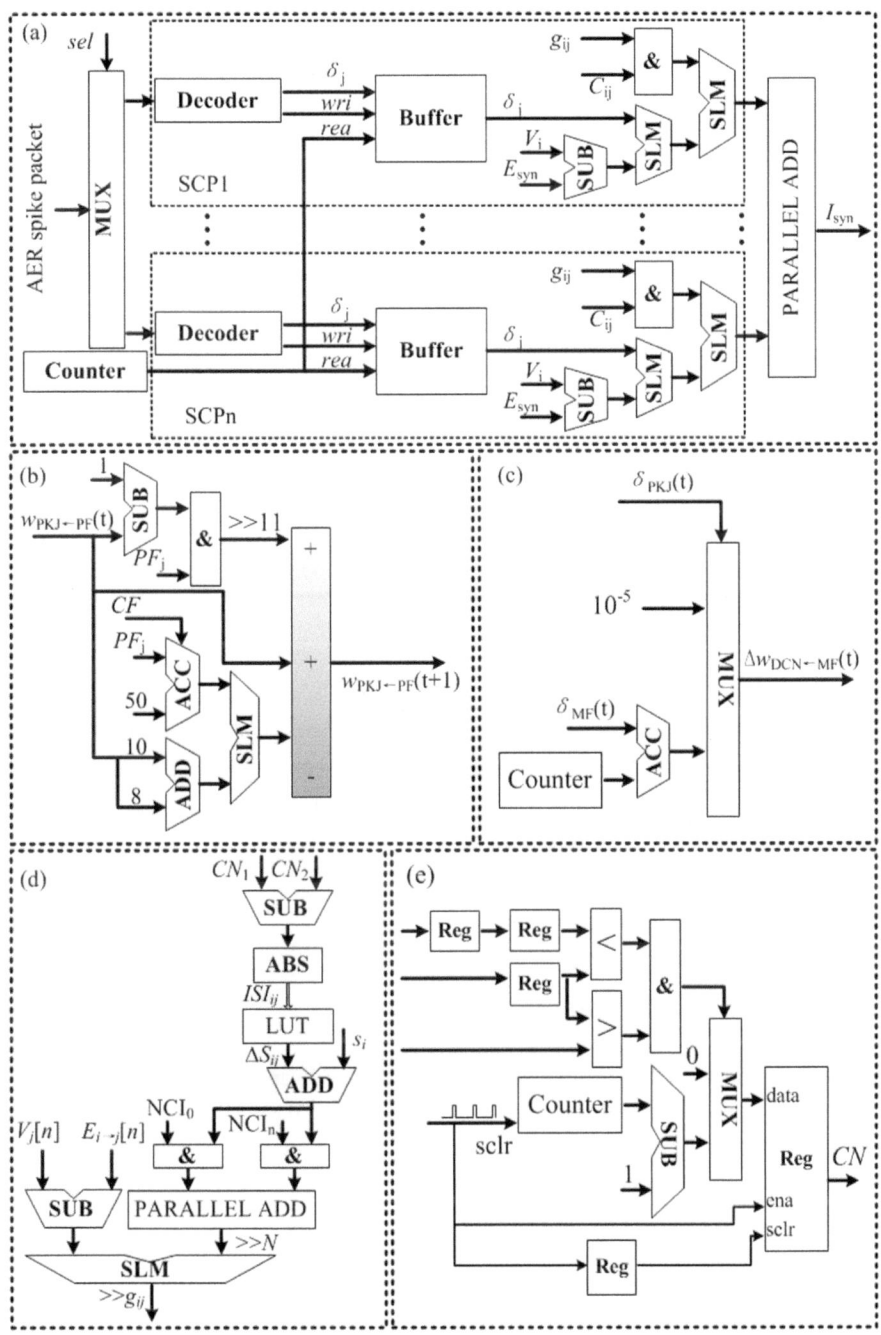

Fig. 5.10 Digital neuromorphic realizations of cerebellar synaptic plasticity. **a** Digital architecture of the silicon synapse unit, which uses a parallel computing architecture. **b** Synaptic weight calculation from PF to PKJ. **c** Synaptic weight changed from VN to MF. **d** STDP learning calculation from PKJ to VN [32]

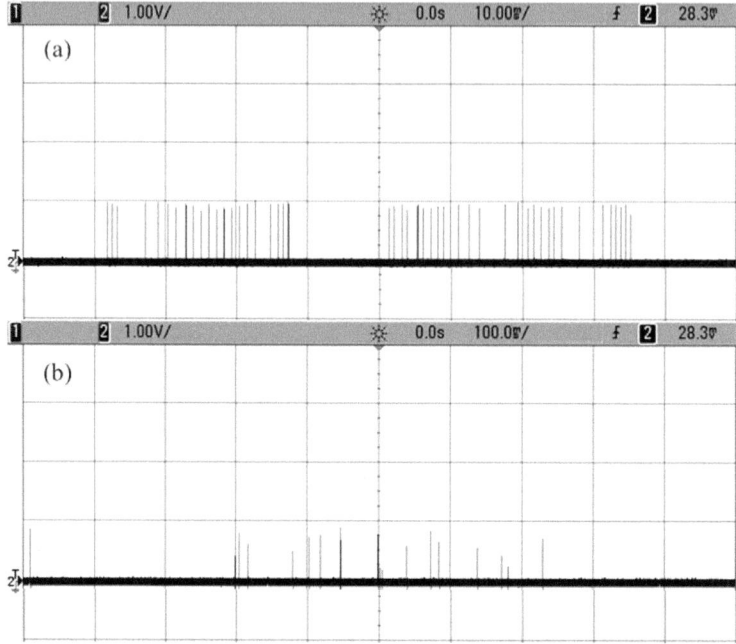

Fig. 5.11 Real-time spike activity of the cerebellar morphology system presented on an oscillo-scope. **a** Input discrete spikes from MFs. **b** Discrete spikes are output from the CMAC network [32]

$$
\begin{cases}
RSME = \sqrt{\dfrac{1}{N}\displaystyle\sum_{i=1}^{N}\left(x_{sof}(i) - x_{har}(i)\right)^2} \\[2mm]
MAE = \max\left|x_{sof}(i) - x_{har}(i)\right| \\[2mm]
CORR = \dfrac{\mathrm{cov}\left(x_{sof}, x_{har}\right)}{\sigma\left(x_{sof}\right)\sigma\left(x_{har}\right)} \\[2mm]
ERRTT = \left|\dfrac{\Delta T_{har} - \Delta T_{sof}}{\Delta T_{sof}}\right|
\end{cases}
\tag{5.2}
$$

where $x_{sof}(i)$ and $x_{har}(i)$ represent the software and hardware computation results at the i_{th} iteration, respectively. The variables T_{har} and T_{sof} are the spike time intervals for the hardware and software results, respectively. $CORR$ is defined as the ratio of the product of the covariance and variance of two datasets.

$$
\begin{cases}
\mathrm{cov}\left(x_{sof}, x_{har}\right) = \displaystyle\sum_{i=1}^{N}\left(x_{sof}(i) - \bar{x}_{sof}\right)\left(x_{har}(i) - \bar{x}_{har}\right) \\[2mm]
\sigma(x) = \sqrt{\displaystyle\sum_{i=1}^{N}\left(x(i) - \bar{x}\right)^2}
\end{cases}
\tag{5.3}
$$

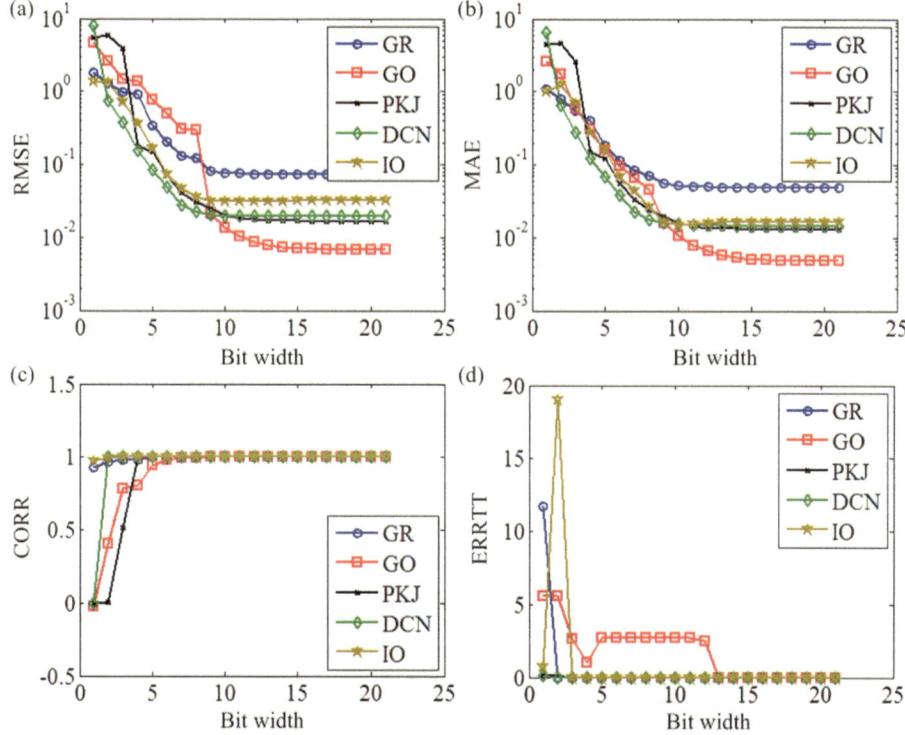

Fig. 5.12 Accuracy analysis of digital neuromorphic calculations for each nucleus in the neuromorphic cerebellar model. Different types of nuclei were considered, including GR, GO, PKJ, VN and IO. Here, the analysis of the impact of bit width is shown on **a** RMSE, **b** MAE, **c** CORR and **d** ERRTT [32]

The parameters x_{sof} and x_{har} represent the average values $x_{sof}(i)$ and $x_{har}(i)$, respectively. As anticipated, the error evaluation results presented in Fig. 5.13 underscore that computational accuracy can be significantly enhanced by increasing the bit width within the digital neuromorphic cerebellar network. To assess the system's performance, a thorough throughput comparison was conducted between the CerebelluMorphic system and three state-of-the-art cerebellar digital neuromorphic systems [3, 24, 28]. These comparisons involved configuring the cerebellar morphology to route neural events across the architecture at various levels of synaptic events, ranging from 20 to 180 million Synaptic Operations (SynOps). Figure 5.13a illustrates a comparison of system throughput with normalized traffic patterns, while Fig. 5.13b and c depict comparisons of system throughput with 10% and 30% hotspot traffic, respectively. Additionally, Fig. 5.13d presents a comparison of system throughput with the tornado traffic pattern. Notably, as the rate of injected synaptic events increases, the throughput of the cerebellar morphology surpasses that of the other two systems significantly. At a SynOps event rate of 160 million, the cerebellar morphology achieves throughput values that are 4.09 and 1.18 times greater than the other systems in normalized traffic patterns, 4.70 and 1.26 times greater in 10% hotspot traffic patterns, 3.27 and 1.55 times greater in 30% hotspot traffic patterns, and 1.78 and 1.33 times greater in tornado traffic patterns. This substantial improvement can be attributed to the BFT-based architecture of the CMD neuromorphic system. The enhanced throughput attained demonstrates that the cerebellar system can effectively handle a significantly larger information load over time compared to the other two neuromorphic systems.

Cerebellar PKJ cells exhibit intricate intrinsic biological behaviors and possess the capacity to integrate numerous synaptic inputs. As the sole output from the cerebellar cortex, gaining insights into the dynamics of PKJ cells holds paramount importance in comprehending cerebellar function. Synchronization and resonance dynamics serve as fundamental mechanisms for encoding and transmitting neural information. The orchestration of neuronal activities is intricately linked to neural correlations within group coding, with synchronization representing a crucial manifestation of this correlation among neurons. Resonance dynamics, on the other hand, elucidates the firing output response to an input signal—a phenomenon that has been the subject of study in biological nervous systems for many years. Dynamic analyses of PKJ cells are conducted in the context of active plasticity. To investigate the synchronization of the PKJ population, network synchronization criteria are defined as follows:

$$x_s = \frac{\left(\langle E(t)^2 \rangle_t - \langle E(t) \rangle_t^2\right)}{\frac{1}{N} \sum_{i=1}^{N} \left(\langle E_i(t)^2 \rangle \quad \langle E(t) \rangle^2\right)} \tag{5.4}$$

where $E(t)$ represents the mean spike events of the PKJ neurons and N represents the total number of neurons in the PKJ. $E(t)$ is defined as

$$E(n) = (1/N) \sum_{i=1}^{N} E_i(n) \tag{5.5}$$

As shown in Fig. 5.14a and b, an increase in the synaptic weight $w_{bs \to PKJ}$ from BS to PKJ cells increases the level of synchronization in the PKJ population, while an increase in the weight $w_{grc \to PKJ}$ from GrC to PKJ decreases the dynamics of PKJ synchronization. Furthermore, as the oscillation frequency of the MFs increases, synchronicity will increase, and a larger synaptic weight from GrC to PKJ will cancel this effect. The weights from BS to PKJ are low and the weights from GrC to PKJ are high, and the network synchronization of the PKJ units is not directly affected by the mossy fiber (MF) firing rate. To quantitatively describe the dynamic behavior of the nervous system, the linear response criterion x_{lr} is defined as

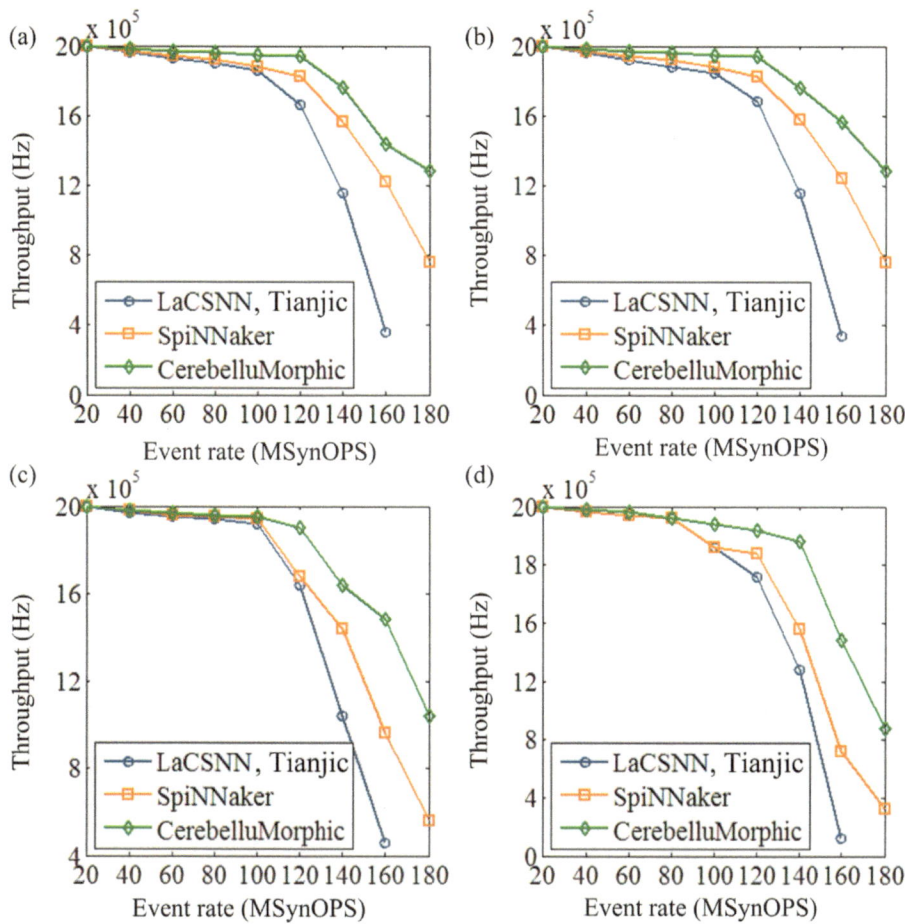

Fig. 5.13 Systematic evaluation and comparison of cerebelluMorphic under different conditions. Three structures of the neuromorphic system were considered, including LaCSNN/Tianjic, SpiN-Naker and CerebelluMorphic. **a** Throughput in normalized traffic patterns. **b** 10% and **c** 30% hotspot traffic patterns, and **d** tornado traffic patterns versus event rate display [32]

$$x_{lr} = \sqrt{\left(x_{lr}^{\sin}\right)^2 + \left(x_{lr}^{\cos}\right)^2} \tag{5.6}$$

where $x_{lr}^{\sin}$ and $x_{lr}^{\cos}$ are

$$\begin{cases} x_{lr}^{\sin} = \frac{2}{Tt} \sum_{n=1}^{Tt} E(n)\sin(\omega n) \\ x_{lr}^{\cos} = \frac{2}{Tt} \sum_{n=1}^{Tt} E(n)\cos(\omega n) \end{cases} \tag{5.7}$$

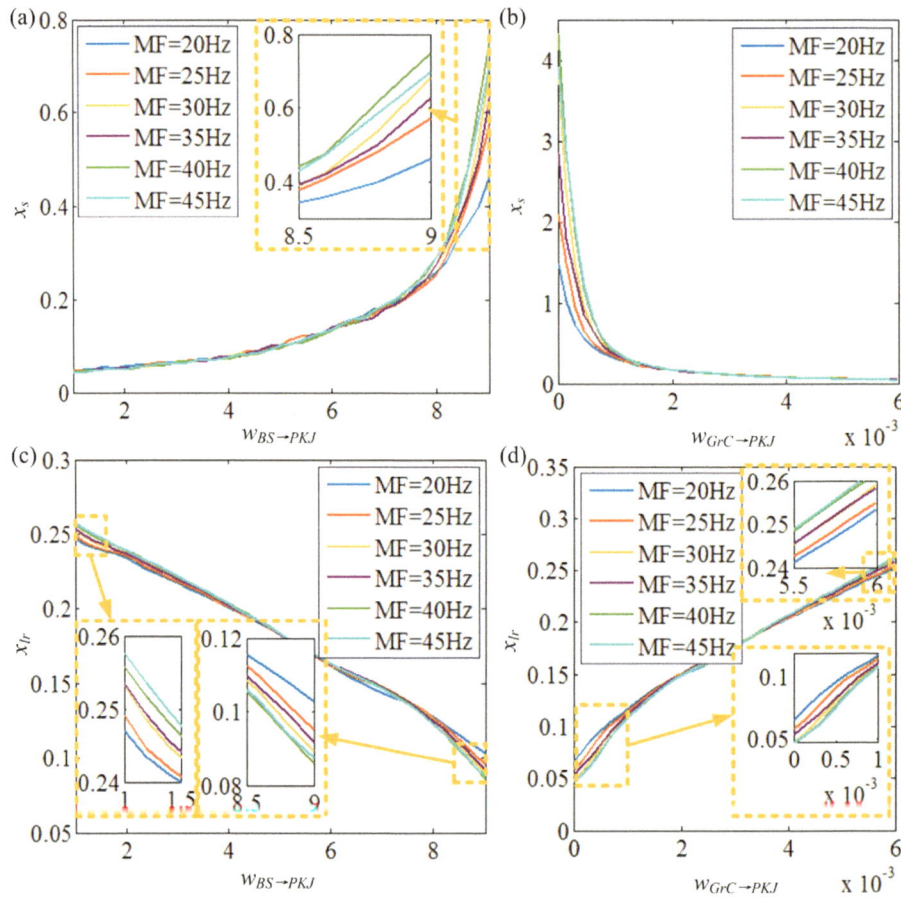

Fig. 5.14 Dynamic analysis of PKJ population. The different firing rates of MF are taken into account, which means that the neuromorphic model is stimulated at different levels. Here, different levels of MF were used to study the xs of PKJ neurons as a function of **a** $w_{bs \to PKJ}$ and **b** $w_{PF \to PKJ}$. In addition, different levels of MF were used to study the x_{lr} of PKJ neurons as a function of **c** $w_{bs \to PKJ}$ and **d** $w_{PF \to PKJ}$ [32]

where ω is the angular frequency of the oscillation. The linear response criterion x_{lr} reflects the resonance dynamics of the neural population, as shown in Fig. 5.14c and d. As the synaptic weight decreased from BS to BS, the resonance status of the PKJ population is enhanced. Interestingly, different levels of $w_{bs \to PKJ}$ and $w_{grc \to PKJ}$ adversely affect resonance dynamics as resonance increases. At low levels of $w_{bs \to PKJ}$, an increase in the MF firing rate will improve resonance within the PKJ population, while at high levels of larger $w_{bs \to PKJ}$ resonance will be reduced. The enhancement of MF firing can inhibit the resonance of PKJ neurons at low $w_{grc \to PKJ}$ and improve it at high $w_{grc \to PKJ}$.

(2) Neuromorphic Learning of PKJ and VN Cells

To investigate the learning-induced alterations in PKJ cells and their corresponding VN responses, the MF input undergoes oscillations spanning cycles 1 to 300, as depicted in Fig. 5.15a. As the cycle count increases, the maximum firing rate exhibits moderate variations, ranging from 88.5 to 75.62 Hz, while the minimum firing rate experiences a substantial decrease from 83.59 to 34.7 Hz. Consequently, the modulation in PKJ cell firing can be attributed to the decline in the minimum firing rate, aligning with the observed changes in PKJ neuron firing rates during OKR adaptation [13]. Figure 5.15b illustrates the firing activity of VN cells. The firing dynamics of VN neurons are phase-modulated by MF oscillations, driven by the modulation of PKJ cell inhibition. The maximum firing frequency increases from 60.37 to 109.6 Hz, and the minimum firing rate of VN neurons transitions from 23.17 to 42.43 Hz. To investigate VN modulation of the MF signal, the gain ratio is defined as the modulation of VN spike activity per cycle divided by the modulation observed during the first cycle. Due to the inhibition of IO information, neuromorphic learning varies with MF oscillation, ultimately resulting in a gain ratio of 1.791, as depicted in Fig. 5.15c. The distribution of synaptic weights for active GrC neurons is portrayed in Fig. 5.15d. At the onset and conclusion of the modulation period, synaptic weights are uniformly distributed within the range of 0.175 to 1.0. However, during the midpoint of the MF period, characterized by the most substantial MF and climbing fiber (CF) inputs, most synaptic weights are concentrated between 0.175 and 0.5. Therefore, the regulation of PKJ cell firing is attributed to the spatial distribution of synaptic weights between parallel fibers (PF) and PKJ cells, in conjunction with the feedforward inhibition from BC neurons.

(3) Dynamic Response Analysis in Neuromorphic GrC Layer

To delve into the underlying mechanisms governing the GrC response to temporal oscillations, our model is utilized to simulate an OKR adaptation experiment using retinal slip signal input. As portrayed in Fig. 5.16a, this OKR adaptation experiment incorporates MFs and CFs for real-time oscillatory retinal slip information transmission. This setup is based on a conceptualization of the neural circuits involved in OKR adaptation, as proposed by previous experimental studies [13]. GrC groups are sequentially activated, initiating with a sinusoidal oscillation cycle. A Poisson spike, oscillating sinusoidally at 0.5 Hz, is input to the MFs in accordance with prior biological research [13]. The phenomenon of long-term depression (LTD) imparts a sinusoidal spatial distribution to the synapses of PF-PKJ cells, resulting in a gradual increase in sinusoidal modulation of PKJ cell responses. This pattern aligns with findings from previous experimental studies [13]. Figure 5.16b reveals that, at the onset and conclusion of the MF signal oscillation period, the MF firing rate is low, leading to uniform, random spiking among GrCs. However, as

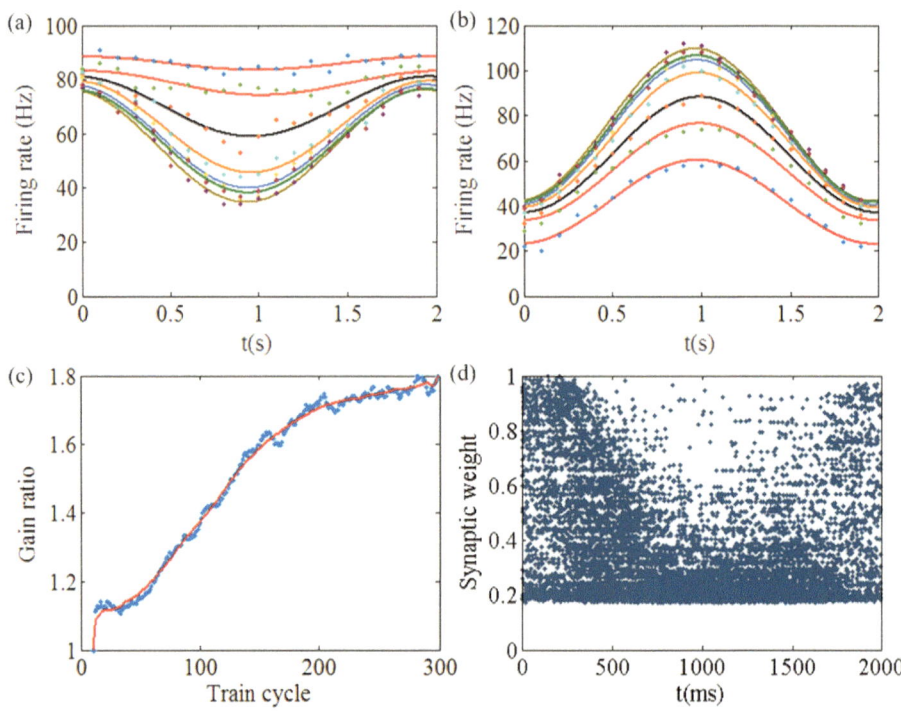

Fig. 5.15 Neuromorphic learning in OKR adaptation experiment. The color curves from the top to the bottom sequentially represent learning-induced firing rate changes of **a** PKJ cells and **b** VN cells in the 1st, 50th, 100th, 150th, 200th, 250th and 300th cycles of MF signal oscillation. **c** Gain change rate according to the MF train. Here, the blue dots are discrete data of the gain ratio, and the red solid line represents the fitted curve. **d** Synaptic weight distribution between PKJ cells and active GrC cells after 300 MF oscillation cycles [32]

the MF transmission rate escalates, GrCs become activated at a rate substantially proportional to the transmission activity of the MFs. This observation suggests that GrCs can effectively transmit MF amplitude information to PKJ cells. Building upon this mechanism, the PKJ unit can acquire scalar information, encompassing timing and gain signals, along with the complete waveform processed by the CFs, thus unifying gain and timing control. Due to the random cyclic connections between GrC and GoC cells, individual GrC neurons exhibit diverse temporal spike activities, as depicted in Fig. 5.17a. This phenomenon underscores that the active GrC neuron population gradually changes over time. Our results further demonstrate that the firing rate of different GrC neurons fluctuates over time in response to MF signals, contributing to the temporal dynamics of the cerebellum. To assess this property, a similarity index $S(t)$ between active GrC neurons is calculated, as presented in Fig. 5.17b. The decreasing trend of $S(t)$ with increasing t indicates infrequent temporal variation in the GrC layer and a one-to-one correspondence between the active GrC population and the oscillation time step of the MF signal. Despite the temporal dynamics within the GrC layer, the generation of time-fluctuating spikes remains reproducible under MF signal oscillations. The reproducibility index $R(t)$, quantifying the similarity between two spike patterns of all GrC neurons for consecutive cycles, is illustrated in Fig. 5.17c. $R(t)$ values increase toward 0.9 at the cycle's outset and gradually decline to approximately 0.8. This trend indicates that GrC neuron spike activity exhibits high reproducibility within the MF oscillation cycle. The GrC layer's role involves generating the same sequence of active GrC neurons across different trials and cycles, ensuring the reliable transmission of MF signals. As dynamic network activity in a cyclic structure relies on external inputs and initial states, the GrC layer resets its internal state at each cycle's inception. Consequently, a gradual increase in the MF signal is sufficient to facilitate this internal state reset.

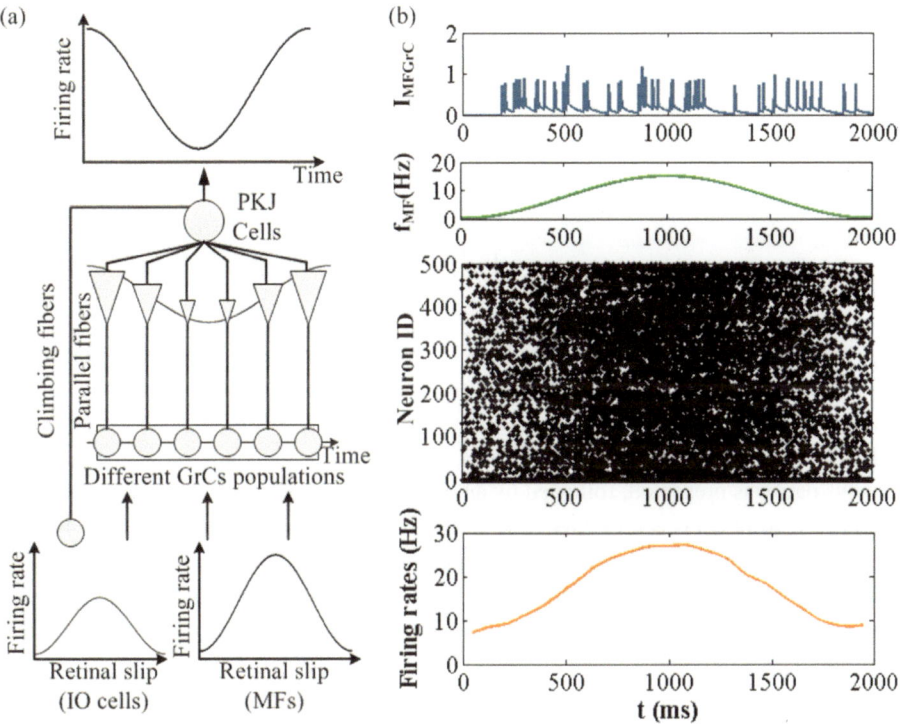

Fig. 5.16 Dynamics of GrC in response to MF input information oscillating sinusoidally at 0.5 Hz. **a** Simulation results of OKR adaptability experiment with retinal slip signal input using the SNN model. **b** Spike pattern and firing rate of 500 GrC during the MF signal oscillation period. The black dots represent spike firings. $I_{MF \rightarrow GrC}$ and f_{MF} in the first two figures represent the synaptic current from MF cell to GrC cell and the mean firing frequency of MF cells, respectively [32]

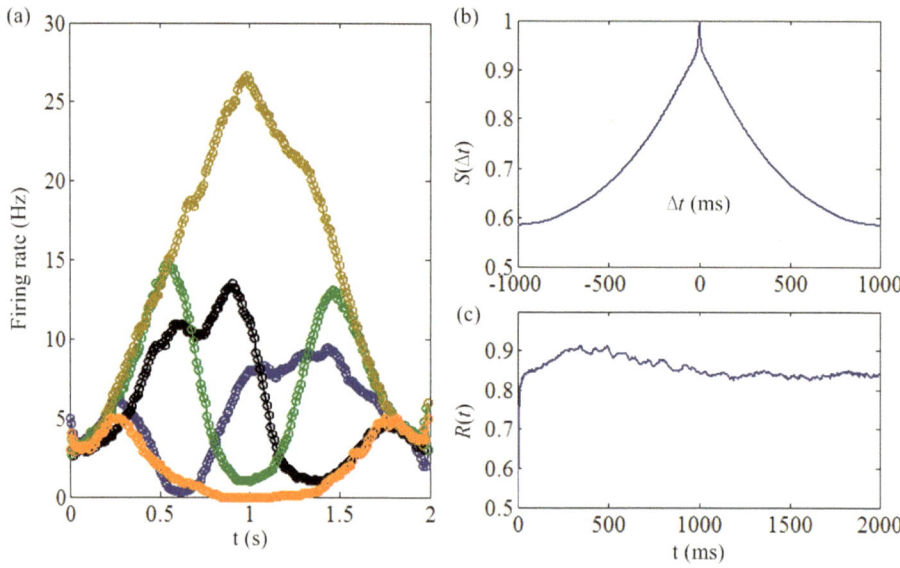

Fig. 5.17 Dynamic activity of GrC neurons in response to sinusoidal oscillatory MF inputs. **a** Average firing rate of five representative GrC neurons. **b** Similarity index $S(t)$ of the spike pattern in the GrC layer. **c** Repeatability index $R(t)$ of the spike pattern of all GrC neurons in two consecutive MF oscillation periods [32]

(4) Robustness analysis of the model

To assess the robustness of the large-scale CerebelluMorphic model, the classical conditional paradigm of delayed blinking is employed, dividing it into two sessions, each consisting of 100 trials. This design is illustrated in Fig. 5.18a. In each session, there was an acquisition phase spanning 80 trials, during which the conditioned-unconditioned stimulus pair was presented, followed by an extinction phase with 20 trials where only the conditioned stimulus was presented. The interspike interval (ISI) was set to 300, 400, and 500 ms, while the conditioned stimuli had durations of 400, 500, and 600 ms, equivalent to the ISI plus the unconditioned duration plus an additional 100 ms. A 100 ms pause was introduced between consecutive trials to silence the network. Conditioned stimuli were input via MFs, while unconditioned stimuli were delivered through IO neurons. In Fig. 5.18b–d, the first 100 trials correspond to the acquisition phase, and the subsequent 100 trials represent the extinction phase. The %CR denotes the probability of VN neurons producing a conditioned response (CR) during stimulation. Our results reveal that the %CR of the network remains robust even with increased ISI durations.

Due to the regular structure of SNNs [15–18], the cerebellum has served as an inspiration for numerous theoretical models, focusing on the combination properties of SNNs. Large-scale cerebellar networks exhibit the capacity to generate biodynamic behaviors,

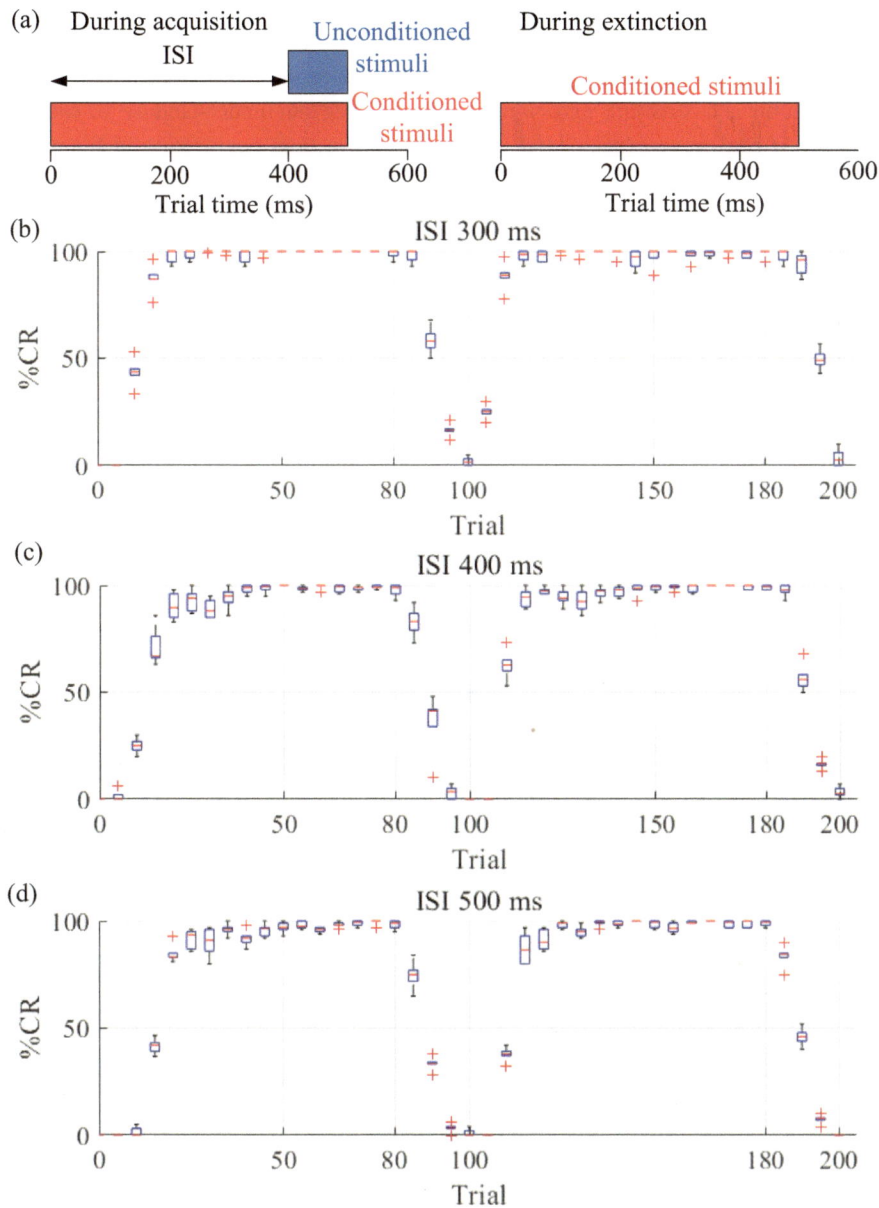

Fig. 5.18 Test results about robustness of the model. Behavioral CR results under different blinking classical conditions. **a** Blink classic conditioning protocol. **b–d** % CR with different ISI values from 300 to 500 ms [32]

including motor learning and memory consolidation. As a vital component of the mammalian brain, the cerebellum efficiently executes highly intricate motor learning tasks in parallel. Simulating cerebellar dynamics on large hardware and applying it to replicate brain function and structure can significantly enhance our understanding of the brain. Moreover, it can prove instrumental in addressing practical engineering challenges and developing intelligent machines.

A novel, highly biologically relevant neuromorphic model is developed inspired by cerebellar supervised learning. Real-time CerebelluMorphic offers a valuable approach for studying very gradual neural processes and facilitating interactions with the external environment, whether for robot control or brain-computer interfaces. The model's improvements enhance both its biological plausibility and computational learning capabilities within the cerebellar loop. Previous research has demonstrated that considering the differential fraction of multiple synaptic sites can replicate the more intricate dynamics of supervised learning compared to solely considering a single site of synaptic plasticity [19].

Furthermore, in contrast to prior models [14, 20, 21], a novel neuromorphic approach is introduced, which is aimed at capturing biologically significant mechanisms in SNNs. GPU simulation represents another potent technique for real-time cerebellar modeling, enabling the scaling up of network sizes to encompass up to 1 million neurons [22, 23]. Neuromorphic computing is grounded in a non-von Neumann architecture inspired by the brain's computational prowess. The brain has naturally evolved to process neurosensory information in a highly parallel and asynchronous manner, aligning with the fundamental computational principle of neuromorphic hardware. Our proposed neuromorphic approach leverages an event-driven 3D BFT neuromorphic architecture, capable of accommodating 3.5 million neurons across six nucleus types and three plasticity sites. This architectural choice enables the real-time replication of biologically plausible mechanisms, a crucial requirement for emulating the supervised motor learning function of the biological cerebellum.

To summarize, the development of the large-scale CerebelluMorphic model is rooted in a bottom-up modeling approach that explicitly incorporates a neuromorphic computational architecture. The key contributions of this endeavor include: (1) Provision of a theoretical and computational foundation for unraveling the motor learning mechanism, encompassing multiple plasticity rules in cerebellar learning. This model extends the capabilities of supervised learning beyond the confines of the Marr-Albus-Ito theory. (2) Introduction of a fresh perspective in the form of reverse engineering for understanding large-scale brain cognition. (3) Establishment of a novel engineering framework tailored for real-time motor learning. This framework can be harnessed to implement diverse computational intelligence models, such as cerebellar architectures and their motor learning, while ensuring biologically sound learning and computation.

5.4   BiCoSS Neuromorphic System

5.4.1   System Design

As described in [31] and [33], BiCoSS comprises seven subsystems interconnected by an enhanced tree structure within the network-on-chip architecture, significantly bolstering system scalability. In this configuration, a neuron serves as the fundamental computational core, while a neural network acts as the basic computational unit. Each of these computational units is composed of multiple computing cores, the exact number contingent upon the unit's maximum computing capacity. Illustrated in Fig. 5.19, each subsystem is composed of four neuron network units and one routing unit, with each of these units implemented using FPGA technology. The improved tree structure's nodes are effectively realized by employing four computing units through a hybrid interconnection scheme. This approach mitigates the need for frequent cross-level communication between computing units, which can arise when using a simplistic tree topology structure. Consequently, it helps reduce unnecessary propagation delays and resource consumption. Furthermore, the four computing units within each node of the tree structure utilize a hybrid interconnection pattern and are interconnected with other nodes in the improved tree structure via routing units. In comparison to previous topological designs, this structure offers greater support for the expansion of multi-core parallel computing systems.

Figure 5.20 illustrates the physical layout of the BiCoSS system. Within each computational unit of the BiCoSS system, there is one FPGA, two synchronous dynamic random access memorys (SDRAMs), and one serial configuration chip EPCS128 dedicated to program storage. SDRAM serves as the repository for intermediate variables and network parameters required for network computations, while the serial configuration chip EPCS128 stores programs to ensure their persistence in case of power loss. For the LIF or Izhikevich neuron model, each neuron network unit employs an on-chip network structure known as IBFT, comprising 16 nodes. This utilization of on-chip network technology enhances communication efficiency, system scalability, and overall flexibility within each computing unit of the multi-core system. Additionally, the system is equipped with GPIO communication ports designed for interfacing with external devices such as cameras, mechanical arms, digital-to-analog conversion modules, analog-to-digital conversion modules, and other input/output peripherals, enabling seamless interaction with the external environment. In the event of an external analog signal input, the analog signal is first converted into a digital signal through the analog-to-digital conversion module. Subsequently, the digital signal is transmitted to a universal IO pin on the FPGA chip via the communication port, thus facilitating the reception and processing of external signals. Conversely, when there's a need to output an analog signal, the FPGA's output signal can be routed through the digital-to-analog conversion module connected to the communication port. This process converts the digital signal into an analog form for output. Digital

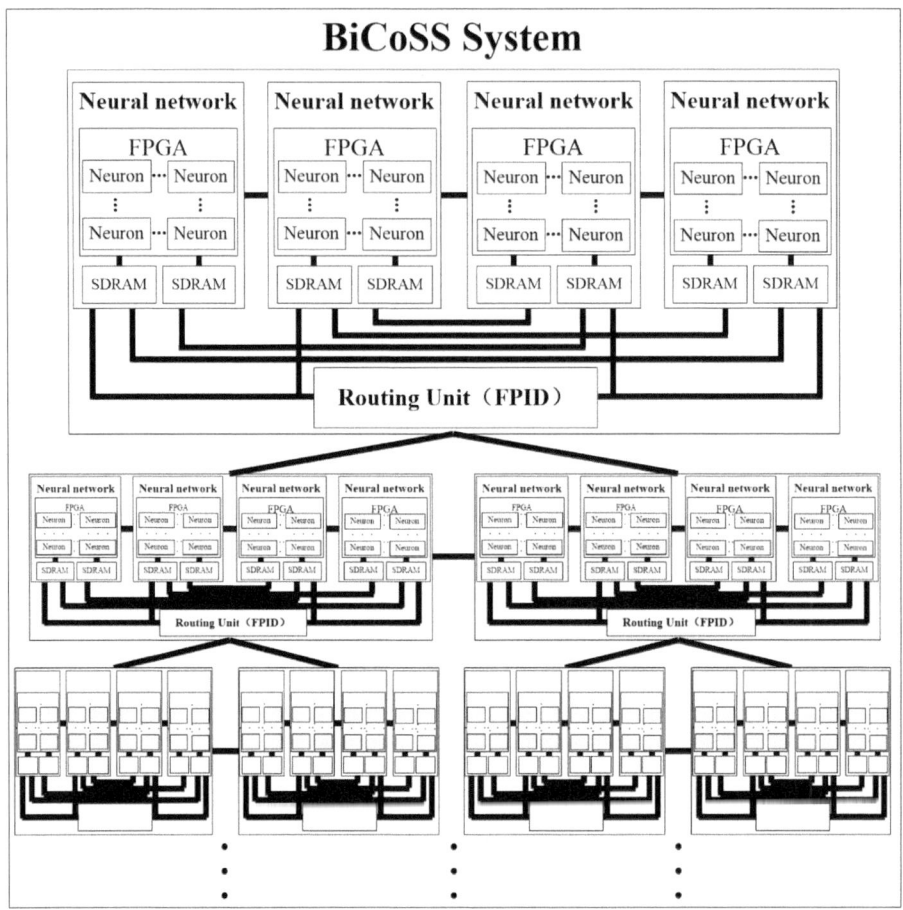

Fig. 5.19 BiCoSS architecture [33]

signals, on the other hand, can be directly sent through the communication port. The availability of these communication ports allows for real-time application development, including brain-inspired perception and decision-making.

The intricate computational architecture of BiCoSS is meticulously depicted in Fig. 5.21. Within the realm of neural network computation, BiCoSS harnesses a neuron computation unit, comprising the neuron computation module, synapse computation module, configuration unit module, and routing module. As illustrated in Fig. 5.21a, the neuron computing module dutifully conveys the synapse variable to the routing module, facilitating the transmission of essential synapse variable information to other neuron network units. Simultaneously, the membrane potential variable is dispatched to the synapse calculation module, where synaptic computations take place. Furthermore, the configuration

Fig. 5.20 BiCoSS system physical diagram [33]

unit module is responsible for configuring parameters within the routing module, while the routing module is entrusted with the task of receiving external AER spike data packets from adjacent nodes within the on-chip network. The routing module adeptly determines the direction of data transmission based on a programmable routing table found within the configuration unit. Figure 5.21b delves into the detailed computational architecture of the neuron computation module and the synapse computation module. For the variables V and X_i, these components encompass the respective neuron model pipeline calculation module and RAM storage array. The pipeline calculation module works tirelessly to update the neuron state in real time, while the RAM array serves as the repository for storing intermediate variables associated with neuron computations. This arrangement facilitates time division multiplexing operations on the neuron network, with BiCoSS performing this operation a total of 9000 times on the neuron calculation module. Additionally, the STDP compute unit diligently updates and outputs synapse weights in real time to the synapse computing unit. The synapse computing unit, in turn, receives both the AER spike data packet and the membrane potential variable denoted as V, subsequently engaging in computations. The ultimate result of these intricate operations is the computation and subsequent output of the membrane potential variable and the synapse variable.

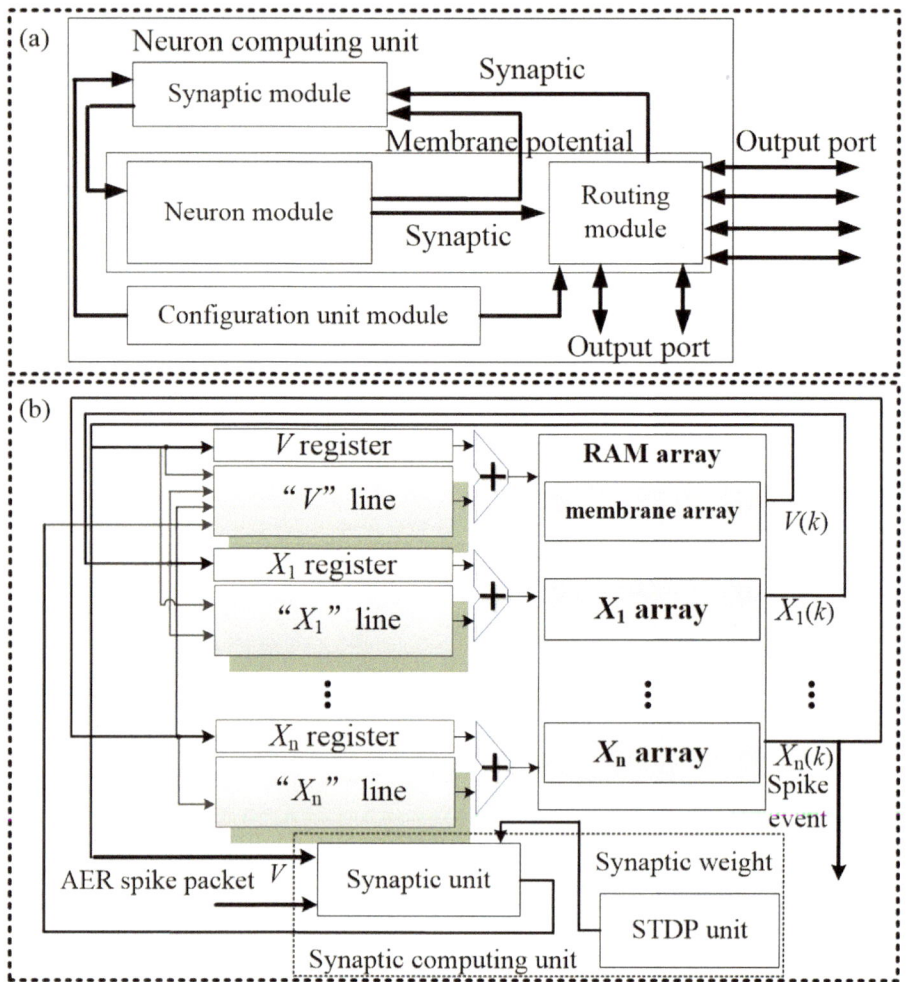

Fig. 5.21 Neural network computational architecture of the BiCoSS system. **a** Neuron computing unit top-level architecture. **b** Detailed architecture of neuron computation module and synapse
computation module [33]

Figure 5.22 provides an intricate glimpse into the architectural intricacies of the neuron and synapse computation modules within the BiCoSS system. Figure 5.22a delves
into the architecture of the cerebellar neuron computation module, exemplified by the
LIF neuron model. A resource-efficient approach is taken to optimize constant multiplication, achieved through a shifter, while variable multiplication is tactfully handled by
the logic unit. Figure 5.22b unveils the detailed computational architecture of the synapse

computation unit, showcasing a parallel structure with n synapse computation submodules. The read address of the buffer is denoted as "*rea*". A stream of AER spike packets is seamlessly introduced to the multiplexer and meticulously selected via a conventional counter. Further down the processing chain, the decoder skillfully deciphers the synaptic variable information and the timestamp data embedded within the AER spike packet. The timestamp data assumes the crucial role of serving as the write address for the buffer. Within each parallel synapse calculation module, the network connection matrix is dutifully stored in a read-only memory (ROM). A parallel adder then steps into action, summing up the outputs generated by the n parallel synapse calculation modules. Subsequently, the updated result is seamlessly channeled into the neuron calculation module, where further computations are executed with precision.

Selecting the appropriate neural network model is a pivotal challenge in the realm of large-scale brain simulation. This choice effectively delineates the composition of the neural network, the operational dynamics of its constituents, and the intricacies of their mutual interactions. Guided by insights from biological neural networks, the foundational elements of neural networks primarily comprise neurons and synapses. The neural network model plays a fundamental role in elucidating how distinct neurons and synapses interconnect and collaborate during the information processing that mirrors the functioning of the human brain. Tailored to the specific computational tasks characteristic of human brain cognition, the BiCoSS system, illustrated in Fig. 5.23a, charts the firing times of the neuron network on the abscissa, juxtaposed with the serial numbers assigned to neurons that exhibit firing behavior within the network. This temporal sequence of neural activations orchestrates the intricate cognitive processes akin to those within the human brain. In regard to synaptic modeling, the BiCoSS system employs synapses inspired by the intricacies of the human brain's neural network. These synapses are characterized as follows:

$$I_{ij}^{syn}(t) = g_{ij}s_j[V_{syn} - V_i(t)] \tag{5.8}$$

where the variables are defined as follows: C_{ij} represents the connection matrix, g_{ij} signifies the coupling strength, V_i pertains to the membrane potential of both the pre-synaptic and post-synaptic neurons, V_{syn} represents the reversal potential of the synapse, and s_j denotes a synaptic variable. The plasticity mechanism observed in the human brain is intimately tied to the brain's capacity for learning, imbuing neural networks with the ability to adapt over time. This adaptation manifests as changes in the synaptic weights within the neural network. In the BiCoSS system, an online unsupervised learning approach is employed, incorporating the spike-timing-dependent plasticity (STDP) learning rule, renowned for its high biological plausibility. The STDP rule is articulated as follows:

$$g_{ij} = g_{ij} + \Delta g_{ij} \tag{5.9}$$

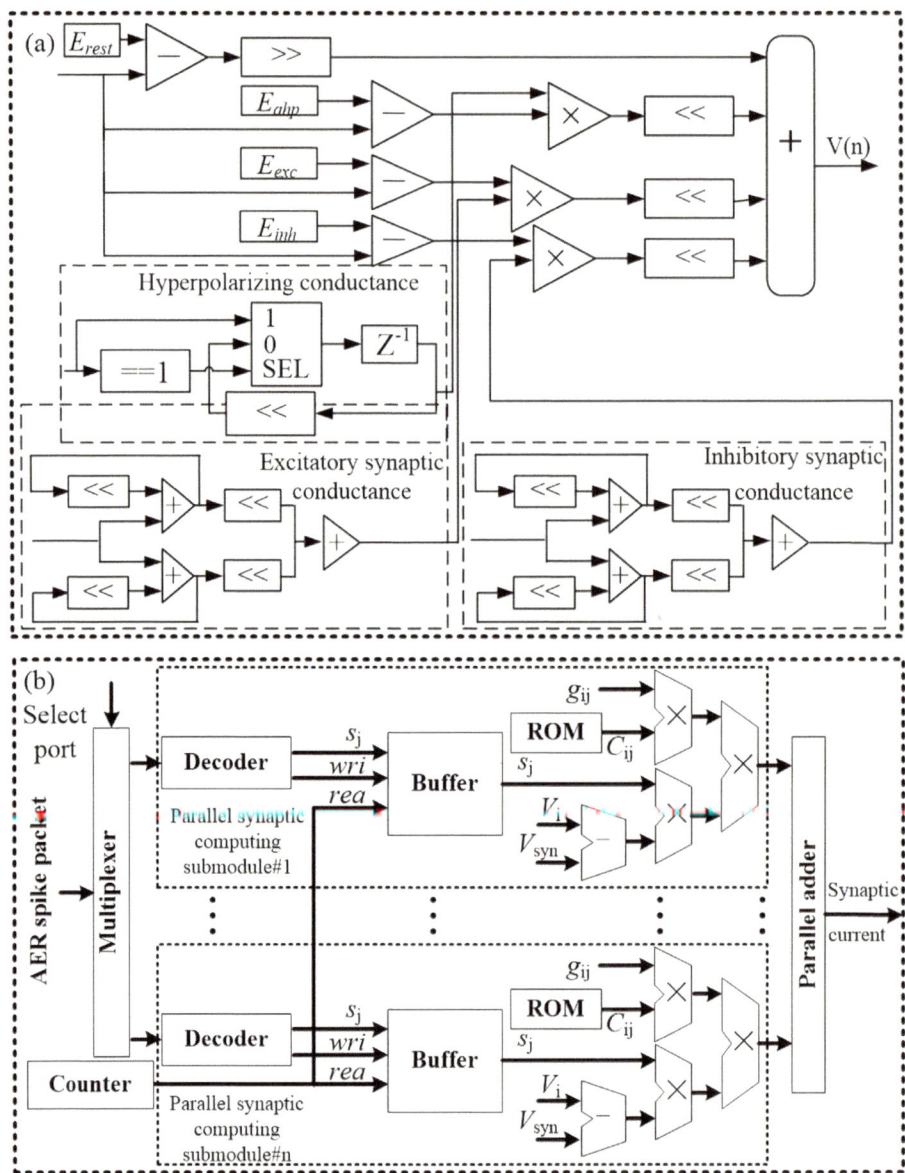

Fig. 5.22 BiCoSS system neuron and synapse computing module architecture. **a** Digital implementation of neuron computing module. **b** Digital implementation of the synapse computation module [33]

Fig. 5.23 Biologically
inspired firing behavior and
synaptic plasticity of neuronal
networks. **a** Raster plot of
neuronal networks.
b Dynamics of synaptic
plasticity [33]

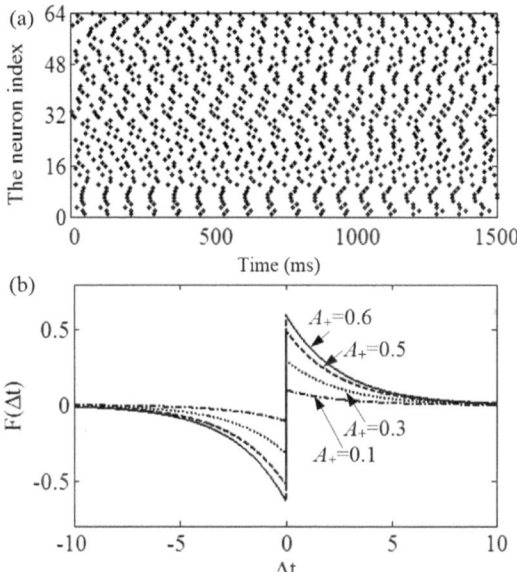

$$\Delta g_{ij} = g_{ij} F(\Delta t) \tag{5.10}$$

$$F(\Delta t) = \begin{cases} A_+ \exp(-|\Delta t|/\tau_+) & \text{if } \Delta t > 0 \\ -A_- \exp(-|\Delta t|/\tau_-) & \text{if } \Delta t < 0 \\ 0 & \text{if } \Delta t = 0 \end{cases} \tag{5.11}$$

where g_{ij} and Δg_{ij} represent the synaptic strength and synaptic strength change value
between two neurons, respectively. $\Delta t = t_i - t_j$ represents the time difference between
pre-and post-synaptic neuron firing, A_+ and A_- limit the maximum value of synaptic
correction, τ_+ and τ_- represent the time window parameters in the process of synaptic strength. Figure 5.23b visually elucidates the dynamics of synaptic plasticity. When
the firing activity of the presynaptic neuron precedes that of the postsynaptic neuron,
the resulting curve resides in the first quadrant, indicative of an augmentation in synaptic coupling strength. Conversely, if presynaptic neurons fire subsequent to postsynaptic
neurons, the curve occupies the third quadrant, signaling a diminishment in synaptic coupling strength. Notably, the rate of change of the F-function is influenced by distinct
STDP accommodation rates, with an escalating trend in the F-function's alteration as the
accommodation rate increases.

The transmission of information in the human brain relies on the firing behavior of
neurons. The neural information processing mode, based on AER, draws inspiration from
the information processing within biological neural networks. In this mode, when a neuron generates an action potential, it emits a firing signal. This firing signal then triggers
an address coding circuit to encode the event occurrence address. The neural signal's

attributes are defined by factors such as the peak-peak interval or firing frequency of neural firing. Importantly, firing events are actively triggered, meaning neurons that do not generate firing events do not produce output signals. This approach effectively eliminates information redundancy, reduces the volume of firing data, minimizes system overhead, lowers power consumption, and enhances system computational efficiency. Real-time firing events can be output without significant contention, enabling real-time computing. In cases of simultaneous neural firing transmission, potential data transmission conflicts are managed through communication arbitration by a control module.

Figure 5.24a illustrates the format of the firing information transmitted by the BiCoSS system, utilizing a 28-bit packet that comprises 8-bit AER data, 5-bit chip address data, 3-bit layer address data, 6-bit node address data, and a 6-bit timestamp. The 5-bit chip address encodes the 28 FPGA chips responsible for computing the neural network. The 3-bit layer address (with values 000, 001, 010, 011) designates the current layer number within the BFT structure. The 6-bit node address (ranging from 000,000 to 001,111) represents the neuron computing nodes, with codes 0 to 15, and can be further expanded to accommodate 64 neuron computing nodes. The 6-bit timestamp determines the calculation timing of the AER data packet and can introduce a delay of 0 to 63 clock cycles for computation. This data packet representation facilitates arbitrary connections among 114,688 neurons. In Fig. 5.24b, the router within the BiCoSS system comprises six input ports for connecting to other routing nodes or neural computing units. Each input port is equipped with a virtual channel featuring six FIFOs. These virtual channels are overseen by a virtual channel arbiter, which employs a counter to sequentially select FIFOs within each virtual channel. During the routing of firing information, when the first-layer router receives firing event information from the neural computing unit, the firing information wrapper module initiates a packaging process. The firing information wrapper module is situated after the dummy channel, allowing the neural computing unit's firing behavior output to be processed into an AER neural firing packet. The AER neural firing packet consists of 28 bits and comprises five sections: AER data, chip address, layer address, node address, and timestamp.

The firing information wrapper draws upon data stored in the configuration processor, which can be reprogrammed on-site as needed to modify stored information based on the number of neurons and connection details. The configuration processor holds chip address, layer address, node address, and timestamp data. Chip and layer addresses are represented as 3-bit fixed-point numbers, while node addresses are expressed as 6-bit fixed-point numbers. Timestamp data also employs a 6-bit fixed-point representation. Incoming events are buffered in memory and processed upon the arrival of the timestamp. Table 5.2 provides the address codes pertinent to the current router.

After the nerve firing event is encapsulated into the AER nerve firing packet, the routing logic module processes the AER nerve firing packet based on the routing algorithm as shown in Fig. 5.24c. The cross-interconnect module is implemented by a multiplexer and controlled by a cross interconnect arbiter. The routing algorithm of BiCoSS system is

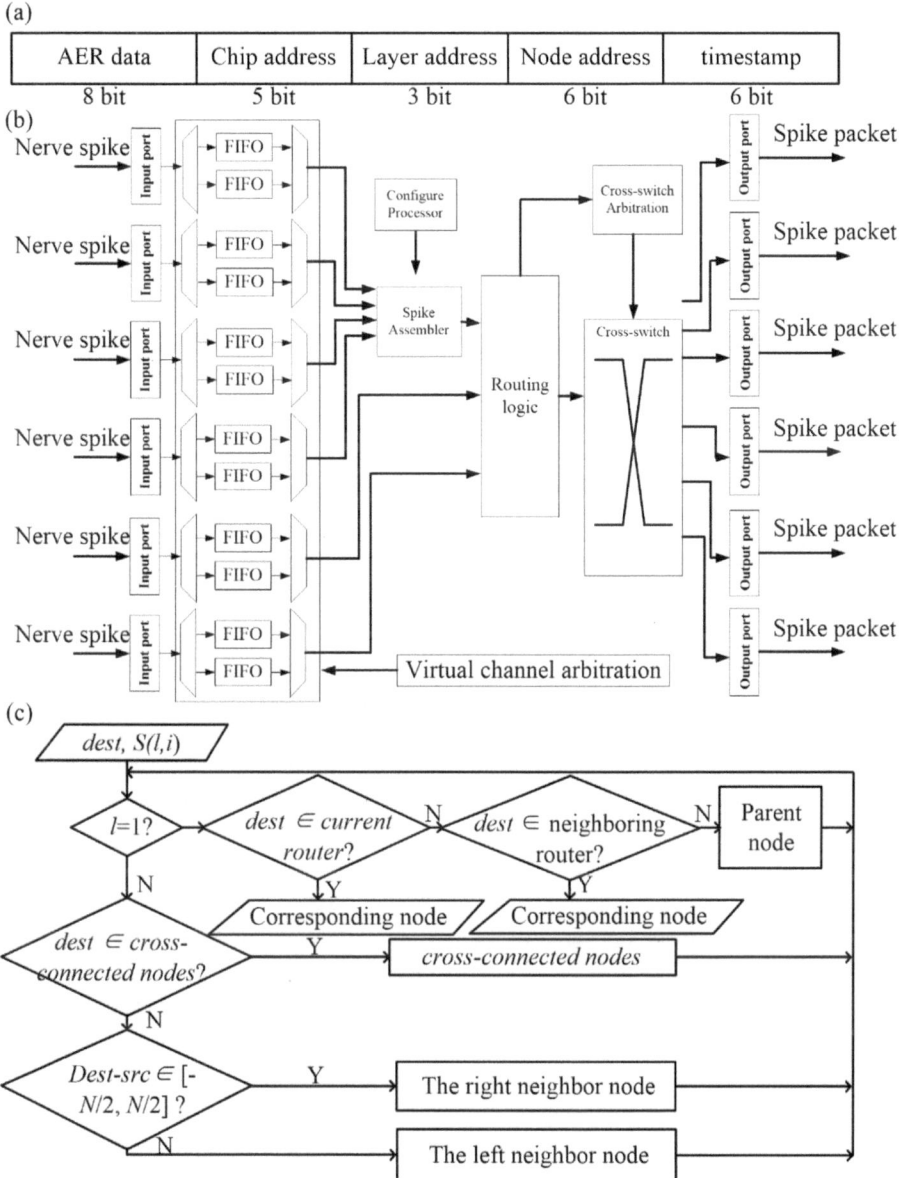

Fig. 5.24 Neural firing information routing for the BiCoSS system. **a** AER packets. **b** Routing architecture of system. **c** Routing algorithms [33]

Table 5.2 Current router-related address code

Current node	Child node	Neighbor node	Parent node	Compute unit
001,0000	–	001,0100	010,0000	0000;0001, 0010;0011
001,0100	–	001,0000	010,0000	0100;0101, 0110;0111
001,1000	–	001,1100	010,1000	1000;1001, 1010;1011
001,1100	–	001,1000	010,1000	1100;1101, 1110;1111
010,0000	001,0000; 001,0100	010,1000	–	–
010,1000	001,1000; 001,1100	010,0000	–	–

a deterministic non-shortest path length routing algorithm. The node code of the router is compared with the code of the destination node. In the algorithm, *dest* denotes the destination node, and the node address of the current route is denoted by $S(l, i)$. The addresses of the source and destination nodes are denoted by *src* and *dest*, respectively. The number of the neural computing units is N. Start by entering the addresses of the source and destination nodes (*src, dest*) and the address of the current router (l, i). If $l = 1$, the values of the lowest range *lowRange* and the highest range *highRange* of the current router are calculated. If *dest* belongs to this range, the nerve firing packet will be transmitted to the destination node. If not, that neural fire packet is transmit to any parent node of the current routing node and the route algorithm is restarted. If *dest* belongs to the cross-connected router, the neural firing packet is passed through the cross-connected link to the corresponding routing node and the routing algorithm is restarted. If $dest - src < N/2$ and $dest - src > -N/2$, the neural firing packet will be delivered to the correct neighbor node, otherwise it will be delivered to the left neighbor node, and the routing algorithm will be restarted and continued.

The BiCoSS system leverages Intel Corporation's Cyclone EP4CE115 series FPGA chips, a hierarchical heterogeneous architecture, and a tree-based structure to manage chip cascading and system expansion. Each chip employs an IBFT-type topological structure to enable system-on-chip functionality, while on-chip and inter-chip communication rely on a neural firing information routing mode based on address event expression. This approach significantly enhances calculation efficiency, reduces computational overhead and power consumption, and bolsters system scalability. The BiCoSS system specifically employs Intel's Cyclone IV series FPGA chip, EP4CE115F29C7N, featuring 115K vertically arranged logic elements, 4Mbit embedded memory, and 528 IO pins. Each chip is connected to two SDRAMs and a serial configuration device, EPCS128, effectively extending the core system to a six-layer PCB. To support simulations of neural networks, the serial configuration chip, EPCS128, stores programs to ensure data integrity when the system is powered off. Without this feature, the large-scale neural network would need to be re-uploaded chip by chip each time the system restarts. The BiCoSS system's seven subsystems are directly interconnected through the PCB backplane, with extra pins available on each subsystem, serving as external interfaces for digital-to-analog conversion, vision equipment, robot control, system expansion, and more.

To address the resource limitations of FPGAs relative to network scale, the BiCoSS system incorporates time-sharing multiplexing technology. It efficiently utilizes the same computing unit to process multiple channels of data during different time intervals. In simulating neural networks, the state variables computed by the neuron calculation unit are employed as output and stored in on-chip RAM. The system maintains continuous excitation, sequentially inputting the state information from the previous step of each neuron into the neuron computing unit. After updating the state of the time-shared neuron, the computation proceeds to the next step. This approach effectively manages resource constraints while maintaining simulation performance. Assuming that the storage depth of the RAM memory of the network state storage unit is N_{neu}, the total number of neurons contained in the network is N_{total}, and considering the time complexity, it can be described as follows

$$\begin{cases} N_{total} = n \cdot N_{neu} \\ N_{Npipe} \cdot N_{neu.} + N_{Spipe} \cdot N_{syn} \leq 1/f_{\max} \\ N_{syn} = m \cdot N_{neu} \end{cases} \tag{5.12}$$

where $f_{\max}$ is the maximum calculation frequency that can be realized by the FPGA, and N_{Npipe} is the pipeline depth of the neuron calculation unit. N_{Spipe} is the pipeline depth of the synaptic current calculation unit, and m is the number of the parallel synaptic current calculation unit. n is the number of the parallel neuron calculation unit.

In a brain simulation system, the utilization of hardware resource grows in tandem with the expansion of the functional network. Figure 5.25a offers a visual representation of the relationship between on-chip storage resource utilization, the bit width of neuronal computations, and the frequency of multiplexing neurons. Meanwhile, Fig. 5.25b illustrates that the network's update rate escalates as the number of SCCUs increases. This implies that augmenting the count of SCCUs facilitates the simulation of a larger population of neurons within a network possessing a predetermined connection topology. Conversely, reducing the number of neurons in the network enhances the overall computational speed of the system. The network can employ the minimum number of neurons required to achieve the desired dynamic behavior effectively.

5.4.2 Experiments

To assess and validate the BiCoSS system's performance, four critical aspects are considered: computational efficiency, power consumption, communication efficiency,

Fig. 5.25 Performance
analysis of the model
implementation. **a** Relationship
between on-chip memory
resource overhead and neuron
computation bit width, and the
number of neuron reuse.
b Relationship between
network update speed and
SCCU and N_{total} [33]

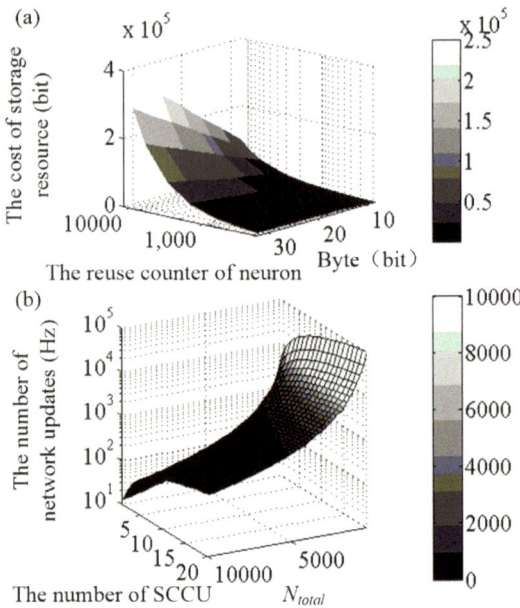

and scalability. The system's performance is evaluated by simulating both biologically plausible neuron models and biological neural networks featuring STDP learning
mechanisms.

To better illustrate the computational efficiency of the BiCoSS system, it's compared with three alternative platforms: CPU, GPU, and a multi-core bus platform. The
computational efficiency is quantified using a cost function, Q, defined as:

$$Q = t_{com}/t_{bio} \tag{5.13}$$

where t_{com} and t_{bio} represent the computation times of the reference biological neural
network and the platform being assessed, respectively. The CPU platform employs an
Intel dual-core 2.4GHz CPU, the GPU platform uses an NVIDIA GTX 280 GPU, and
the multi-core bus platform features an EP3SE340 FPGA chip. The simulation employs
a biological small-world neuron network based on the LIF neuron model. Figure 5.26a
examines neuron counts ranging from 10^3 to 4×10^6. As the neuron count increases,
the computational efficiency of the BiCoSS system maintains a higher value, while the
other two alternative computing platforms exhibit an exponential decline in efficiency.
This difference is attributed to BiCoSS's adoption of a fully parallel non-von Neumann
computing architecture, significantly enhancing computational efficiency.

In terms of power consumption, the BiCoSS system comprises 35 Intel EP4CE115
FPGA chips, consuming 10.419 W with a power density of 35.4 mW/cm^2. In comparison,

Fig. 5.26 BiCoSS system
performance analysis.
a Computation efficiency.
b Average delay comparison.
c Comparison of network sizes
with the same number of nodes
[33]

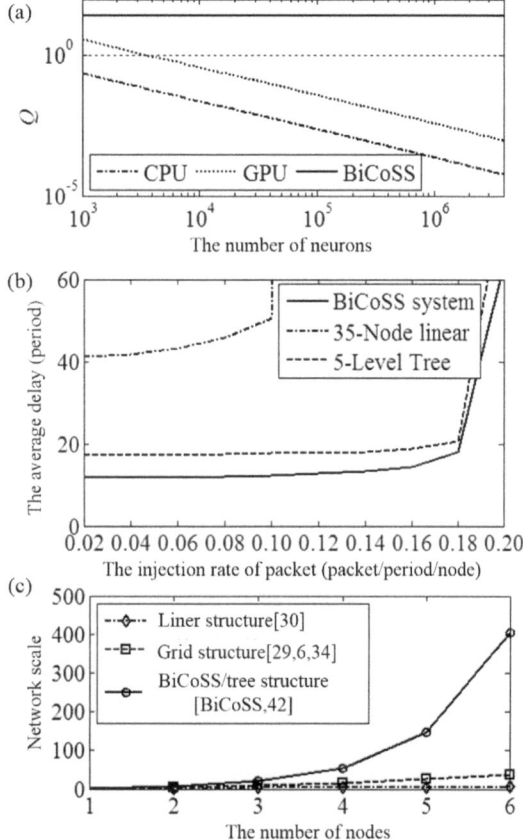

traditional CPU power densities range from 50 to 100 W/cm^2, while GPUs consume 50–300 W with power densities of 10–100 W/cm^2. BiCoSS boasts a power density 2.8 k times lower than that of GPUs. Regarding communication efficiency, Fig. 5.26b compares the delay of the BiCoSS system with that of traditional linear and tree structures. When the packet rate is below 0.18 packet/cycle/node, the BiCoSS system's structural load delay is notably lower than that of the traditional linear structure (exemplified by Neurogrid) and even lower than the traditional tree structure (as represented by HiAER). The average delay slightly increases with rising packet injection rates. The inflection point for linear topology is at 0.1 packet/cycle/node, while the traditional tree structure and the BiCoSS structure reach their inflection points at 0.18 packet/cycle/node. Given that the BiCoSS structure boasts higher node connectivity compared to the traditional tree structure, its growth rate post-inflection point is marginally slower, with the average delay being 68.7% of that in the tree structure.

Historically, various neuromorphic computing platforms have used different multi-chip topological structures to realize and simulate large-scale biological neural networks. These structures include the linear layout used by Neurogrid and the grid structures employed by BrainScaleS, Truenorth, and SpiNNaker. In terms of scalability, as shown in Fig. 5.26c, the BiCoSS system adopts an enhanced tree structure that offers significantly greater scalability compared to traditional linear and grid structures. This design choice bolsters the system's scalability, facilitating the expansion of large-scale human brain neuron networks.

The neural network unit within the BiCoSS system is implemented using a BFT structure, known for its favorable system delay characteristics. To showcase the delay superiority of the BiCoSS system's neuron network unit, average delays are compared with traditional NoC architectures like mesh, torus, Butterfly, and Flattened Butterfly types, as shown in Fig. 5.27a–d. The synaptic event input rate is defined in Million Synaptic Events/s (MSynE/s). The results illustrate that the average delay of the BiCoSS system's neuron network unit remains relatively stable as the input rate of synaptic events increases. In contrast, the average delay for the other three types of network structures on chips substantially increases with higher input rates of synaptic events, highlighting the BiCoSS system's robust delay characteristics under conditions of elevated synaptic event input rates.

To further highlight the performance advantages of the BiCoSS system, Table 5.3 provides a comparison between BiCoSS and major large-scale brain simulation systems worldwide. Regarding implementation mode, BrainScaleS [10], HiAER [26], and other systems employ analog circuits to realize large-scale neuromorphic computing, which comes with drawbacks such as lower computing precision compared to digital circuits, limited one-to-one correspondence between software instructions and hardware actions, and the need for extensive development cycles when neural network parameters or structures change. In contrast, digital neuromorphic computing systems like Truenorth [1] and SpiNNaker [9], as well as digital-analog hybrid systems like Neurogrid [8], combine the strengths of digital and analog circuits. However, these hybrid systems are sensitive to external environments, exhibit poor stability, involve digital-to-analog conversion delays, and pose challenges in application development and programmability. Unlike analog and hybrid circuits, BiCoSS employs digital implementation for large-scale neuromorphic computing. While digital circuits have higher power consumption, they offer superior computational accuracy, robust processing capabilities, and reconfigurability. The complexity of the neuron model plays a crucial role in ensuring the biological fidelity of neuromorphic computing. The BiCoSS system, with its digital circuitry, can realize any model, including the HH model based on conductance and learning algorithms based on the STDP mechanism inspired by biology. This versatility empowers it to better elucidate the human brain's information processing mechanisms. Regarding computational scale, BiCoSS can handle large-scale neural networks comprising up to 100 million neurons. For a BiCoSS system with N compute nodes, its scalability reaches $4e^N$. It's important to note that while BiCoSS consumes 10.419W, which is significantly less than SpiNNaker's 49W, it still consumes more power compared to systems like Neurogrid (2.7 W), Truenorth (63

Fig. 5.27 BiCoSS system neuron network unit average delay. **a** Average delay comparison of BiCoSS and mesh architectures. **b** Average delay comparison of BiCoSS and Torus type architectures. **c** Average latency comparison between BiCoSS and Butterfly architectures. **d** Average latency comparison of BiCoSS and Flattened butterfly architectures [33]

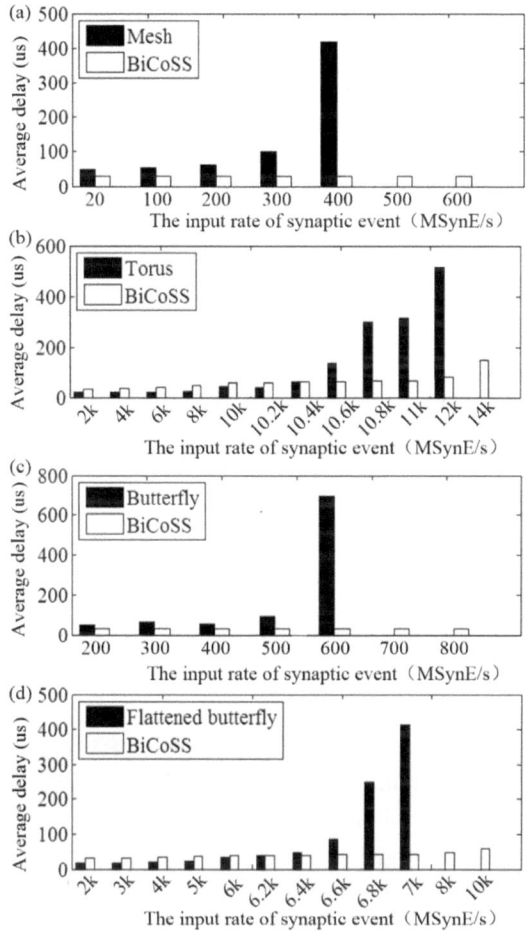

mW), etc. Nevertheless, considering the network scale, the complexity of realizable models, biology-inspired learning rules, and system scalability, the BiCoSS system outshines current representative large-scale neuromorphic computing systems worldwide.

A simple high-level cognitive behavior in the human brain involves millions of neuron spike activities. Computing large-scale neural networks aids in connecting the dynamic characteristics of neural networks with high-level cognitive human brain behaviors. The scale of computation inspired by the human brain enhances its biological plausibility. The BiCoSS system can calculate large-scale neural networks with cognitive functions resembling those in the human brain in real-time. On one hand, increased network scale brings it closer to the computational scale of human brain cognitive behavior, enabling accurate exploration of large-scale cognitive mechanisms. On the other hand, BiCoSS's real-time computational performance can be utilized for interfaces with living organisms

Table 5.3 Comparison with current representative large-scale neuromorphic computing systems

Neuromorphic system	A/D	Model	Learning	Scale	Scalability
BrainScaleS [10]	A	AIF	STDP	4M	N^2
Truenorth [1]	D	LIF	/	1M	N^2
Neurogrid [8]	AD	QIF	/	1M	2^N
SpiNNaker [9]	D	/	STDP	1B	N^2
LaCSNN [24]	D	/	STDP	1M	N^2
BlueHive [25]	D	/	/	64k	N^2
IFAT [11]	A	LIF	/	65k	2^N
HiAER [26]	A	LIF	/	1M	2^N
BiCoSS	D	/	STDP	4M	$4 \cdot 2^N$

or artificial intelligence devices. To further study the mechanisms of human brain cognitive behavior, a deep dive into brain-inspired decision-making, motor control, neural disease mechanisms, and brain-inspired visual target recognition from the neuron level is conducted, as depicted in Fig. 5.28. In this context, the BiCoSS system delves into brain mechanisms across multiple brain regions, realizing large-scale brain-inspired models of various nuclei, including the basal ganglia network responsible for decision-making and behavior selection, the cerebellar model for motor control, the visual cortex for object recognition, and the thalamic cortex, a focal point in movement disorders.

In brain-inspired computing applications such as brain-inspired decision-making, motor control, and the study of movement disorder mechanisms, all 35 FPGAs within the BiCoSS system are utilized to construct a large-scale neuron network with 4 million neurons for the exploration and real-time computation of human brain cognitive mechanisms. Figure 5.29a illustrates the physical diagram of the brain cognitive mechanism computing platform based on the BiCoSS system. The input signal generator, realized by the DE2 development board, generates discrete spike sequences as input, which are transmitted to the BiCoSS system via a two-bit line. The BiCoSS computing system, used to simulate the large-scale neuron network, converts the output into analog signals through a digital-to-analog converter and displays the output waveform on an oscilloscope. Figure 5.29b–e showcases the real-time output of discrete spike signals from basal ganglia GPi neurons, cerebellar GO neurons, and thalamic cortex neurons under decision-making, motor control, normal, and Parkinsonion states. In the Parkinsonion disease state, thalamic cortex neurons transition from regular spiking to burst firing. The displayed real-time discrete spike signals are sampled from randomly selected neurons for observation in specific nuclei within the BiCoSS large-scale network.

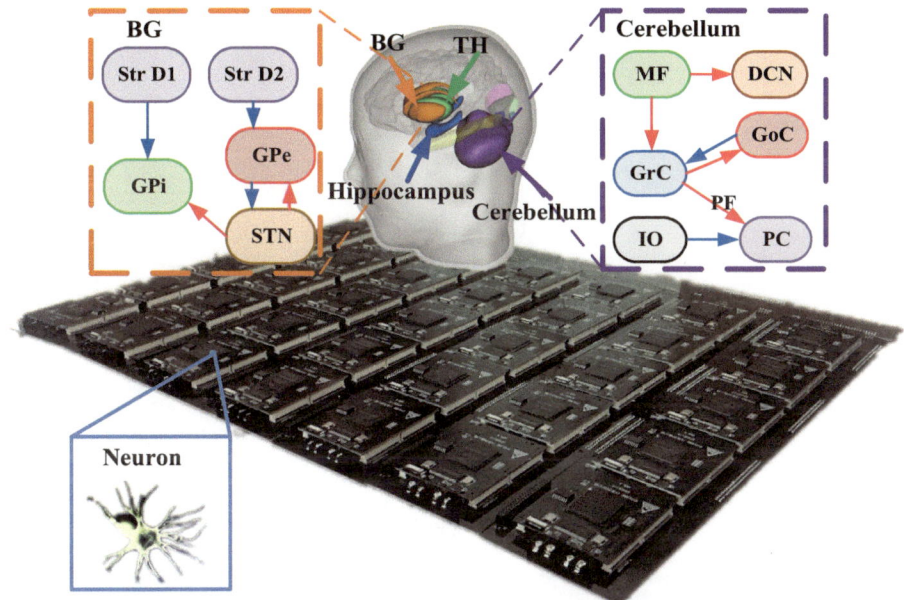

Fig 5.28 Neuromorphic computing for different brain regions based on the BiCoSS platform [31]

(1) Object Recognition Based on Unsupervised Learning Mechanism of Visual Pathway

The task of digit recognition is first performed using the BiCoSS system, using the SNN model proposed by Diehl et al. [27], the model used follows the LIF neuron model dynamics, expressed as follows

$$\tau_m \frac{dV}{dt} = -g_L(V - V_L) - g_E(V - V_E) - g_I(V - V_I) \tag{5.14}$$

where τ_m is the membrane constant, g_E and g_I are the excitatory and inhibitory conductance, V_E and V_I are the back potentials of the excitatory and inhibitory synapses, V_L and g_L are the leakage potential and leakage conductance. When the membrane potential exceeds a threshold, the neuron fires and resets back and goes through a refractory period.

In addition, the neuron model uses a self-regulating mechanism, and excitatory neurons use dynamic threshold V_{th}. It is based on the firing rate of neurons and is expressed as follows

$$V_{th} = V_\theta + \theta_{th} \tag{5.15}$$

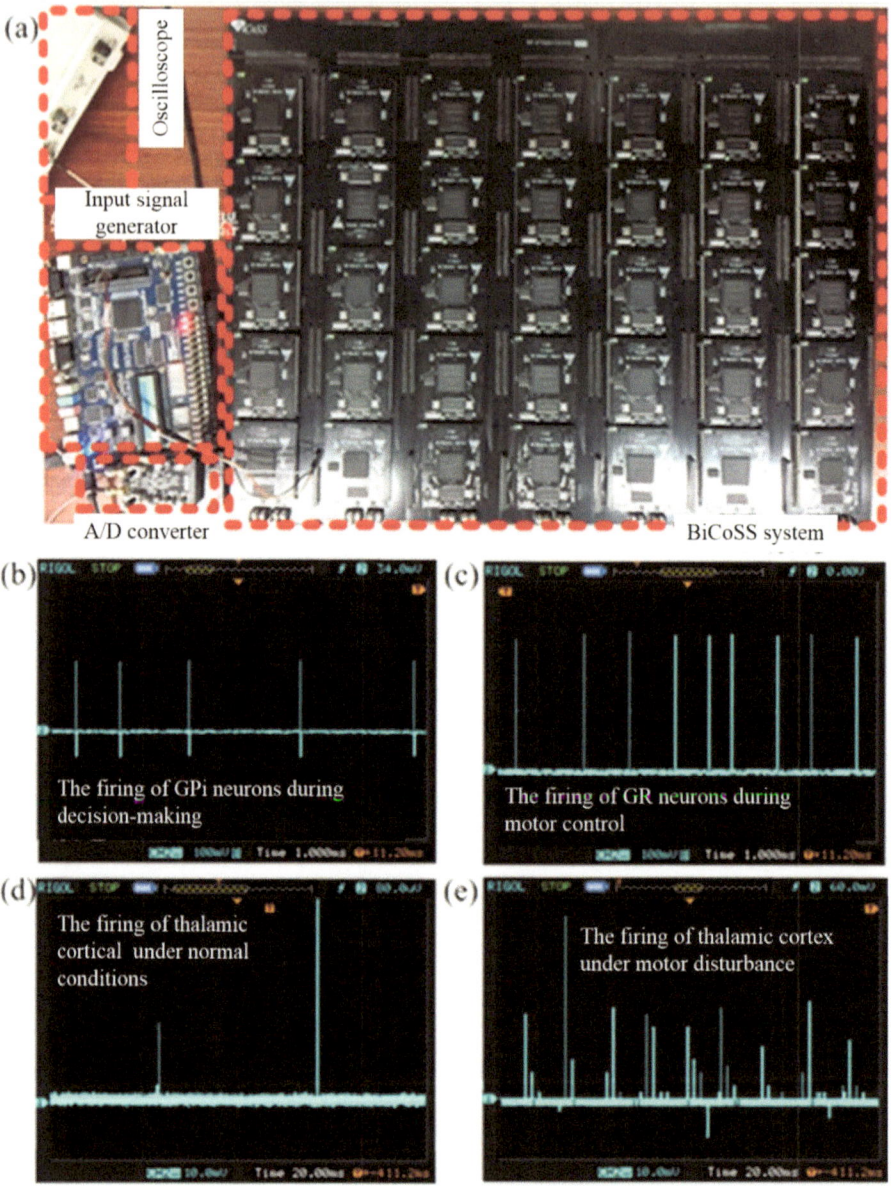

Fig. 5.29 Physical diagram and calculation results of experimental system [31]

$$\tau_{th}\frac{d\theta_{th}}{dt} = -\theta_{th} + \Delta\theta_{th}\delta_{t,t_k^j} \tag{5.16}$$

Besides, conductance-based chemical synapses are used, and the expressions for excitatory and inhibitory synapses are as follows

$$
\begin{cases}
\tau_{g_E}\frac{dg_E}{dt} = -g_E + \sum_{j\in C_E} w_{i,j} \sum_k \delta_{t,t_k^j} \\
\tau_{g_I}\frac{dg_I}{dt} = -g_I + \sum_{j\in C_I} w_{i,j} \sum_k \delta_{t,t_k^j}
\end{cases} \tag{5.17}
$$

where $w_{i,j}$ is the synaptic weight between neurons and t_k^j is the firing time of neuron j. In the absence of presynaptic spike stimulation, the excitatory and inhibitory synaptic conductance of neurons are based on a time constant c and τ_{g_I}. The T-STDP learning rule is used to modify the synaptic weights in real time. The updated expressions of three additional trace variables are as follows

$$
\begin{cases}
\frac{dx_1(t)}{dt} = -\frac{x_1(t)}{\tau_{x_1}}, & \text{if } t = t_{pre}, x_1(t) = 1 \\
\frac{dy_1(t)}{dt} = -\frac{y_1(t)}{\tau_{y_1}}, & \text{if } t = t_{post}, y_1(t) = 1 \\
\frac{dy_2(t)}{dt} = -\frac{y_2(t)}{\tau_{y_2}}, & \text{if } t = t_{post}, y_2(t) = 1
\end{cases} \tag{5.18}
$$

where x_1, y_1 and y_2 respond to pre-and postsynaptic spikes, τ_{x_1}, τ_{y_1} and τ_{y_2} are time constants. The weight change of the synapse may be calculated as follows

$$
\begin{cases}
\Delta w = -A_- y_1(t), & \text{if } t = t_{pre} \\
\Delta w = A_+ x_1(t)y_2(t), & \text{if } t = t_{post}
\end{cases} \tag{5.19}
$$

The parameters $A+$ and $A-$ are the learning rates of the pre-and postsynaptic neurons. It can be noted that the synaptic weights remain within the desired range $[w_{min}, w_{max}]$. In order to verify that application performance of BiCoSS system from different learning rule, Q-STDP learning rules including four trace variables are constructed according to the extended rules of T-STDP, thus the application model of BiCoSS system is extended.

As depicted in Fig. 5.30, an SNN is instantiated by comprising conductance-based LIF neurons and train it with T-STDP for digit recognition. The network consists of three layers: an input layer with 36 Poisson input neurons, a learning layer comprising 100 excitatory neurons, and 100 inhibitory neurons, and an output layer comprising 10 excitatory neurons. The input layer is fully connected to the learning layer. The learning layer itself consists of an excitatory layer and an inhibitory layer, where neurons in the inhibitory layer are interconnected with neurons in other excitatory layers. Lateral inhibition and competitive connections are realized via a winner-take-all (WTA) mechanism at the learning level. Plastic synapses connect the excitatory and efferent layers, with one efferent neuron per digit. The digit represented by the neuron with the highest firing rate signifies the predicted digit. As shown in Fig. 5.31, as the network size increases, the recognition accuracy improves, peaking when the network size reaches 100 neurons. With a noise rate of 16.67%, the recognition effectiveness can reach 96.3% when employing the Q-STDP learning rule. However, when the network size exceeds 100 neurons, overfitting becomes apparent. The FPGA resource utilization for the spiking neuron network is detailed in the table, with the primary on-chip hardware resources being logic resources.

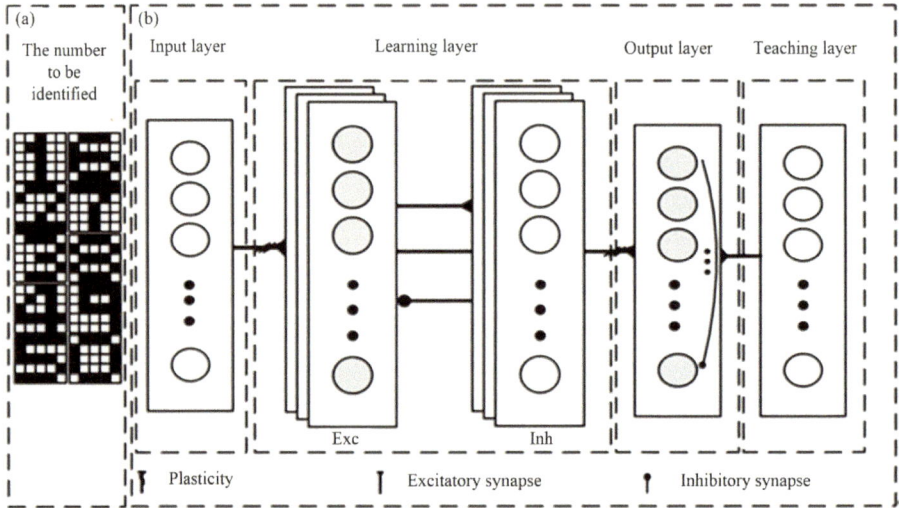

Fig. 5.30 Network structure. **a** Number to be identified. **b** Network connectivity

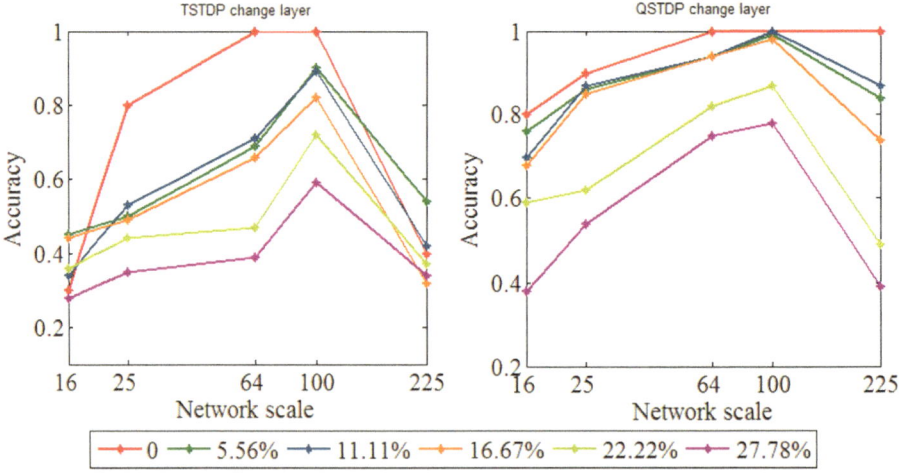

Fig. 5.31 Effect of different synaptic plasticity rules on recognition accuracy. **a** T-STDP. **b** Q-STDP

To demonstrate the system's performance on more complex classification tasks, the complexity of image recognition is expanded to encompass the MNIST (modified national institute of standards and technology) handwritten numeral recognition dataset. The number of neurons in the input layer increased to 784, with 5000 neurons in the excitatory layer, 5000 in the inhibitory layer, and 10 in the output layer. After 2000 training iterations, the test results indicate that the recognition efficiency for MNIST, based on T-STDP and Q-STDP, reaches 93.4% and 94.6%, respectively.

(2) Decision Making and Behavior Selection Based on Reinforcement Learning Mechanism of Basal Nuclei

Among various functional neural networks, the neural network related to cognitive decision-making is a prominent research area. Brain decision-making is an intricate neural process, and the cortex-basal ganglia-thalamus circuit plays a vital role in human cognitive decision-making. Different dopamine levels influence the brain's adoption of diverse decision-making strategies, encompassing conservatism, risk-taking, and randomness. Consequently, the BiCoSS system constructs a large-scale basal ganglia neuron network responsible for human brain decision-making. This network encompasses four nuclei: striatum (Str), GPe, GPi, and STN. Stratum consists of two parts, D1 and D2, and the latter three nuclei are common to both parts, with each part comprising one million neurons arranged in a grid pattern. Apart from the striatum, which simulates firing information in a Poisson sequence, neurons in the other nuclei follow the Izhikevich model with ion channel currents, as defined by the neuron equation below

$$
\begin{cases}
\frac{dV_{ij}^x}{dt} = 0.04\left(V_{ij}^x\right)^2 + 5V_{ij}^x - u_{ij}^x + I_{ij}^x + I_{ij}^{syn} & V_{ij}^x < V_{peak} \\
\frac{du_{ij}^x}{dt} = a\left(bV_{ij}^x - u_{ij}^x\right) & V_{ij}^x < V_{peak} \\
V_{ij}^x = c & V_{ij}^x \geq V_{peak} \\
u_{ij}^x = u_{ij}^x + d & V_{ij}^x \geq V_{peak}
\end{cases}
\tag{5.20}
$$

where V_{ij}^x is the membrane potential, and u_{ij}^x is the membrane recovery variable. I_{ij}^{syn} is the total synaptic current received by the neuron, and I_{ij}^x is the external current of the neuron at the position coordinate (i, j) of neuron x. V_{peak} is the firing threshold of the neuron (30mV in this model), and x can represent any of the three neurons of GPe, GPi and STN.

The synaptic current can be calculated as follows

$$
\begin{cases}
\tau_{Re}\frac{dh_{ij}^{x\rightarrow y}}{dt} = -h_{ij}^{x\rightarrow y}(t) + S_{ij}^x(t) \\
I_{ij}^{x\rightarrow y}(t) = W_{ij}^{x\rightarrow y}h_{ij}^{x\rightarrow y}(t)\left(E_{Re} - V_{ij}^y(t)\right)
\end{cases}
\tag{5.21}
$$

where τ_{Re} is the decay constant of the synaptic receptor, E_{Re} is the synaptic potential of the relevant receptor, $S_{ij}^{x}(t)$ is the firing condition of neuron x at time t. $h_{ij}^{x \rightarrow y}$ is the gating variable of the synaptic current from neuron x to neuron y, $S_{ij}^{x}(t)$ is the synaptic weight of the connection between neuron x and neuron y, V_{ij}^{y} is the membrane potential at (i, j). Re can represent any one of AMPA, GABA and NMDA synaptic receptors. Table 5.4 shows the specific parameter values of different types of cells.

In the basal ganglia network, D1 Str and D2 Str convey inhibitory signals to GPi and GPe, respectively. GPe sends inhibitory signals to STN, which, in turn, transmits excitatory signals to GPi and GPe. The basal ganglia network's ultimate output is provided by GPi. In the presence of high dopamine levels, two different stimulation signals are input into the BiCoSS network, both being Poisson sequences with frequencies of 4 Hz and 8 Hz, respectively. The raster plot for 2,000 neurons from 1 million STN and GPi neurons is displayed in Fig. 5.32. As shown in Fig. 5.32, under high dopamine levels, the basal ganglia network adopts a corresponding selection strategy. Stronger Poisson sequence stimulation accelerates network activity, leading the basal ganglia network to select 8 Hz stimulation as the optimal decision.

(3) Motor Learning and Control Based on Cerebellar Supervised Learning Mechanism

The cerebellum plays a crucial role in human brain motor control, encompassing tasks like maintaining body balance, regulating muscle tone, and coordinating movements. The

Table 5.4 Parameter values for different cells in the basal ganglia model

Parameter	GPe	GPi	STN
a	0.1	0.1	0.005
b	0.2	0.2	0.265
c	−65	−65	−65
d	2	2	1.5
I^{x} (nA)	10	10	30
E_{AMPA} (mV)	0	0	0
E_{NMDA} (mV)	0	0	0
E_{GABA} (mV)	−60	−60	−60
τ_{AMPA} (ms)	6	6	6
τ_{NMDA} (ms)	160	160	160
τ_{GABA} (ms)	4	4	4
$W_{\text{StrD2} \rightarrow \text{GPe}}$	0.8	–	–
$W_{\text{StrD1} \rightarrow \text{GPi}}$	–	1	–
$W_{\text{STN} \rightarrow \text{GPi}}$	–	–	1.15

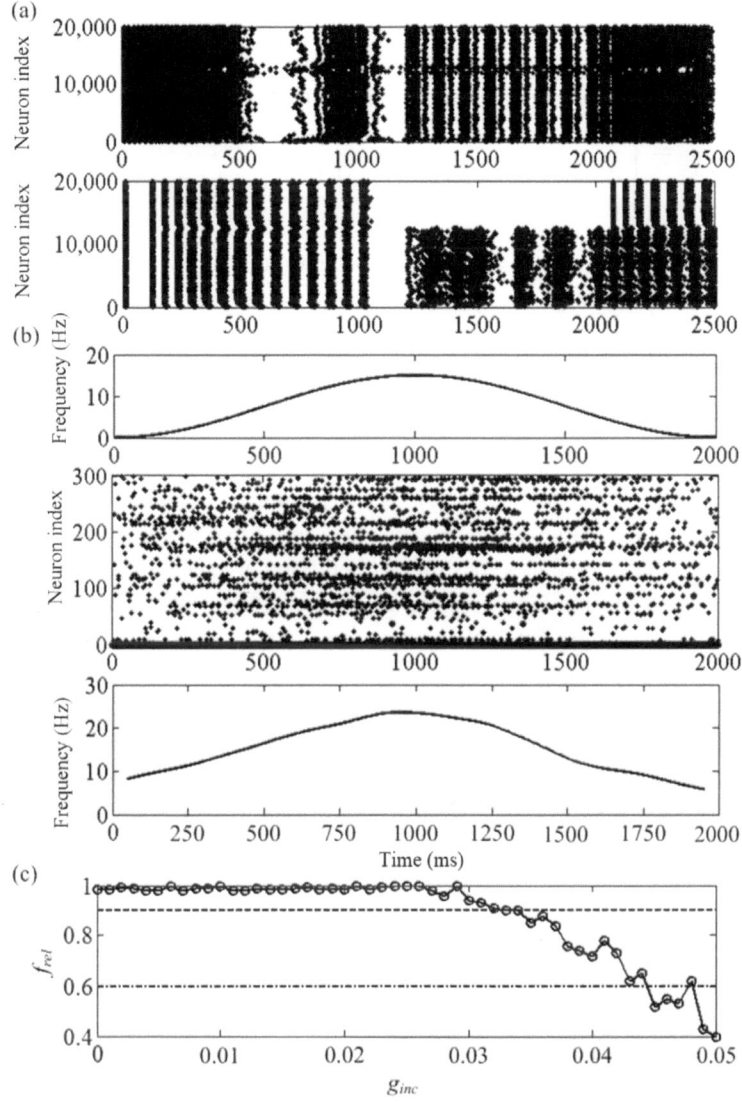

Fig. 5.32 Cognitive calculations based on the BiCoSS system. **a** Raster plot. **b** Firing characteristics of motor learning. **c** Relationship between Dyskinesia state and coupling strength [33]

BiCoSS system establishes a large-scale cerebellar neuron network to explore the mechanisms of human brain motor control. The cerebellar model incorporates four million neurons, including one million GrCs, one million GoCs, half a million Purkinje cells (PKJ), half a million BS cells, half a million IO cells, and half a million VN cells. GR and GO neurons are arranged in a grid pattern. All neuronal cells follow the LIF model with ion channels, as defined by the unified membrane potential equation and synaptic conductance equation provided below

$$
C_m \frac{dV}{dt} =
\begin{cases}
-g_{leak}(V(t) - E_{leak}) - g_{exc}(t)(V(t) - E_{exc}) \\
-g_{inh}(t)(V(t) - E_{inh}) \\
-g_{ahp}(t - t_0)(V(t) - E_{ahp}) + I & V \le V_{th} \\
E_{leak} & V > V_{th}
\end{cases}
\tag{5.22}
$$

$$
\begin{cases}
g_{ahp}(t - t_0) = G_{ahp} \exp\big(-(t - t_0)/\tau_{ahp}\big) \\
g_{exc}(t) = G_{exc1} \exp(-t/\tau_{exc1}) + G_{exc2} \exp(-t/\tau_{exc2}) \\
\quad + G_{exc3} \exp(-t/\tau_{exc3}) \\
g_{inh}(t) = G_{inh1} \exp(-t/\tau_{inh1}) + G_{inh2} \exp(-t/\tau_{inh2})
\end{cases}
\tag{5.23}
$$

where C_m is the membrane capacitance of the neuron, and V is the membrane potential. g_x represents the synaptic conductance, and τ_{ahp} is the conductance delay time. x can be any one of $\{leak, exc, inh, ahp\}$, and V_{th} is the membrane potential threshold. t_0 is the last firing time of the neuron. I is a spontaneous current that is present only in a few cell types. Specific parameter values for different cell types are shown in Table 5.5.

The calculation utilizes the CerebelluMorphic structure with anatomical significance previously proposed [32]. External input is transmitted to GR via Mossy fiber (MF), with GR and GO interconnected. The input information is processed and transmitted to PKJ and BS through PFs. External error signals are input to IO and transformed into learning signals, which are then transmitted to PKJ through CFs. PKJ receives inhibitory signals from BS before passing information to VN, whose output constitutes the cerebellum's final output signal. The vestibulo-oculomotor reflex experiment illustrates the neural response process where retinal images remain stable as the head position changes. During head movement, vestibular stimulation triggers eye movement, ensuring a stable retinal image despite shifting head positions. Figure 5.33a provides an overview of the brain-inspired motor control network structure inspired by the cerebellum, while Fig. 5.33b showcases the trajectory of robot arm control based on neuromorphic computing. These demonstrations highlight how BiCoSS-based neuromorphic computing can achieve cognitive motor control functions akin to those of the cerebellum.

5.5   Summary

This chapter provides an in-depth exploration of several notable large-scale neuromorphic systems, including TrueNorth, SpiNNaker, Loihi, Tianjic, LaCSNN, CerebelluMorphic, and BiCoSS . It discusses these systems from the perspectives of their architectural

Table 5.5 Parameter values of different cells in cerebellar model

Parameter	GR	GO	PKJ	BS	VN	IO
θ (mV)	−35	−52	−55	−55	−38	−50
C (pF)	3.1	28	106	107	122.3	10
G_{leak} (nS)	0.43	2.3	2.32	2.32	1.63	0.67
E_{leak} (mV)	−58	−55	−68	−68	−56	−60
G_{exc1} (nS)	0.1584	36.4	1	1	33	1
G_{exc2} (nS)	0.0216	3.003	–	–	17	–
G_{exc3} (nS)	–	6.097	–	–	–	–
τ_{exc1} (ms)	1.2	1.5	8.3	8.3	9.9	10
τ_{exc2} (ms)	52	31	–	–	30.6	–
τ_{exc3} (ms)	–	170	–	–	–	–
E_{exc} (mV)	0	0	0	0	0	0
G_{inh1} (nS)	0.012	–	1	–	30	0.18
G_{inh2} (nS)	0.016	–	–	–	–	–
τ_{inh1} (ms)	7	–	10	–	42.3	10
τ_{inh2} (ms)	59	–	–	–	–	–
E_{inh} (mV)	−82	–	−75	–	−88	−75
G_{ahp} (nS)	1	20	100	1	50	1
E_{ahp} (mV)	−82	−72.7	−70	−70	−70	−75
τ_{ahp} (ms)	5	5	5	2.5	2.5	10
I (nA)	–	–	250	–	700	–

designs and application examples, revealing the forefront of neuromorphic computing. Specifically, the chapter introduces the LaCSNN, CerebelluMorphic, and BiCoSS systems designed and implemented by the authors, offering a comprehensive understanding of these innovative neuromorphic platforms. It showcases the significant potential of large-scale neuromorphic systems and their various applications in understanding the complexities of brain-inspired intelligence.

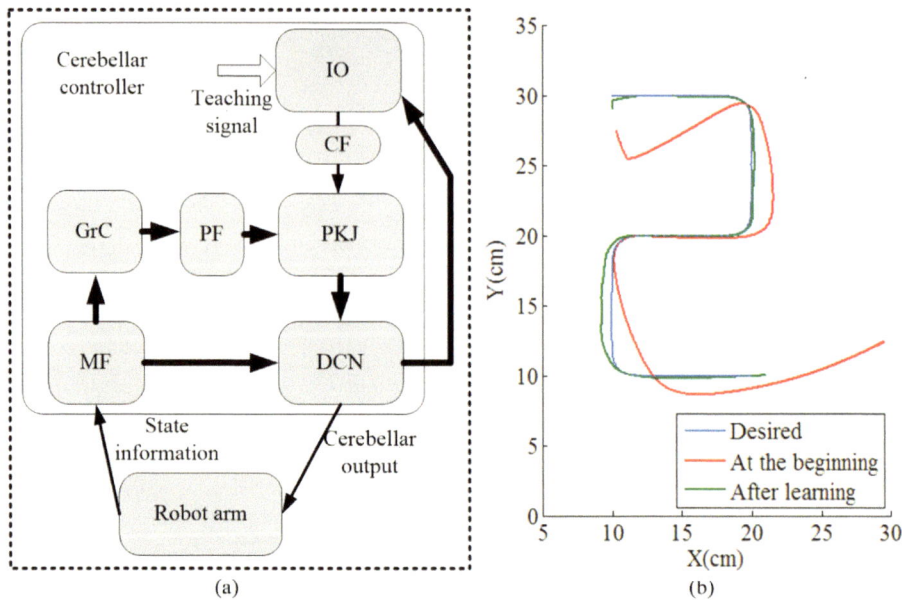

Fig. 5.33 Cerebellar arm control. **a** Cerebellar controller architecture. **b** Manipulator control effect [32]

References

1. Merolla PA, Arthur JV, Alvarez-Icaza R, et al. A million spiking-neuron integrated circuit with a scalable communication network and interface. Science. 2014;345(6197):668–73.
2. Davies M, Srinivasa N, Lin TH, et al. Loihi: a neuromorphic manycore processor with on-chip learning. IEEE Micro. 2018;38(1):82–99.
3. Painkras E, Plana LA, Garside J, et al. SpiNNaker: a 1-W 18-core system-on-chip for massively-parallel neural network simulation. IEEE J Solid-State Circ. 2013;48(8):1943–53.
4. Galluppi F, Lagorce X, Stromatias E, et al. A framework for plasticity implementation on the SpiNNaker neural architecture. Front Neurosci. 2015;8:429.
5. Han J et al. Tianjic: a 65 nm 1.3TOPS/W reconfigurable AI accelerator with always-on timbre recognition for mobile applications. IEEE J Solid-State Circ. 2018;54(1):244–57.
6. Esser Steven K, et al. Convolutional networks for fast, energy-efficient neuromorphic computing. Proc Natl Acad Sci. 2016;113(41):11441–6.
7. Wang Y, et al. An on-line learning spiking neural network implemented on Tianjic chip for dynamic pattern recognition. Front Neurosci. 2017;11:740.
8. Benjamin BV, Gao P, McQuinn E, et al. Neurogrid: a mixed-analog-digital multichip system for large-scale neural simulations. Proc IEEE. 2014;102(5):699–716.
9. Furber SB, Galluppi F, Temple S, et al. The Spinnaker project. Proc IEEE. 2014;102(5):652–65.
10. Petrovici MA, Vogginger B, Müller P, et al. Characterization and compensation of network-level anomalies in mixed-signal neuromorphic modeling platforms. PLoS ONE. 2014;9(10):e108590.

11. Yu T, Park J, Joshi S et al. 65k-Neuron integrate-and-fire array transceiver with address-event reconfigurable synaptic routing. In: Biomedical circuits and systems conference, (BioCAS), 2012. IEEE;2012. p. 21–4.

12. Luque NR, Garrido JA, Naveros F, et al. Distributed cerebellar motor learning: a spike-timing-dependent plasticity model. Front Comput Neurosci. 2016;10:17.

13. Nagao S. Behavior of floccular Purkinje cells correlated with adaptation of horizontal optokinetic eye movement response in pigmented rabbits. Exp Brain Res. 1988;73:489–97.

14. Yamazaki T, Nagao S. A computational mechanism for unified gain and timing control in the cerebellum. PLoS ONE. 2012;7(3):e33319.

15. Dean P, Porrill J, Ekerot CF, et al. The cerebellar microcircuit as an adaptive filter: experimental and computational evidence. Nat Rev Neurosci. 2010;11(1):30–43.

16. Albus JS. A theory of cerebellar function. Math Biosci. 1971;10(1–2):25–61.

17. Marr D, Thach WT. A theory of cerebellar cortex. From the retina to the neocortex: selected papers of David Marr;1991. p. 11–50.

18. Ito M. Long-term depression. Annu Rev Neurosci. 1989;12(1):85–102.

19. Carrillo RR, Ros E, Boucheny C, et al. A real-time spiking cerebellum model for learning robot control. Biosystems. 2008;94(1–2):18–27.

20. Gallimore AR, Kim T, Tanaka-Yamamoto K, et al. Switching on depression and potentiation in the cerebellum. Cell Rep. 2018;22(3):722–33.

21. Lev-Ram V, Mehta SB, Kleinfeld D, et al. Reversing cerebellar long-term depression. Proc Natl Acad Sci. 2003;100(26):15989–93.

22. Yamazaki T, Igarashi J. Realtime cerebellum: A large-scale spiking network model of the cerebellum that runs in realtime using a graphics processing unit. Neural Netw. 2013;47:103–11.

23. Hausknecht M, Li WK, Mauk M, et al. Machine learning capabilities of a simulated cerebellum. IEEE Trans Neural Netw Learn Syst. 2016;28(3):510–22.

24. Yang S, Wang J, Deng B, et al. Real-time neuromorphic system for large-scale conductance-based spiking neural networks. IEEE Trans Cybern. 2018;49(7):2490–503.

25. Moore SW, Fox PJ, Marsh SJT et al. Bluehive-a field-programable custom computing machine for extreme-scale real-time neural network simulation. In: 2012 IEEE 20th international symposium on field-programmable custom computing machines. IEEE;2012. p. 133–40.

26. Park J, Yu T, Joshi S, et al. Hierarchical address event routing for reconfigurable large-scale neuromorphic systems. IEEE Trans Neural Netw Learn Syst. 2016;28(10):2408–22.

27. Diehl PU, Cook M. Unsupervised learning of digit recognition using spike-timing-dependent plasticity. Front Comput Neurosci. 2015;9:99.

28. Pei J, Deng L, Song S, et al. Towards artificial general intelligence with hybrid Tianjic chip architecture. Nature. 2019;572(7767):106–11.

29. Kornijcuk V, Jeong DS. Recent progress in real-time adaptable digital neuromorphic hardware. Adv Intell Syst. 2019;1(6):1900030.

30. Voogd J, Glickstein M. The anatomy of the cerebellum. Trends Cogn Sci. 1998;2(9):307–13.

31. Yang S, Wang J, Hao X, et al. BiCoSS: toward large-scale cognition brain with multigranular neuromorphic architecture. IEEE Trans Neural Netw Learn Syst. 2021;33(7):2801–15.

32. Yang S, Wang J, Zhang N, et al. CerebelluMorphic: large-scale neuromorphic model and architecture for supervised motor learning. IEEE Trans Neural Netw Learn Syst. 2021;33(9):4398–412.

33. Yang S, Hao X, Wang J, Li H, Wei X, Yu H, Deng B. Large-scale brain-inspired computing system BiCoSS: its architecture, implementation and application. Acta Automatica Sinica. 2021;47(9):2154–69.